Andrew Gray

Theory and Practice of Absolute Measurements in Electricity and Magnetism

Volume 2, Part 1

Andrew Gray

Theory and Practice of Absolute Measurements in Electricity and Magnetism
Volume 2, Part 1

ISBN/EAN: 9783337406042

Printed in Europe, USA, Canada, Australia, Japan

Cover: Foto ©berggeist007 / pixelio.de

More available books at **www.hansebooks.com**

THE THEORY AND PRACTICE OF
ABSOLUTE MEASUREMENTS
IN
ELECTRICITY AND MAGNETISM

VOL. II.—PART I.

THE THEORY AND PRACTICE OF

ABSOLUTE MEASUREMENTS

IN

ELECTRICITY AND MAGNETISM

BY

ANDREW GRAY, M.A.

PROFESSOR OF PHYSICS IN THE UNIVERSITY COLLEGE OF NORTH WALES

IN TWO VOLUMES

VOLUME II—PART I

London

MACMILLAN AND CO.

AND NEW YORK

1893

[All Rights Reserved]

RICHARD CLAY AND SONS, LIMITED,
LONDON AND BUNGAY.

PREFACE

THOUGH the object and scope of the present work is set forth in the preface to Volume I., some further explanation is necessary in connection with the part now issued. First, in order to embrace all the topics which it was thought desirable to discuss it has been found necessary to use small type to a greater extent than had been anticipated, and it is hoped that this may not too seriously detract from the usefulness of the book.

The work has been composed in the short intervals of heavy teaching and other college duties, and printed off in instalments as progress was made. The subject also has made immense strides in the interval during which the second volume has been in the press, so that large additions to the subject-matter have come into existence, while various parts of magnetic and electromagnetic theory have been improved and altered in relative importance. Thus if the work were to be rewritten some changes would have to be made, especially in the earlier chapters. Also, had a better estimate of the ground to be covered, and more rapid progress been possible, a larger page would no doubt have been chosen, and the book improved in homogeneity and general balance.

Still, it is hoped that it may prove useful to the student of electrical science, and to the increasing number of electrical engineers who, to the great advantage of science, endeavour to solve the problems which their daily experience brings before them.

Accordingly I have given accounts of experimental investigations, in which I have tried to place before the reader not merely a skeleton of the method followed and the result arrived at, but such a statement in each case as may serve to show the procedure adopted, the difficulties met with, the mode in which they were overcome, the corrections made, and the reduction of the observations to the final result. Thus I have described the researches of Lord Rayleigh, Rowland, J. J. Thomson, and others, on the absolute measurement of currents, the determination of the ohm and of v, the practical methods of testing and measurement devised by Hopkinson, Ayrton, Perry, and other engineers, the magnetic researches of Hopkinson and Ewing, and the discoveries of Hertz in electrical radiation. In all cases it has been my aim, avoiding really unimportant detail, to describe the events of the investigation, and especially the manner in which practical difficulties were met as they arose;

for in this way only is it possible I think to learn fully the true lessons of physical experimental science. This plan of proceeding has however rendered the matter, though compressed as far as possible, almost too extensive for a single volume.

Peculiar difficulties attend the discussion of many parts of the subject of the present volume, especially points of electromagnetic theory. Generally I have tried to elucidate or avoid difficulties I had myself felt, and to give some idea of the later developments of electromagnetism.

In Chapter VI. is given a fairly full discussion of the calculation of constants of coils, &c., in which I have anxiously endeavoured to avoid error in the statement of results. As however the proofs were not read or the work verified by any one but myself, I fear there may be some slips left undetected. In this Chapter I have included some results of my own as to special arrangements, principally of single-layer coils, which I hope may prove useful. The Chapter concludes with an account of Lord Rayleigh's and Lord Kelvin's treatment of the effective resistance and inductance of conductors carrying alternating currents, together with a sketch of Lord Rayleigh's dynamical theory of this subject. I regret that by an oversight the valuable experiments of Prof. Hughes have not been mentioned in this connection.

In the notation adopted for electric and magnetic

quantities I have followed the proposal of Heaviside to use Clarendon type for directed quantities, but have done so in general without regard to whether the scalar magnitude merely or the complete notion was concerned in the statement or result. As however I have not ventured as yet to employ the vector analysis, this cannot, I think, in the present work lead to confusion. Clarendon type seems much to be preferred to the German capitals which have been so much used in this connection, and to be better also than the plain block type with which some writers have endeavoured to carry out Heaviside's suggestion. Block letters have however been here used for scalar quantities such as energy, total magnetic induction through a coil, &c., which continually occur. Thus for example I have adopted **B** for magnetic induction itself, and B for total magnetic induction, which in practical work has so much more frequently to be considered. Some inconsistencies in notation may be met with; these are in the main due to the peculiar circumstances of composition, and the large demands made by the great number of sets of quantities to be symbolized.

In the correction of the proofs of the first five chapters I received valuable assistance from my friend George A. Gibson, Esq., M.A., but in some of the more extensive later chapters I have had no help whatever. Hence although the most anxious care has been taken, there are no doubt many errors and misprints left uncorrected.

For notes of these I shall be grateful to any reader who will kindly send them to myself or to Messrs. Macmillan & Co.

In an Appendix are given short accounts of Zonal Spherical Harmonics, and the Theory of Errors of Observation. Some notes on points which arose in the course of the work, and a few tables complete the volume.

In the first volume a much larger number of typographical and other errors than I had hoped existed, have been found, chiefly by myself in the first instance. I have to thank those readers who have kindly forwarded lists; and I take the present opportunity of sending out a sheet which I hope contains all the errata of any importance.

A. GRAY.

UNIVERSITY COLLEGE OF
   NORTH WALES, BANGOR.
      Nov. 1892.

TABLE OF CONTENTS

CHAPTER I

MAGNETIC THEORY

SECTION I

MAGNETS. MAGNETIC POTENTIAL. POTENTIAL ENERGY OF A MAGNET

	PAGE
Hypothesis of imaginary magnetic matter	1
Unit quantity of magnetism or magnetic pole	2
Dimensional formula of magnetic pole	3
Realization (approximate) of unit pole	3
Magnetic field ; uniform field	4
Magnetic axis of a magnet	5
Magnetic moment	5
Uniform magnetization ; magnetic filament	7
Non-uniform magnet	8
Small magnet ; potential and force	8
Equation of lines of force	9
Description of lines of force	9
Description of equipotential surfaces and lines	11
Analogues of lines of force and equipotential surfaces	13
Magnetic filament	14
Potential	14
Equation of lines of force	15
Description of lines of force and equipotential curves	16
Magnetic potential of any magnet	19
Intensity of magnetization defined	19
Solenoidal distribution of magnetism	20
Potential energy of a magnet	20
Potential energy of a magnetic doublet	21

Potential energy of a finite magnet 21
    Expression in terms of components of field and magnetization intensities 22
    Expression in terms of magnetic moment 22
    Expansion in spherical harmonics 23
Centre and principal axis of a magnet 24

SECTION II
MAGNETIC INDUCTION. VECTOR POTENTIAL. MAGNETIC ENERGY

Induced magnetization 25
Force in interior of magnetized body 25
Surface distribution in cavity 26
Force within cavities of different shapes 26
    Cylindrical cavity 26
    Crevasse 27
    Spherical cavity 27
Magnetic induction, and magnetic force 27
    Components 29
    Solenoidal condition 29
    Variation of induction at surface 29
    Variation of force at surface 29
    Surface integral of induction 30
Vector potential of magnetic induction 33
    Components of induction 33
    Specification of vector potential 33, 34
Energy of a magnet 34
    Two distributions 35
    Intrinsic and mutual energies 37
Magnetic energy generally 37

SECTION III
APPLICATIONS OF GENERAL THEORY. MAGNETIC SHELLS. LAMELLAR DISTRIBUTION. UNIFORMLY MAGNETIZED ELLIPSOID

Magnetic shell 38
    Potential 39
    Potential and force due to closed shell 39

TABLE OF CONTENTS

	PAGE
Reckoning of solid angles	40
Solid angle at opposite sides of shell	40
Lamellar magnetization	41
Condition	41
Potential due to lamellar magnet	41
Continuity of potential at surface	42
Action of lamellar magnet	43
Energy of simple shell	43
Mutual energy of two shells	43
Vector potential of magnetic shell	44
Energy of lamellar distribution	44
Complex shell	45
Potential due to circular magnetic shell	45
Spherical harmonic expansions	46, 47
Mutual energy of two circular shells	48
Expression in elliptic integrals	49
Uniform magnetization : synthesis	51
Ellipsoid	52
Prolate ellipsoid	54
Oblate ellipsoid	55
Sphere	55

Section IV

INDUCED MAGNETIZATION

Wever's theory of magnetic induction	56
Coercive force	57
Æolotropic sphere in uniform field	57
Superpositions of magnetizations	58
Magnetization determined by quadric surface	59
Principal axes of magnetization	60
Principal magnetic susceptibilities	60
Principal magnetic inductive capacities	60
Couple on sphere	61, 62
Stable and unstable equilibrium	61
Magnetic induction through sphere	61
Isotropic sphere	62

	PAGE
Æolotropic ellipsoid in uniform field	63
Components of magnetized force	63
Couple in ellipsoid	64
(1) Weakly magnetic ellipsoid	64
(2) Strongly magnetic ellipsoid	64
Homogeneous isotropic ellipsoid	65
Induced magnetization	65
Diamagnetism	65
Apparent diamagnetism	65
Magnetic "poles"	66

CHAPTER II

DETERMINATION OF THE HORIZONTAL COMPONENT OF THE EARTH'S MAGNETIC FIELD

Determination of H	70
Method of Gauss	70
Magnetometer	70
Adjustment	71
Deflecting magnets	72
Deflection experiments	72
"End-on" position	72
"Side-on" position	74
Oscillation experiments	75
Theoretical results	77
Correction for inductive magnetization	78
Account of actual experiments	79
Deflection experiments	80
Observation of deflections	82
Oscillation experiments	83
Reduction of observations	84
Corrections	
Distribution of magnetism	85
Alteration of moment	86
Induction	86
Variations of H	88
Effect on induction-correction of varying thickness of magnet	89
Effect of length and hardness on induction-correction	91
Elimination of magnetic distribution	92

TABLE OF CONTENTS xv

	PAGE
Magnetic survey	93
Comparison of moments of large magnets	94
Stroud's compound magnetometer	94
Theory and use of instrument	97

CHAPTER III

THEORY OF ELECTROMAGNETIC ACTION

SECTION I

ACTIONS BETWEEN CURRENTS AND MAGNETS

Oersted's experiment	101
Equivalence of a current and a magnetic distribution	103
Ampère's theorems	104, 105
Magnetic shell equivalent to a current	107
Work done in carrying magnetic pole round current	107
Path of pole and circuit interlaced	107, 108
Magnetic field of current in long straight conductor	109
Lines of force	109
Deduction of law of force	110
Potential of current in long straight conductor	112
Work done in carrying pole round current	113
Line integral of magnetic force expressed in terms of components of current	114
Equations of currents	116
Effect of nature of medium	116
Action of magnetic system on current	117
Case of permeability different from unity	118
Equations of electromagnetic force	119

SECTION II

ACTION OF CURRENTS ON CURRENTS

Mutual action of two circuits	120
Deducible from equivalent shells	121
Ampère's electromagnetic experiments	121
Conclusions	124, 125

Mutual action of two current-elements 125
    Ampère's formula 129
Weber's electromagnetic experiments 131
Forces between straight conductors 133
Applications of Ampère's formula 134
    Action of thin solenoid on current-element 134
        Components of action of closed circuit on current-element 135
        Directrix of electromagnetic action 136
        Component action in any plane 136
        Action of small circuit on current-element 137
        Action of solenoid 138
Actions of solenoids 139
    Singly infinite solenoid equivalent to magnetic pole . . 140
    Turning moment on finite conductor 140
    Solenoid and uniform magnet compared 141
Turning moment on conductor in field of linear distribution of
    magnetism 142
Reaction of current-element on magnetic system 143
Magnetic field-intensity due to current-element 143
Unit current, second definition 144

CHAPTER IV

GENERAL THEORY OF CURRENT INDUCTION AND ELECTROMAGNETIC ACTION

Section I

LAGRANGE'S DYNAMICAL METHOD

Laws fulfilled by natural system 145
System of currents or of currents and magnets a dynamical system 146
Motion of a system of mutually influencing particles 146
    Degrees of freedom 147
    Lagrange's method 147
    Generalized coordinates 147
    Generalized signification of force 147
    Kinetic energy in generalized coordinates 148
    Time integral of a force 149
    Impulsive forces 149
    Work done in originating a given motion 150

	PAGE
Generalized components of momentum	150
Kinetic energy in terms of momenta	151
Velocities in terms of momenta or impulses	151
Lagrange's equations	153
Meaning of terms in Lagrange's equations	154
Hamilton's equations	155
Application of Lagrange's equations to electricity and magnetism	155
Electrical coordinates	156
Ignoration of coordinates	157
Electrokinetic energy	158
Generalised components of electromagnetic momentum	158
Dynamical theory of two circuits	159
Electromotive forces of induction	160
Dynamical illustrations of current induction	160
Maxwell's mechanical model	161
Dynamical analogues of "make" and break	163
Dynamical analogue of high resistance or air-break in secondary	163
Lord Rayleigh's mechanical model	165
Self-induction	166
Mutual induction of two circuits	167, 169
Analysis of energy of system of two circuits	168
Relation of inductive E.M.F.'s to energy of system	169
Electrokinetic energy and magnetic energy	170
Introduction of magnetic theorems	170
Expressions for inductances	171
1. Mutual inductance	171
2. Self inductance	172
Induction in terms of components of vector potential	173
Components of vector potential for any current system	174
Characteristic equations of vector potential	175
Examples of dynamical theory	176
Two indeformable circuits	176
Electrokinetic energy	176
Electromagnetic work done	176
Total energy spent in any change	176
Ampère's theory of magnetism	177
Current and magnetic shell	177
Two magnetic shells	179
Currents in two fixed mutually influencing circuits	179
Differential equations	179

	PAGE
Integral equations	181
Total flow of electricity at make	181
Total flow of electricity at break	182
Alternate current transformer without iron	183
Current in primary	184
Current in secondary	184
Definition of Impedance	185
Virtual resistance and inductance of primary and secondary	185
Theory of multiple arc in alternating circuit	186
Discharge of condenser through inductive resistance	188
Oscillatory discharge	188

CHAPTER V

GENERAL ELECTROMAGNETIC THEORY

SECTION I

INDUCTIVE ELECTROMOTIVE FORCES IN A CONTINUOUS MEDIUM. ELECTROMAGNETIC THEORY OF LIGHT.

Inductive electromotive forces	190
Electromotive force due to motion of circuit	190
Equations of electromotive force	191
Cases of simple induction	192
1. Rails and slider	192
2. Metal disk spinning in its own plane in magnetic field	193
Maxwell's electromagnetic theory of light	194
Total current	195
Displacement current	195
Conduction current	195
Differential equations of wave propagation	196, 197
Solution	198
Wave length and velocity of propagation	199
Electric displacement and magnetic induction in electromagnetic wave	200
Heaviside and Hertz's electromagnetic equations	201
Conditions fulfilled by the electric and magnetic field intensities	201
Derivation of equations of wave propagation	202
Analogy between electric and magnetic force	202

TABLE OF CONTENTS

	PAGE
Waves in insulating æolotropic medium	203
Principal specific inductive capacities	204
Relation of velocity of propagation to direction of ray	205
Double refraction	205
Ratio of units	206

SECTION II

ELECTROSTATIC AND ELECTROKINETIC ENERGY OF THE MEDIUM.—MOTION OF ENERGY IN THE ELECTROMAGNETIC FIELD

Energy of electromagnetic field	207
Electrostatic energy	207
Electrokinetic energy	207, 208, 209
Localization of electrokinetic energy	211
Dissipation of energy in a magnetic cycle	211
Motion of energy in electromagnetic field	213
Examples of flow of energy	216
1. Long straight wire carrying current	216
2. One condenser discharging into another	217
3. Condenser discharging through wire	218
4. Voltaic cell with plates connected by wire	219
Seat of electromotive force of cell	220
5. Generator and motor in same circuit	220
6. Thermoelectric circuit of two metals	221
7. Circuit of two metals with voltaic cell on one	221
8. Thermoelectric circuit of iron and two metals of lead type	222
Bearing of theory on Thomson effect	223
Apparent paradox	223
Velocity of electromagnetic wave deduced from theory of flow of energy	223

SECTION III

MAGNETO-OPTIC ROTATION

Faraday's magneto-optic experiment	225
Law of the phenomenon	226
Verdet's constant	226
Non-magnetic turning of plane of polarization	226

TABLE OF CONTENTS

	PAGE
Comparison of phenomena	226
Physical interpretation of points of difference	227
Non-magnetic turning explainable by helical structure	227
Magnetic turning by rotatory motion of molecules	229
Diamagnetism not a differential effect	230
Dynamical theory of circular polarization	230
Motion in plane polarized ray compounded of opposite circular motions	231
Gyrostat	233
Examples of precessional motion	234
Reactional couple due to motion of axis of gyrostat	234
Gyrostatically loaded medium	235
Wave of transverse displacement	235
Differential equations of motion	235
Solution	236
Relation of rate of turning to wave-length	236
Gyrostatic molecule	237
Magneto-optic rotation according to electromagnetic theory of light	238
Hall's phenomenon	241
Rowland's theory of Hall effect	243
Kerr's magneto-optic and electro-optic effects	244

CHAPTER VI

SECTION I

MAGNETIC ACTION OF CIRCUITS AND SHELLS

Solid angle subtended by circle	245
Potential due to circular current	246
Potential due to circular surface distribution of magnetism	246
Potential due to disk of magnetic matter	247
Magnetic forces due to circular magnetic shell	248
Field of circular current	249
Couple on magnetic needle in field of thin circular coil	250
Correction for coil-section	252, 257
Gaugain's galvanometer	253
Helmholtz's double-coil galvanometer	254
Couple on needle	254
Field of thin double coil	255
Correction for finite cross-section of coil	257, 258

TABLE OF CONTENTS

	PAGE
Four-coil galvanometer	258
Three-coil galvanometer	259
Long coil of single layer of wire	259
Potential and force at centre	261
Potential at internal point	263
Long coil of several layers	263
Potential, &c., of circular current	264
Series for, at near and distant points	264, 265
Mutual action of two circular conductors	266
Electrokinetic energy	267
Attraction between two parallel circular currents	268
Action between two coils of finite cross-section	269
Electrokinetic energy	269
Attraction and turning moment	269
Electrodynamometer with two double coils	270
Mutual energy of coil systems	271
Turning moment on either double coil	272
Mutual energy of two long single-layer coils	272
Expression in zonal harmonics	272
Integration of terms of expansion	273
Integrated expression for mutual energy	274
Mutual action of two long single-layer coils	275
Simplification by adoption of particular arrangement	275
Application to standard electrodynamometer	276
Absolute galvanometer	276
Mutual induction of two single-layer coils	277
Mutual induction of two multiple-layer coils	277
Simple uniform solenoid	278
Approximate realization by helix	278
Closed solenoid with circular axis	278
Ring electromagnet	278
Solenoid enclosing different media	281
Magnetic "current" and magnetic "resistance"	282
Magnetomotive force	282
Magnetic *reluctance*	283
Total induction in ring electromagnet	283
Induction in straight solenoid	284
Cylindric conductor replaceable by axial filament	285
Magnetic force within hollow conductor	286
Magnetic force within substance of cylindrical conductor	286
Self-inductance due to magnetization of conductor	287

Section II

CALCULATION OF COEFFICIENTS OF INDUCTION

	PAGE
Self-inductance of two parallel conductors	288, 293
Geometric mean distance of two coplanar areas	290
Geometric mean distance of two circular areas	290
Energy of opposite currents in parallel tubular conductors	291
Geometric mean distances	294
Point from circular area	294
Finite annulus from internal point	295
Finite annulus from itself	296
Point from straight line	297
Boundary of rectangle from centre	297
Two parallel lines from one another	298
Straight line from itself	299
Straight line from parallel rectangle	300
Two parallel rectangles from one another	302
Square from itself	303
Self-inductance of coil of large radius	303
Mutual inductance of two close coaxial coils of large radius	304
Mutual inductance of two coaxial circles	306
Coil of maximum self-inductance	308
Mutual inductance of circle and coaxial helix	309
Expression in series of definite integrals	310
Computation of integrals	311, 312
Result of calculation in actual case	314
Mutual inductance of two coaxial nearly equal circles	315
Expression in elliptic integrals	315
Elliptic integral expansions	315, 320
Elliptic integral expression corrected for cross-section	321
Maxwell's expansion	322
Maxwell's theorem of total induction through circle	323
Self-inductance of coil	325
Weinstein's formula	325

Section III

INDUCTION IN CONDUCTORS CARRYING VARYING OR ALTERNATING CURRENTS

	PAGE
Varying current in cylindrical conductor	325
Current density at any point	326
Total current	327
Lord Rayleigh's calculation for alternating currents	327
Expression for non-inductive E.M.F.	328
Development in series	328
Effective resistance and inductance	329
Expression in terms of Bessel's functions	330
Lord Kelvin's calculation of effective resistance	331
Differential equations	331
Mean square of current	331
Effective resistance	333
Solution for right circular cylinder	333
Investigation of effective inductance	334
Cylindrical conductor of large radius	336
Modified differential equation	336
Effective resistance	337
Lord Rayleigh's calculation for two parallel plane strips	338
Differential equation	339
Effective resistance and inductance	340
Comparison of slow and rapid alternations	340
Thickness of effective surface stratum	341
General dynamical theory of effective resistance and inductance	342
Thomson's and Bertrand's general dynamical theorems	342
General expressions for effective resistance and inductance	345
Analogue of Thomson's theorem	345
Application to electrical problems	345
1. Primary and secondary circuits	345
2. System of primary, secondary, tertiary, &c., circuits	346

ERRATA.

Page 23, line 16 from top, *for* "$\frac{1}{2}(5\mu^2 - 3\mu)$" *read* "$\frac{1}{2}(5\mu^3 - 3\mu)$."

Page 28, line 5 from foot, *for* "every medium" *substitute the words* "at each point, in every medium, depending on the state of the medium at the point considered."

Pages 48, 49, prefix the sign minus to the expressions on the right of the four equations giving values of E.

Page 51, line 8 from foot, *after* "distribution" *insert* "when $\rho = 1$."

Page 52, line 2 from top, *for* "resultant" *read* "total component."

Page 169, line 12 from foot, *for* "$\gamma_1^2 L_1^2 + \gamma_2^2 L_1^2$" *read* "$\gamma_1^2 L'_1{}^2 + \gamma_2^2 L'_2{}^2$."

Page 169, in marginal at foot, *for* "rekinetic" *read* "kinetic."

Page 201, line 12 from top, *for* "(9)" *read* "(6)".

Page 270, in top marginal, *for* "in" *read* "on."

Page 277, line 3 from top, *for* $\dfrac{a^2}{2}\dfrac{x_2}{r_2}$ *read* $\dfrac{2}{a^2}\dfrac{x_2}{r_2}$.

Page 279, line 7 from top, *after* "is" *insert* "$r + x$, and the line integral of magnetizing force round it is."

Page 281, last line, *for* "4π" *read* "$8\pi^2$."

Page 282, line 15 from top, *for* "4π" *read* "$8\pi^2$."

ABSOLUTE MEASUREMENTS

IN

ELECTRICITY AND MAGNETISM

VOLUME II

ADDITIONAL ERRATA.—PART I.

P. 199, line 9 from foot, *for* "$\cos(pz - nt)$" *read* "$\cos(qz - nt)$."
P. 201, line 12 from top, *for* "(9)" *read* "(6)."

and shall therefore not devote space to the description of the ordinary phenomena of attraction and repulsion between permanently or inductively magnetized bodies. We here propose to give such an outline of theory as may suffice to render intelligible the various methods of magnetic measurement, and clearly define the quantities which are determined by these methods.

It can be shown that magnetic phenomena are capable of being accounted for by supposing the magnetized body or system to be the seat of a distribution of what   Hypothesis of Imaginary Magnetic Matter.

ERRATA.

Page 23, line 16 from top, *for* "$\frac{1}{2}(5\mu^2 - 3\mu)$" *read* "$\frac{1}{2}(5\mu^3 - 3\mu)$."

Page 28, line 5 from foot, *for* "every medium" *substitute the words* "at each point, in every medium, depending on the state of the medium at the point considered."

Page 279, line 7 from top, *after* "is" *insert* "$\frac{a^2 r_2}{r} + x$, and the line integral of magnetizing force round it is."

Page 281, last line, *for* "4π" *read* "$8\pi^2$."

Page 282, line 15 from top, *for* "4π" *read* "$8\pi^2$."

ABSOLUTE MEASUREMENTS
IN
ELECTRICITY AND MAGNETISM

VOLUME II

CHAPTER I
MAGNETIC THEORY

SECTION I
MAGNETS   MAGNETIC POTENTIAL   POTENTIAL ENERGY OF A MAGNET

IN dealing with magnetism we shall again suppose the reader to be acquainted with the elementary facts, and shall therefore not devote space to the description of the ordinary phenomena of attraction and repulsion between permanently or inductively magnetized bodies. We here propose to give such an outline of theory as may suffice to render intelligible the various methods of magnetic measurement, and clearly define the quantities which are determined by these methods.

It can be shown that magnetic phenomena are capable of being accounted for by supposing the magnetized body or system to be the seat of a distribution of what Hypothesis of Imaginary Magnetic Matter.

Hypothesis of Imaginary Magnetic Matter. Sir William Thomson has called imaginary magnetic matter. This matter is of two kinds, each of which repels matter of its own kind, and attracts matter of the other kind. If two portions of this matter be supposed concentrated at points in a uniform medium, the force between them is directly as the product of the quantities, and inversely as the square of the distance between them. Both kinds of matter are always present in the distribution in equal amounts, but the distributions may be different in the two cases. It is to be carefully observed, however, that so far as our knowledge goes, no such matter exists. The hypothesis of its existence serves merely to fix the ideas, and afford to them a convenient, but only provisional, mode of expressing the polarity of a magnetized particle.

We shall, following the ordinary convention, call the magnetism of the same kind as that of the extremity of a magnet which points north positive, and the opposite kind negative. The positive direction of magnetic force will then be that in which a positive magnetic pole tends to move.

Unit Quantity of Magnetism or Unit Magnetic Pole. Unit quantity of this magnetic matter (or magnetism as we shall call it) is defined as that quantity which concentrated at a point, at unit distance from an equal quantity of the same kind, also concentrated at a point, is repelled with unit force, when the medium in which both quantities are placed is air. This definition of unit quantity of magnetism, or *unit magnetic **pole*** as it is sometimes called, is that on which the electromagnetic system of units is founded, and corresponds exactly to the definition of unit quantity of electricity given in page 4 of **Vol. I., which forms** the basis of the electrostatic system. **When the** distance between the points is **1** centimetre, and **the** quantities are such that the **force between** them **is 1** dyne, each quantity **is** 1 C.G.S. unit of magnetism, **or,** as it is sometimes put, is unit magnetic *pole* in **the** C.G.S. system of **units.**

If m denote a quantity of magnetism, which, placed **at a point** distant L **units from an equal quantity of the** same **kind, is** repelled **with a force of** F **units, we have** $m^2 = FL^2$, and therefore the **dimensional formula** * $[m]$ of quantity of magnetism is $[F^{\frac{1}{2}}L]$, or $[M^{\frac{1}{2}}L^{\frac{3}{2}}T^{-1}]$. **This** is the same dimensional formula as that of quantity **of** electricity in **the** electrostatic system. This dimensional formula, **and others which** follow, are to be taken **as** provisional. We shall **find** reason hereafter to introduce the dimensions of **magnetic inductive** capacity into **the** formula.

<small>Dimensional Formula of Magnetic Pole.</small>

The poles referred to in **this** definition **are purely ideal, for we** cannot **isolate a** quantity of either kind of magnetism **from** the opposite kind; **but** we can by proper arrangements obtain an approximate realization of the **definition.** Suppose we **have** two long, very thin, straight steel bars, which are uniformly and longitudinally magnetized; they may **be** taken as having poles at their extremities; in fact, the distribution of magnetism in **them is such** that the **magnetic effect** of either bar, at all points external to its **own** substance, would **be** perfectly represented **by a certain** quantity **of one** kind of magnetism **placed at** one extremity **of the bar,** and an

<small>Approximate Realization of Unit Magnetic Pole.</small>

---

* Vol. I. p. 191.

equal quantity of the **opposite kind of** magnetism placed **at the** other extremity. We may imagine, then, these two **bars** placed with their lengths in one line, and like poles turned towards one another, and at unit distance apart. If their lengths be very great compared **with this** unit distance, say 100 or 1000 times as **great, the** farther poles will have no effect on the others comparable **with the** repulsive action of these on **one** another. But there will be an inductive action between the two near poles which will tend to diminish their mutual repulsive force, and this **we** cannot in practice get **rid of.** The magnitude of this inductive effect is, however, **less** for hard steel than **for soft** steel, and we **may** therefore imagine the steel **of the** magnets such **that** the action of one on the other does not appreciably affect the distribution of magnetism in either. If, then, two **equal** like poles repel one another with unit force, **each,** according to the definition, has unit strength.

Magnetic Field.   The whole space surrounding a distribution of magnetism is called **the** magnetic field of the distribution, and the intensity of the field at any point is measured by the force which unit quantity of magnetism, or unit pole, would experience if placed at the point. A magnetic field intensity is therefore **a** directed quantity. If its value at a point P in the field be **H**, the force F on a quantity m of magnetism placed at P will be m**H**. Hence the dimensional formula [**H**] of **H** is $[F/m]$ or $[M^{\frac{1}{2}}L^{-\frac{1}{2}}T^{-1}]$.

Uniform Field.   If **H** be the **same in** magnitude (and therefore also in direction) **at each point in** the field, the field is said to be uniform. Since there is as much magnetism of one kind in a magnetic distribution as of the other kind, a magnetized body, placed in a uniform field, will, if not in equilibrium, experience only a couple, and will, if not resisted, turn round until a certain determinate direction in the magnet is parallel to the direction of the magnetic force in the field. This direction in the body is called the magnetic axis.

For example, if a magnet be suspended so as to be free from the action of all except the magnetic force of the earth, it is found to experience no sensible force of translation as a whole, but takes up a position of directional equilibrium; that is, there is a direction round which if the magnet be turned through any angle it remains in equilibrium in the new position. This direction is that of the magnetic axis of the magnet. *Magnetic Axis of a Magnet.*

The magnet is also in equilibrium if turned through 180° round an axis at right angles to the magnetic axis. Any angular displacement of the magnet not compounded of the two which have just been specified will leave it under the influence of a couple the moment of which depends on (1) the magnet itself, (2) the angle which the new direction of the magnetic axis makes with its direction of stable equilibrium, (3) the intensity of the magnetic field.

In general, for a magnet placed in a uniform magnetic field of intensity **H** so that its axis makes an angle θ with its position of stable equilibrium, that is with the direction of the force, the moment of the couple is **MH** sin θ where **M** is a quantity depending on the magnet, and called its *magnetic moment*. *Magnetic Moment of a Magnet.*

If we assume that the magnet has zero potential energy when its axis is at right angles to the lines of force, its potential energy E in the given position is plainly given by the equation

$$E = \int_{\pi/2}^{\theta} \mathbf{MH} \sin \theta d\theta = - \mathbf{MH} \cos \theta \quad \ldots \quad (1)$$

We shall see below (p. 22) that this is the value of the work done in bringing any magnet into a uniform field, and placing it with its axis inclined at an angle θ to its position of stable equilibrium. For certain simple cases such as symmetrical bar-magnets, &c., it is clear that this is the physical meaning of the potential energy defined with reference to the position of zero potential energy above chosen.

If the components of the magnetic force $\mathbf{H}$ referred to three rectangular axes, one, say that of x, drawn in the true north direction, another, that of y, drawn east, and the third, that of z, drawn downwards, be a, β, γ respectively, and the direction cosines of the magnetic axis referred to the same axes be l, m, n, the equation for E becomes

$$E = - \mathbf{M} (la + m\beta + n\gamma) \quad \ldots \quad (2)$$

For the moment K of the couple tending to bring the magnetic axis into coincidence with the direction of the resultant force we have

$$K = \mathbf{M}\{(m\gamma - n\beta)^2 + (na - l\gamma)^2 + (l\beta - ma)^2\}^{\frac{1}{2}} \quad \ldots \quad (3)$$

The component N of this couple round the axis of z is given by

$$N = \mathbf{M} (l\beta - ma) \quad \ldots \quad (4)$$

If the angle which the magnetic force makes with a horizontal plane, or its *dip*, be ζ, and the angle between a north and south vertical plane, and a vertical plane through the direction of the magnetic force, or the azimuth of the latter plane, be ϕ, and the corresponding angles for the magnetic axis be η and ψ, we have plainly

$$a = \mathbf{H} \cos \zeta \cos \phi, \; \beta = \mathbf{H} \cos \zeta \sin \phi, \; \gamma = \mathbf{H} \sin \zeta,$$
$$l = \cos \eta \cos \psi, \; m = \cos \eta \sin \psi, \; n = \sin \eta.$$

The preceding equations become

$$E = - \mathbf{MH}\{\cos \zeta \cos \eta \cos (\phi - \psi) + \sin \zeta \sin \eta\} \quad . \quad (5)$$

$$N = \mathbf{MH} \cos \zeta \cos \eta \sin (\phi - \psi) \quad \ldots \quad (6)$$

Uniform magnetization has been referred to in p. 3 above, and we shall now consider it a little more fully. A uniformly magnetized magnetic filament is an infinitely thin bar (not necessarily straight nor of uniform cross-section), so magnetized that its action at any external point can be represented by a certain quantity of one kind of magnetism concentrated at one extremity of the bar, and an equal quantity of the opposite magnetism concentrated at the other extremity. Such a filament, if divided across, would be converted into two uniformly magnetized filaments, and each of these in turn into two such filaments if divided, and so on. In short, each small element of the filament is to be supposed magnetized in the same way as the whole bar, so that, when the elements are united, the action of the polarity of any end of an internal element is annulled by the equal and opposite action of that end of the next element which is in contact with it. Thus the equal and opposite polarities of the ends of the complete filament are left unbalanced.

Uniform Magnetization.

Magnetic Filament.

We may suppose, to make this clearer, that each small element of the filament has equal and opposite distributions of magnetic matter over its two ends, so that the total quantity on two end faces in contact is zero. Of course this is only a way of figuring the distribution to the mind; what we really have is no doubt something very different from an actual distribution of matter.

Distribution of Magnetic Matter equivalent to Uniform Filament.

Any uniformly magnetized bar may be supposed made up of uniformly magnetized filaments put together with their ends in the surface of the bar. We

Uniform Magnet.

8   MAGNETIC THEORY

have in this case a surface distribution of magnetism only.

Non-Uniform Magnet. A non-uniformly magnetized bar may be regarded as one in which the polarities of the elements in contact do not counteract one another; in this case we have, besides the end distributions (which are generally opposite but not necessarily equal), a diffused distribution of magnetism throughout the substance of the bar.

Potential and Force due to small Magnet. This subject becomes much more intelligible when considered mathematically. We shall investigate first the potential and force due to an infinitely short and uniformly magnetized filament, and then consider the general case of a magnet made up of such elements. The magnetic filament is its own magnetic axis, and its magnetic action may be supposed due to equal and opposite quantities of magnetism placed at its two extremities. For brevity we shall call this elementary magnet in what follows a *magnetic doublet*. Its magnetic moment we define as the product of either of these quantities of magnetism into the distance between the extremities, and for our present purpose we shall suppose this product finite. Denoting by δx the length of the filament, which we take in the plane of the paper and parallel to the axis of x, with its centre at the origin of coordinates, we have for the coordinates of its extremities $-\frac{1}{2}\delta x, \frac{1}{2}\delta x$. The potential at a point in the plane of the paper the coordinates of which are ξ, η, due to unit quantity of positive magnetism at the origin, is $(\xi^2 + \eta^2)^{-\frac{1}{2}}$. Hence if m be the moment of the short magnet, and the positive magnetism correspond to the point $\frac{1}{2}\delta x$, the potential V of the two equivalent point distributions is given by

$$V = \frac{m}{\delta x}\left\{\frac{1}{\{(\xi - \frac{1}{2}\delta x)^2 + \eta^2\}^{\frac{1}{2}}} - \frac{1}{\{(\xi + \frac{1}{2}\delta x)^2 + \eta^2\}^{\frac{1}{2}}}\right\}$$

$$= \frac{m\xi}{(\xi^2 + \eta^2)^{\frac{3}{2}}} \quad \ldots \ldots \quad (7)$$

This may be written in either of two other equivalent forms, viz.:—

$$V = -m\frac{d}{d\xi}\frac{1}{(\xi^2 + \eta^2)^{\frac{1}{2}}} = \frac{m\cos\theta}{r^2} \quad \ldots \quad (8)$$

where θ is the angle between the axis of the magnet and the line drawn from the centre to the point (ξ, η) and r is the length of that line.

The components X, Y, of magnetic force at the point ξ, η, are given by differentiation of (7), and are

$$\left. \begin{array}{l} X = -\dfrac{dV}{d\xi} = -\dfrac{m(\eta^2 - 2\xi^2)}{(\xi^2 + \eta^2)^{\frac{5}{2}}} \\ Y = -\dfrac{dV}{d\eta} = \dfrac{3m\xi\eta}{(\xi^2 + \eta^2)^{\frac{5}{2}}} \end{array} \right\} \quad \ldots \ldots (9)$$

It is easy to verify that these values of X, Y satisfy the differential equation

$$\frac{dX}{d\xi} + \frac{dY}{d\eta} + \frac{Y}{\eta} = 0 \quad \ldots \ldots (10)$$

which is the well-known form which Laplace's equation [(11) Vol. I. p. 10] takes in the case of a force system symmetrical round an axis. It is to be noted that in (10) the coordinate η is the distance of the point considered from the axis of symmetry taken as axis of x, and that therefore Y in (10) above represents $(Y^2 + Z^2)^{\frac{1}{2}}$, where Y, Z, are taken as the component forces along two other axes at right angles to one another and to that of x.

To find the equation of the lines of force we have [(3 *bis*) Vol. I. p. 7] for any one line $X/d\xi = Y/d\eta$. Hence by (9) the differential equation in its simplest form is

$$3\xi\eta \cdot d\xi + (\eta^2 - 2\xi^2)\, d\eta = 0 \quad \ldots \ldots (11)$$

Equation of Lines of Force due to a small Magnet.

This equation may be integrated either by the ordinary method of separation of the variables, or by restoring the omitted common factor $1/r^5$, and remembering that by (10), η is an integrating factor of the equation thus modified. The integral is

$$\frac{\eta^2}{(\xi^2 + \eta^2)^{\frac{3}{2}}} = \frac{1}{c} \quad \ldots \ldots (12)$$

in which c is a parameter constant for any one line, but variable from one line to another.

This equation may obviously be written in the form

$$r = c \sin^2\theta \quad \ldots \ldots (13)$$

Graphical Description of Lines of Force.

which is very convenient for the graphical description of the curves. For let O, Fig. 1, be the position of the small magnet,

MAGNETIC THEORY

Graphical Description of Lines of Force. OX the direction of its axis, OY an axis in the plane of the paper at right angles to OX. From O as centre and with c as radius describe a semicircle upon the axis of x. Then draw any line OA intersecting the semicircle in A. From A let fall a perpendicular on OY meeting it in B, and from B a perpendicular to OA intersecting it in P. P is a point on the line of force whose parameter is the value of c chosen. The oval curve in Fig. 1 represents a complete line of force successive points on which were found in

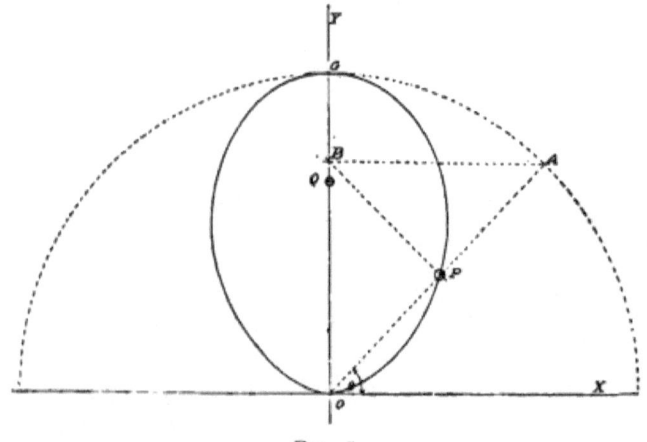

Fig. 1.

this way. It will be seen that the curve, as is also evident from its equation, is symmetrical about its maximum radius vector, OG, which lies along OY, and is equal in length to c. Points on the curve near G cannot be found with accuracy by the method just described,* but this part of the curve can be filled in with sufficient accuracy, by drawing a circular arc from the centre of curvature for the point G. The radius of curvature for any point is easily found from (13) and is

$$c \sin \theta (\sin^2\theta + 4\cos^2\theta)^{3/2} / 3 (\sin^2\theta + 2\cos^2\theta).$$

For the point G this is $c/3$, and the centre is on OG.

---

* This elegant method of describing these curves is due to Mr. John Buchanan, B.Sc. *Nature*, vol. xxi. p. 371.

EQUIPOTENTIAL LINES FOR SMALL MAGNET

Fig. 2 shows **lines of force for** different values **of the** parameter c. The points F_1, F_2 &c. are the points **of maximum** radius of curvature for the several curves.

The direction of the magnetic **force at** any point may easily be **obtained in the** following manner. **If** ϕ be the angle between the radius vector and the tangent **to the curve** at P (Fig. 1) we have $\tan \phi = rd\theta/dr = \frac{1}{2} \tan \theta$, by (13). Hence the following **construction**. Draw from the point of trisection of OP nearest O a perpendicular **to** OP; then if M be **the** point in which **this** perpendicular cuts **the** axis of the **magnet** PM is the direction **of** the line of force at P.

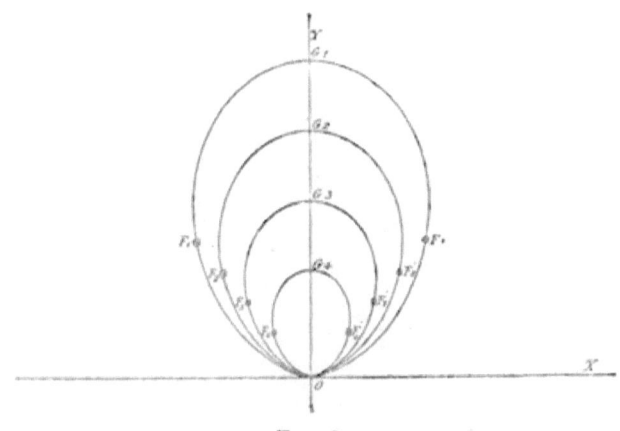

Fig. 2.

This construction gives also the magnitude **of the force at** P, for by (9) we get $X^2 + Y^2 = m^2(4\xi^2 + \eta^2)/(\xi^2 + \eta^2)^4$, and this is easily proved to be $m^2 \cdot PM^2/(OM \cdot OP^3)^2$. Hence **the magnitude** of the force is $m \cdot PM/(OM \cdot OP^3)$.

The equipotential curves **in the** plane of the paper **are obtained** by putting V = const. in (7) or (8). It is easy to **verify by (9) and** (12) that these curves cut the lines of force, as they **ought, at right angles**. They may be constructed graphically in **the following** manner. Draw with $\frac{1}{2}\sqrt{m/V}$ as radius, from a centre on the **axis of** x, a circle (Fig. 3) passing through the position, O, of the **centre of the magnet**. Then draw any line from O to meet this circle in A. The length of this line is $\sqrt{m/V} \cdot \cos\theta$ if θ be the angle which OA makes **with** OX. **Lay** off a distance OB, along the

Description of Equipotential Lines for small Magnet.

Description of Equipotential Lines for small Magnet.

axis of x equal to OA, and on the other segment of the diameter describe a semicircle and draw to it a tangent from O. The length of this tangent is the length of the radius vector r, which

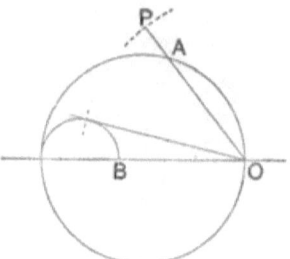

Fig. 3.

laid off from O along OA will give P a point on the curve. Or the construction may sometimes be more conveniently performed as follows: Lay off the length $OB = OA$ as in Fig. 4,

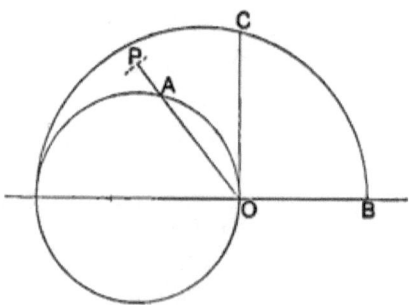

Fig. 4.

and describe a circle on the line made up of OB and the diameter of the former circle. The length of the tangent OC gives the distance OP. The curves, like the lines of force, are symmetrical about the axes of x and y and all pass through the origin. Both sets of curves are shown in Fig. 5.

The lines of force and equipotential surfaces due to a small magnet coincide for all external points with those of a uniformly magnetized sphere, a case approximately realized when a ball of iron is placed in a uniform magnetic field, and also with those of a conducting or dielectric sphere placed without charge in a uniform field of electric force. [See Vol. I. p. 123.] They further correspond exactly to the lines of flow and equipotential

Analogues of Lines of Force and Equipotential Surfaces of small Magnet.

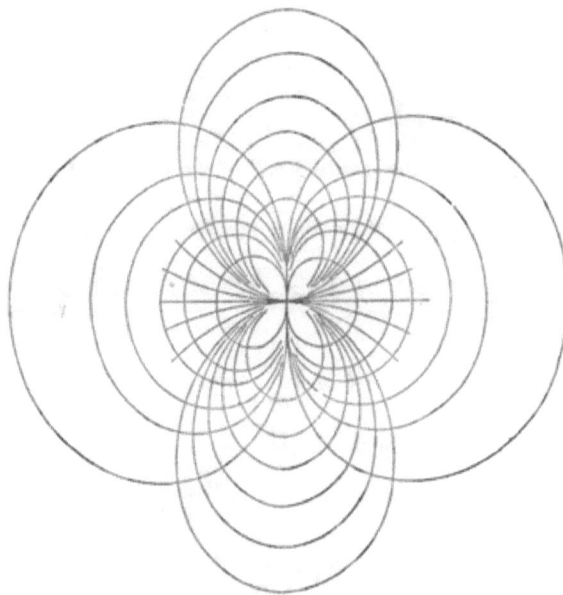

Fig. 5.

surfaces within a large mass of a frictionless incompressible fluid, kept flowing continuously in steady motion through an infinitely short, straight, narrow tube. We have discussed them with so great fulness, on account of their theoretical importance. We shall consider them again later in connection with inductive magnetism, and with the investigations of Hertz on the radiation of electric and magnetic energy.

Let us now consider a magnetic filament regarded as made up of an infinite number of infinitely short magnets placed end to end. Let x, y, z be the coordinates of the centre of one of these

Magnetic Filament.

Magnetic Filament. elementary magnets, ds its length, dm its moment, λ, μ, ν the cosines of the angles which the axis of the element measured in the direction along the filament, from the negative extremity to the positive, makes with the axes: the potential dV at a point $(\xi, \eta, \zeta,)$ external to the filament, produced by the element, is by (8) given by

$$dV = \frac{dm}{r^3}\{\lambda(\xi - x) + \mu(\eta - y) + \nu(\zeta - z)\} \quad . \quad . \quad (14)$$

since $\{\lambda(\xi - x) + \mu(\eta - y) + \nu(\zeta - z)\}/r$ is now the value of $\cos \theta$. But if I denote the magnetic moment of the element per unit of length, taken positive when the direction of the axis, as specified above, is from the negative end of the element to the positive, or, which is the same, if I denote the quantity of magnetic matter on the positive end of the axis, we have

$$dV = \frac{I}{r^3}\{(\xi - x) dx + (\eta - y) dy + (\zeta - z)dz\} \quad . \quad (15)$$

where dx, dy, dz, are the projections of ds on the axes. But since $r^2 = (\xi - x)^2 + (\eta - y)^2 + (\zeta - z)^2$ we can write this equation in the form

$$dV = I \frac{d}{dr}\frac{1}{r} dr.$$

Hence integrating by parts we get

$$V = \frac{I_2}{r_2} - \frac{I_1}{r_1} - \int \frac{1}{r}\frac{dI}{ds} ds . \quad . \quad . \quad . \quad (16)$$

where I_2 I_1 r_2, r_1 are the values of I and r at the positive and negative ends respectively. If I be uniform along the filament the last term vanishes, and (16) becomes

$$V = I\left(\frac{1}{r_2} - \frac{1}{r_1}\right). \quad . \quad . \quad . \quad . \quad (17)$$

or the potential is that due to the two end distributions alone, as stated above, p. 7.

Potential of a Magnetic Filament. Since the potential of a quantity of magnetism $dI/ds \cdot ds$ at the distance r is $dI/ds \cdot ds/r$ the interpretation of the third term in (16) is that $-dI/ds$, if not zero, is the linear density of magnetism diffused throughout the filament. Hence in the general case the total potential is that due to the end distributions together with that produced by the diffused distribution.

LINES OF FORCE OF A MAGNETIC FILAMENT

The equation of the lines of force due to a uniformly magnetized filament is of interest, and may be easily found in a variety of ways. The most elegant is perhaps the following. It is evident that the system of lines is symmetrical about the straight line joining the ends A, B, of the filament. Describe circles from A, B (Fig. 6), as centres with any radii the sum of which is

Equation of Lines of Force of a Magnetic Filament.

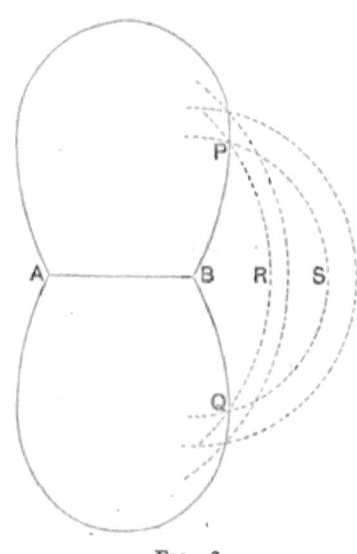

FIG. 6.

greater than the distance A, B. They will intersect in two points which will be points on two lines of force having the same parameter, but on opposite sides of the axis. The circles may be regarded as the intersection with the plane of the paper of two spheres having A, B, as centres, and intersecting in a circle through which pass all lines having a certain parameter. Now considering the total flux of magnetic force (that is, the surface integral of normal magnetic force) in the same direction through any surface having this circle as boundary, and the two centres on the same side of it, it is clear that it may be taken as that due to the quantity of magnetism $-I$ at A, outwards through the segment PRQ of the sphere described from A, and bounded by the circle of intersection, together with that due to $+I$ at B taken outwards through the corresponding segment PSQ of the other sphere. If the angles PRQ, PSQ be respectively $2\theta_1$, $2\theta_2$, these fluxes are respectively $-2\pi I(1-\cos\theta_1)$ and $2\pi I(1-\cos\theta_2)$. Hence the total flux is $2\pi I(\cos\theta_1 - \cos\theta_2)$. Now let two other spheres be described in the same way; then if the flux through a corresponding surface bounded by the circle of intersection is the same as that just found, the two circles of intersection may be supposed joined by a surface generated by the revolution of a line of force round AB as an axis. Hence the equation of a line of force is

$$\cos\theta_1 - \cos\theta_2 = c \quad \ldots \quad (18)$$

where c is a parameter varying from one curve to another.*

Description of Lines of Force for Uniform Filament. To construct the lines of force in this case we may proceed as follows:—Describe a circle on AB (Fig. 6 †) as diameter, and lay off a distance AM such that $AM = c \cdot AB$. Then draw any line from A to cut the

---

* See also Chap. III. below.

† This Fig. is taken by permission from *Constructive Geometry of Plane Curves*, by T. H. Eagles, M.A. (London, Macmillan & Co.). The method of construction here adopted is that given in the same work.

DESCRIPTION OF MAGNETIC CURVES

circle in Q, and lay off Aq along AB equal to AQ. From B as centre with radius Mq describe a circle cutting the former circle in R. Hence since $\cos BAP + \cos ABR = AQ/AB + BR/AB = (Aq + qM)/AB = c$,

Description of Lines of Force for Uniform Filament.

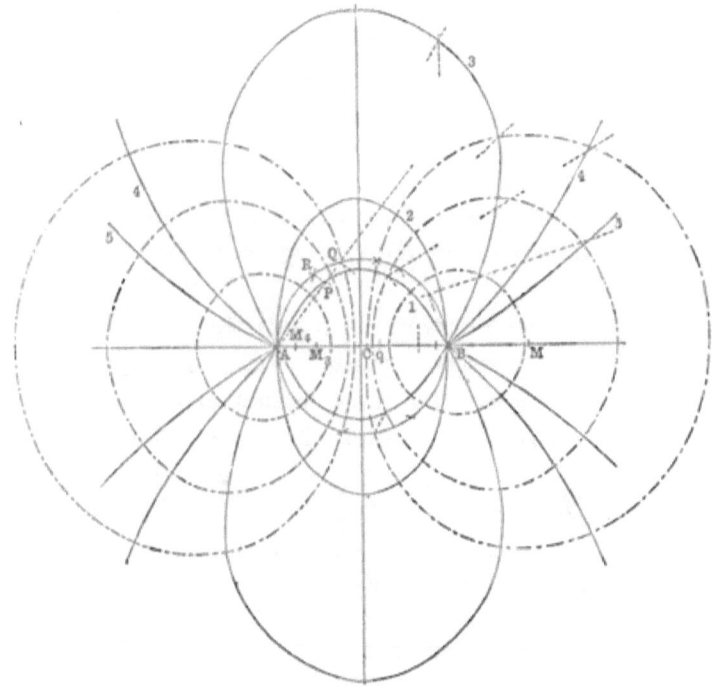

Fig. 6*.

the point in which AQ and BR intersect is a point on the curve. The curve in the vicinity of A or B must be drawn from a knowledge of its inclination θ to the axis of x. This is given by the equation $\cos\theta = c - 1$.

The cut shows curves numbered 1, 2, 3, 4, 5, drawn for the corresponding values of c, $\frac{3}{2}$, 1, $\frac{1}{2}$, $\frac{1}{4}$, $\frac{1}{8}$. When $c = 2$, AB (the axis) is the curve, when $c = 0$, the productions of the axis to the right of B and the left of A are the curves.

Description of Equipotential Curves for Uniform Filament.

The dotted curves intersecting at right angles the lines of force in Fig. 6* are the equipotential curves, which are given by the equation

$$\frac{1}{r} - \frac{1}{r'} = c \quad \ldots \ldots \quad (19)$$

where c is a parameter (the potential per unit of magnetic matter at A or B) varying from one curve to another.

That the dotted curves are intersected at right angles by the lines of force is easily verified by considering that if $f(r, r') = 0$ be the equation of a curve, and lengths df/dr, df/dr' be laid off along r and r', the resultant of these lines is in the direction of the normal. We have from (19) $df/dr = -1/r^2$, $df/dr' = 1/r'^2$. Hence laying off $1/r^2$ from a point on the curve along r towards A, and $1/r'^2$ from the same point along r' in the direction from B, we find that the normal to the equipotential curve is in the direction of the resultant force due to the equal quantities of opposite kinds of magnetic matter at A and B respectively.

Magnetic Potential.

We shall now find the potential at any point in a magnetic field produced by a body magnetized in any given manner. As we shall see later, we are led by magnetic phenomena to suppose a magnetized body made up of an infinitely large number of infinitely small magnetized molecules, each of which may be considered a magnetic doublet, as defined above. We shall also suppose that the magnetic axes of these molecules have in each small element of the body a common direction, of which the cosines for a given element are λ, μ, ν, and which varies from point to point continuously in the body. This is called the direction of magnetization at the element.

We consider an element of the body, in shape a rectangular parallelepiped with its edges parallel to the axes, large enough to contain a very great number of molecules, but not so large that the direction of magnetization varies in it to a sensible extent. Let n be the number of molecules in the element, m their average magnetic moment, then the magnetic moment of the element is nm, and it may be regarded as a small magnet of this moment, with its centre at the point x, y, z, and its axis in the direction λ, μ, ν. For nm we shall write $I dx dy dz$, where I is the magnetic moment of the element per unit of volume, or, as it is usually called, the intensity of magnetization at the element, and $dx dy dz$ is the volume of the element.

Definition of Intensity of Magnetization.

By (14) above we have for the magnetic potential produced by the element at the point (ξ, η, ζ) the expression $I dx dy dz \{(\xi - x)\lambda + (\eta - y)\mu + (\zeta - z)\nu\}/r^3$, where $r^2 = (\xi - x)^2 + (\eta - y)^2 + (\zeta - z)^2$. Writing $I\lambda$, $I\mu$, $I\nu = A, B, C$, so that A, B, C, are what are called the components of magnetization at the point x, y, z, and integrating throughout the body we get for the total potential V at (ξ, η, ζ) the equation :

Calculation of Magnetic Potential due to Given Magnetic Distribution.

$$V = \iiint \frac{1}{r^3} \{A(\xi - x) + B(\eta - y) + C(\zeta - z)\} dx dy dz . \quad (20)$$

This may be written in the form :

$$V = \iint \frac{A}{r} dy dz + \iint \frac{B}{r} dz dx + \iint \frac{C}{r} dx dy$$
$$- \iiint \frac{1}{r} \left(\frac{dA}{dx} + \frac{dB}{dy} + \frac{dC}{dz}\right) dx dy dz . \quad (21)$$

in which the first three integrals are confined to the surface and are reckoned in the following manner. Taking the first of the three, conceive a prism of cross-section $dy dz$, and length parallel to the axis of x, drawn in the body. The area $dy dz$ is the projection at right angles to the axis of x of the element dS of the surface intercepted at either end by the prism; and the element of the integral corresponding to the negative or left-hand end of the prism is to be taken negative, the element for the other end positive. Now if l_1, l_2, be the x direction cosines of the normals drawn outwards from the surface elements at these ends respectively, dS_1, dS_2 the corresponding areas, we have $dydz = l_2 dS_2 = -l_1 dS_1$; so that the elements of the integral are $A_2 l_2 dS_2/r_2 + A_1 l_1 dS_1/r_1$. Hence we may write the first integral in the form $\iint Al/r \, dS$, in which the integration is to be extended over the whole surface. The other double integrals may be similarly transformed, and we get

$$V = \iint \frac{1}{r}(Al + Bm + Cn)\,dS$$

$$- \iiint \frac{1}{r}\left(\frac{dA}{dx} + \frac{dB}{dy} + \frac{dC}{dz}\right)dx\,dy\,dz \quad \ldots \quad (22)$$

in which l, m, n are the direction cosines of the normal to an element dS of the surface of the body.

Interpretation of Expression for Magnetic Potential. Clearly we may interpret the quantity $Al + Bm + Cn$ as a surface density σ of magnetic distribution, equal at each surface element to the normal component of intensity of magnetization.

The expression $-(dA/dx + dB/dy + dC/dz)$ is interpretable in the same way as the volume density ρ of a distribution of magnetism throughout the substance of the body.

From these expressions we get by direct integration, as we clearly ought, the total magnetism of the body equal to zero.

It is almost needless to say that these results are consequences of our suppositions as to the structure of the magnetized body, and that the interpretations just stated are to be regarded merely as convenient modes of expressing the outcome of the analysis. If, however, as seems certain, the magnetized body be made up of polarized molecules of some kind, the surface and body distributions found, will correspond to unbalanced surface and body polarities respectively.

Solenoidal Distribution of Magnetism defined. If the potential at (ξ, η, ζ) due to the body is expressed by the surface integral alone, then

$$\frac{dA}{dx} + \frac{dB}{dy} + \frac{dC}{dz} = 0 \quad \ldots \quad (23)$$

A distribution of magnetism fulfilling this condition is said to be *solenoidal*.

Potential Energy of a Magnet. We have now to consider the potential energy of a magnet situated in a magnetic field. By this we mean the work which has been done against magnetic forces, in bringing the magnet into the field and placing it in the given position. The potential energy of unit quantity of negative magnetism at a point P, at

POTENTIAL ENERGY OF MAGNET

which the potential is V, is of course simply $-V$; and hence that of a unit of positive magnetism at a point at distance ds from this point is $V + dV/ds \cdot ds$. The potential energy of a magnetic doublet with its extremities at these points is therefore $m\,dV/ds$, where m is the moment of the doublet. If the direction cosines of the axis of the doublet be λ, μ, ν, we have of course

Potential Energy of a Magnetic Doublet.

$$m\frac{dV}{ds} = m\left(\lambda \frac{dV}{dx} + \mu \frac{dV}{dy} + \nu \frac{dV}{dz}\right) \quad \ldots \quad (24)$$

Now, as above (p. 18), we may regard this small magnet as a magnetic molecule of a body of finite size, and take a parallelepiped of the body, large enough to contain a great number of such molecules, but not so large that the direction of magnetization, that is the common direction of the axes of the molecules, varies to a sensible extent. The potential energy of the element will be proportional to the number of such molecules contained in the element. Hence by the expression above, the potential energy dE of an element of volume $dx\,dy\,dz$, is given by

Calculation of Potential Energy of Finite Magnet.

$$dE = \mathbf{I}\left(\lambda \frac{dV}{dx} + \mu \frac{dV}{dy} + \nu \frac{dV}{dz}\right) dx\,dy\,dz,$$

where $\mathbf{I}$ denotes the intensity of magnetization as defined above. Writing as before $A, B, C = \lambda\mathbf{I}, \mu\mathbf{I}, \nu\mathbf{I}$, and integrating throughout the magnet we get

$$E = \iiint \left(A \frac{dV}{dx} + B \frac{dV}{dy} + C \frac{dV}{dz}\right) dx\,dy\,dz \quad \ldots \quad (25)$$

Integrated by parts this becomes

$$E = \iint V(A\,dy\,dz + B\,dz\,dx + C\,dx\,dy)$$
$$- \iiint V\left(\frac{dA}{dx} + \frac{dB}{dy} + \frac{dC}{dz}\right) dx\,dy\,dz. \quad \ldots \quad (26)$$

The triple integration is taken throughout the space occupied by the magnet; the double integrations give, when l, m, n, are put for the direction cosines of the normal to an element dS of the surface, an integration over the whole surface of the magnet, so that

$$E = \iint V(lA + mB + nC)\,dS$$
$$- \iiint V\left(\frac{dA}{dx} + \frac{dB}{dy} + \frac{dC}{dz}\right) dx\,dy\,dz \quad \ldots \quad (27)$$

By the interpretations, stated on p. 20 above, of the quantities in brackets, this may be written

$$E = \iint V\sigma dS + \iiint V\rho\, dxdydz \quad \ldots \ldots \ldots (28)$$

which is the energy equation (31) given on p. 32 of Vol. I., except that the factor $\frac{1}{2}$ in that equation does not appear, since the field in the present case is independent of the distribution brought into it.

Potential Energy in terms of Components of Field and Magnetization Intensities.

It is to be noted that V is the potential due to the magnetic system producing the field, and that therefore $-dV/dx$, &c, are the components a, β, γ, parallel to the axes of the magnetic field intensity due to this system. Hence we may write (25) in the form

$$E = -\iiint (Aa + B\beta + C\gamma)\, dxdydz \quad \ldots \ldots (29)$$

In the case of a uniform field intensity for every part of the magnet this becomes

$$E = -(a\mathbf{M}_1 + \beta\mathbf{M}_2 + \gamma\mathbf{M}_3). \quad \ldots \ldots (30)$$

where $\mathbf{M}_1, \mathbf{M}_2, \mathbf{M}_3$, denote the integrals $\iiint A\,dxdydz$, &c. Now we can find three quantities p, q, r, fulfilling the equation

$$p^2 + q^2 + r^2 = 1,$$

such that

$$\mathbf{M}_1 = p\mathbf{M}, \quad \mathbf{M}_2 = q\mathbf{M}, \quad \mathbf{M}_3 = r\mathbf{M},$$

so that we have

$$E = -\mathbf{M}(pa + q\beta + r\gamma) \quad \ldots \ldots (31)$$

Potential Energy in terms of Magnetic Moment.

$\mathbf{M}$ is what has been defined above (p. 5) as the magnetic moment of the magnet and p, q, r, are the direction cosines of its axis. If $\mathbf{H}$ be the resultant magnetic intensity of the field, its direction cosines are $a/\mathbf{H}, \beta/\mathbf{H}, \gamma/\mathbf{H}$, and (31) may be written

$$E = -\mathbf{MH}\left(p\frac{a}{\mathbf{H}} + q\frac{\beta}{\mathbf{H}} + r\frac{\gamma}{\mathbf{H}}\right) = -\mathbf{MH}\cos\theta \quad . \quad (32)$$

the equation already obtained (p. 6 above).

POTENTIAL ENERGY OF MAGNET

It is instructive and useful to consider as a particular case the potential energy of a magnet in the field due to a single magnetic pole, as this gives the potential of the magnet at the point at which the pole is situated, and supplies conditions by which the centre and axes of the magnet may be determined. The following treatment of the problem is due to Sir William Thomson.* In this case if PP' be the distance of the point P, (ξ, η, ζ), at which the pole is situated, from the point P', (x, y, z), of the magnet, we have for the potential V at P', due to the unit pole at P the value $1/PP'$.

Potential Energy of a Magnet in presence of a Unit Magnetic Pole.

$$V = \frac{1}{PP'} = \frac{1}{r}\left(Z_0 + Z_1\frac{r'}{r} + Z_2\frac{r'^2}{r^2} + \&c.\right) \quad \ldots \quad (33)$$

where r, r' ($r' < r$) are the distances OP, OP' from the origin of coordinates to the points P, P', and Z_0, Z_1, Z_2, &c. are zonal surface harmonics † of the orders specified by their suffixes, and having their pole at P. If μ be the cosine of the angle POP', then

Expansion in Spherical Harmonics.

$$Z_0 = 1, \quad Z_1 = \mu, \quad Z_2 = \tfrac{1}{2}(3\mu^2 - 1), \quad Z_3 = \tfrac{1}{2}(5\mu^2 - 3\mu), \&c.$$

Substituting these values of Z_0, Z_1, &c. in (33), then putting for μ its value $(\xi x + \eta y + \zeta z)/rr'$, and differentiating we evaluate dV/dx, dV/dy, dV/dz. Using these in (25) and putting

$$P_1 = \iiint Ax\,dx\,dy\,dz, \quad P_2 = \iiint By\,dx\,dy\,dz, \quad P_3 = \iiint Cz\,dx\,dy\,dz$$

$$Q_1 = \iiint(Bz + Cy)dx\,dy\,dz, \quad Q_2 = \iiint(Cx + Az)dx\,dy\,dz,$$

$$Q_3 = \iiint(Ay + Bx)dx\,dy\,dz, \text{ we get}$$

$$E = \mathbf{M}(p\xi + q\eta + r\zeta)\frac{1}{r^3}$$
$$+ \{\xi^2(2P_1 - P_2 - P_3) + \eta^2(2P_2 - P_3 - P_1) + \zeta^2(2P_3 - P_1 - P_2)$$
$$+ 3(Q_1\eta\zeta + Q_2\zeta\xi + Q_3\xi\eta)\}\frac{1}{r^5} + \&c. \quad \ldots \quad (34)$$

---

* See Reprint of *Papers on Electrostatics and Magnetism*, 2nd Ed., p. 373.

† For the theory of Spherical Harmonics which we shall frequently have to employ in what follows, the student may consult Thomson and Tait's *Nat. Phil.* Vol. I. Part I., or Ferrers's *Spherical Harmonics*. A clear and brief account of the subject is given in Minchin's *Statics*, Vol. II. 3rd Edition. A short explanation, covering the theorems used in this work, is given in an Appendix to the present volume.

MAGNETIC THEORY

Simplification by Change of Origin.

The quantities P_1, P_2, &c. are functions of the coordinates x, y, z, and of A, B, C; hence we may change the origin to another point (x', y', z'), and take the direction of the axis of the magnet as that of x, so that $p = 1$, $q = 0$, $r = 0$. This makes

$$\iiint A dx dy dz = \mathbf{M}, \quad \iiint B dx dy dz = 0, \quad \iiint C dx dy dz = 0.$$

We have then new values P', &c. given by the equations

$$P'_1 = P_1 - \mathbf{M} x', \quad P'_2 = P_2, \quad P'_3 = P_3,$$
$$Q'_1 = Q_1, \quad Q'_2 = Q_2 - \mathbf{M} z', \quad Q'_3 = Q_3 - \mathbf{M} y'.$$

Hence, if the new origin be taken so that

$$x' = \frac{2 P_1 - P_2 - P_3}{2\mathbf{M}}, \quad y' = \frac{Q_3}{\mathbf{M}}, \quad z' = \frac{Q_2}{\mathbf{M}}.$$

we get $\quad P'_1 = \tfrac{1}{2}(P_2 + P_3), \quad Q'_2 = 0, \quad Q'_3 = 0,$

and (34) takes the simplified form (accents omitted)

$$E = \mathbf{M}\frac{\xi}{r^3} + \tfrac{3}{2}\frac{(P_2 - P_3)(\eta^2 - \zeta^2) + 2Q_1\eta\zeta}{r^5} + \&c. \quad . \quad (35)$$

in which ξ, η, ζ have of course the proper values for the new origin.

Centre and Principal Axis of Magnet. Secondary Axes of Magnet.

The origin thus found is called the centre of the magnet, and the definition enables us to specify the position of the magnetic axis, as well as its direction. The magnetic axis is sometimes called the principal axis of the magnet.

If we turn the axes of y and z round that of x, through the angle $\tfrac{1}{2}\tan^{-1} Q_1/(P_2 - P_3)$, (35) takes the form

$$E = \mathbf{M}\frac{\xi}{r^3} + \tfrac{3}{2}\frac{R(\eta^2 - \zeta^2)}{r^5} + \&c. . \quad . \quad . \quad . \quad (36)$$

where R is the quantity which replaces $P_2 - P_3$. These directions of the axes of y and z are called the secondary axes of the magnet.

In the case of symmetry round the axis of x, the second term of the expression on the right of (36) is zero, since then whatever magnetization at right angles to the axis there be throughout the body, it must be such that the coefficient R vanishes identically. To a close approximation therefore for a unit pole placed at a point (ξ, η, ζ), the distance r of which from

MAGNETIC INDUCTION

the origin is considerably greater than that of any part of the magnet from the same point, the mutual potential energy is $M\xi/r^3$. *Potential &c. at External Point due to Magnet.*

Since the potential energy is mutual, the equations (34), &c. found for E, give the potential energy of the unit pole in the field due to the magnet, that is the potential due to the magnet at the point (ξ, η, ζ).

Section II.

MAGNETIC INDUCTION. VECTOR POTENTIAL. MAGNETIC ENERGY.

When a substance capable of being magnetized is placed in a magnetic field, it becomes magnetic, and a definite relation in general exists between the magnetization produced at each part and the field intensity. *Induced Magnetization.* The determination of this relation has been, especially in late years, the subject of much careful investigation, and we shall return to it later. Here we deal mainly with the theory of certain given cases of magnetization. In general, in the substances with which we have to deal in practice, the magnetization is in the direction of the magnetic force, and we shall first consider this case.

We have first to define what we mean by the force in the interior of a magnet. Here, as in all other cases, the magnetic force is that which would be exerted on a unit magnetic pole if placed at the point, and, since we could make no experiment as to the internal state of the body, except within a cavity hollowed out within it, we imagine a small portion of the magnetized body excavated so as to give a space in which the force might be measured. The formation of this cavity leaves unbalanced the *Force in the Interior of a Magnetized Body.*

magnetism on the extremities of the molecules which abut against its surface. We shall suppose it formed without disturbing the magnetization of the rest of the body, and since we cannot divide a magnetic molecule the signs of the surface distributions will be perfectly definite. Thus for a crevasse * cut at right angles to the direction of magnetization there is positive magnetism on the face next the negative end of the magnet, and negative magnetism on the opposite face. On a surface the normal to which, drawn into the cavity, is inclined at an angle ϵ to the direction of the intensity of magnetization **I**, taken as positive when drawn in the magnet from the negative pole to the positive, the density of distribution is $\mathbf{I}\cos\epsilon$, and is positive therefore when ϵ is acute, and negative when ϵ is obtuse.

Surface Distribution in Interior of a Cavity.

The force within the cavity depends upon the shape and dimensions of the cavity, and upon the position of the pole within it. In the first place we shall consider a cylindrical cavity of finite length and diameter, cut with its axis in any given direction, in a uniformly magnetized body. If the intensity of magnetization of the body be **I**, and θ be the angle which the axis of the cylinder makes with the direction of **I**, we have for the density of the distribution on the curved surface of the cavity the value $\mathbf{I}\sin\theta$ at points in a plane through the axis parallel to the direction of magnetization. In another plane through this axis making an angle ϕ with the direction of magnetization the density is $\mathbf{I}\sin\theta\cos\phi$. Now the force which this distribution exerts at right angles to the axis on a pole placed at the centre of the axis, is, if $2l$ be the length of the cylinder, and $2r$ its diameter,

Force within Cavities of Different Shapes.
1. *Cylindrical Cavity inclined to Direction of Magnetization.*

$$2\mathbf{I}\sin\theta\int_{-\frac{\pi}{2}}^{\frac{\pi}{2}}\int_{-l}^{l}\cos^2\phi\,\frac{r^2}{(r^2+x^2)^{\frac{3}{2}}}d\phi dx = 2\pi\mathbf{I}\sin\theta\,\frac{l}{(r^2+l^2)^{\frac{1}{2}}}.$$

---

* A narrow cavity with parallel plane faces, every dimension of which is great in comparison with the width of the cavity.

The ends of the cylinder give a resultant force along the axis of amount $4\pi I \cos\theta \{1 - l/(r^2 + l^2)^{\frac{1}{2}}\}$; and the total force within the cylinder at the centre of the axis is the resultant of these two components. Hence if l be great in comparison with r, the force is at right angles to the axis and of amount $2\pi I \sin\theta$. Hence if $\theta = \pi/2$, the force is $2\pi I$. If l be small in comparison with r, the force is $4\pi I \cos\theta$.

Particular Cases: Axis of Cylinder parallel to Magnetization.

If $\theta = 0$, so that the axis of the cylinder is parallel to I, the force becomes $4\pi I\{1 - l/(r^2 + l^2)^{\frac{1}{2}}\}$, and is therefore $4\pi I$ or zero, according as l is small or great in comparison with r. Also* the force is $4\pi I$ in any narrow crevasse bounded by planes at right angles to I, and is plainly zero in any elongated narrow cavity with its length parallel to I.

Crevasse.

In the important case of a spherical hollow the surface distribution follows the law of variation from point to point of a material distribution formed by placing two spheres of equal uniform densities $+\rho$ and $-\rho$ in coincidence, and displacing the positive sphere in the direction of I through a small distance δx. We may suppose ρ very great, and δx very small, so that $\rho \delta x = I$. The potential due to the inner nucleus of the positive sphere at a point distant r from the centre is $\frac{4}{3}\pi I r^2/\delta x$. The potential due to the outer shell is $2\pi I(R^2 - r^2)/\delta x$. Hence the whole potential is $2\pi I(R^2 - \frac{1}{3}r^2)/\delta x$. The potential at the same point due to the negative sphere is plainly $-2\pi I(R^2 - \frac{1}{3}r^2)/\delta x + \frac{4}{3}\pi I r dr/dx$. Hence the total potential is $\frac{4}{3}\pi I x$. The force within the spherical hollow produced by the surface magnetization is therefore in the direction of magnetization, and equal to $\frac{4}{3}\pi I$.

2. Spherical Cavity.

In the case of a non-uniformly but continuously magnetized body these cavities have only to be taken small enough to enable the average value of I over each to be used in the values of the force.

The cases most important for our present purpose are (1) the comparatively long narrow cylinder, (2) the short comparatively wide cylinder, both with axes parallel to I. In each case the force within the hollow, due to the surface distribution upon it, must be increased by the resultant force at the point due to the distribution producing the magnetic field and to the rest of the

Magnetic Force (H), and Magnetic Induction (B).

---

* See Vol. I. p. 58.

magnetic distribution of the magnet. If we call this force **H**, the force within the cavity is in case (1) simply **H**, in case (2) it is $\mathbf{H} + 4\pi\mathbf{I}$. **H** is thus the magnetic force within the magnet, apart from any action of unbalanced polarity produced by cutting a hollow in the substance. The other quantity $\mathbf{H} + 4\pi\mathbf{I}$, is called the *magnetic induction* within the magnet. We shall denote it by **B**.

*[margin: Magnetic Force (**H**), and Magnetic Induction (**B**).]*

In the more general case in which **H** and **I** are not in the same direction the equation

$$\mathbf{B} = \mathbf{H} + 4\pi\mathbf{I} \quad \ldots \ldots \quad (37)$$

is true in the vector sense, that is **B** is the resultant of the vectors **H** and $4\pi\mathbf{I}$.

In the case in which the magnetization is induced by the magnetizing force **H**, and has the same direction, if we put $\mathbf{I} = \kappa\mathbf{H}$, we get

$$\mathbf{B} = (1 + 4\pi\kappa)\mathbf{H} \quad \ldots \ldots \quad (38)$$

[margin: Magnetic Inductive Capacity and Magnetic Permeability.]

The multiplier $1 + 4\pi\kappa$ is generally denoted by μ, and called the *magnetic inductive capacity*, or the *magnetic permeability*. The factor κ is called the *magnetic susceptibility*. In general, as we shall see below, it is a function of **H**.

It is clear that as here defined κ is a mere number. The quantity μ is also a mere number when defined by the equation $1 + 4\pi\kappa$. Now μ has a definite value for every medium, and it is possible that that property of the medium (say some form of motion), which makes the magnetic inductive capacity vary from medium to medium, may give to it certain dimensions at present unknown. We may use μ as the absolute magnetic

VARIATION OF **B** AND **H** AT SURFACE OF BODY

inductive capacity depending on this property, that is the **magnetic** inductive capacity with reference to an absolutely unmagnetizable medium as standard, and regard its dimensions as unknown; but we shall in the account of magnetic measurements which follows, in general employ it to denote $1 + 4\pi\kappa$.

Since **B**, **H**, and **I** are vectors in the general case we may replace each by three components along the axes. We have then instead of (37) the equation

$$\left.\begin{array}{l} a = \alpha + 4\pi A \\ b = \beta + 4\pi B \\ c = \gamma + 4\pi C \end{array}\right\} \quad \ldots \ldots \ldots \quad (39)$$

Components of Magnetic Induction.

where $a, b, c, \alpha, \beta, \gamma, A, B, C$, are the components of **B**, **H**, and **I** for the point considered.

It is easy to prove that the magnetic induction fulfils the solenoidal condition. We have from (39)

$$\frac{da}{dx} + \frac{db}{dy} + \frac{dc}{dz} = \frac{d\alpha}{dx} + \frac{d\beta}{dy} + \frac{d\gamma}{dz} + 4\pi\left(\frac{dA}{dx} + \frac{dB}{dy} + \frac{dC}{dz}\right) \quad (40)$$

Solenoidal Condition fulfilled by Magnetic Induction.

Now remembering that the quantity within the brackets may be regarded as a volume density $(-\rho)$ of magnetism, and that by the definition of **H** we must have by the characteristic equation * of electric and magnetic potential

$$\frac{d\alpha}{dx} + \frac{d\beta}{dy} + \frac{d\gamma}{dz} + 4\pi(-\rho) = 0,$$

and therefore

$$\frac{da}{dx} + \frac{db}{dy} + \frac{dc}{dz} = 0. \quad \ldots \ldots \quad (41)$$

In the space surrounding the magnetized body, **B** coincides with **H** in all respects. The transition in value from one side of the surface to the other, takes place differently in the two cases. The normal component of **B** varies continuously from one side of the surface to the other, the tangential component discontinuously; and the reverse is the case with the value of **H**. To

*Variation of **B** and **H** at Surface.*

---

* Poisson's Theorem, Vol. I. p. 13, Eq. (16).

Variation of B and H at Surface.

prove this we have only to notice that by (39) if θ be the angle between the normal to the surface drawn outwards, and the common direction of **B**, **H**, and **I**, we have for the normal component of **B** in the interior

$$\mathbf{B}\cos\theta = (\mathbf{H} + 4\pi\mathbf{I})\cos\theta \quad \ldots \ldots \quad (42)$$

and that if $\mathbf{H}'$ be the magnetic force just outside the surface at the same place, and θ' its inclination to the normal, the characteristic equation of the potential gives, since $\mathbf{I}\cos\theta$ is a surface density of magnetism

$$\mathbf{H}'\cos\theta' = \mathbf{H}\cos\theta + 4\pi\mathbf{I}\cos\theta \quad \ldots \ldots \quad (43)$$

Since **B** and **H** coincide outside the magnet the quantity on the left is the normal component of the magnetic induction. The expression on the right, therefore, shows the normal continuity of **B**, and at the same time the normal discontinuity of **H**.

The tangential component of **B** is $(\mathbf{H} + 4\pi\mathbf{I})\sin\theta$ inside the surface, and $\mathbf{H}\sin\theta$ outside the surface. The latter is the value of the tangential component of the magnetic force on both sides.

Integral of B over any closed surface is zero.

Since the magnetic induction fulfils the solenoidal condition, it follows that the surface integral of magnetic induction taken over any closed surface whatever, whether wholly within or wholly without, or partly within and partly without the magnetized body, is zero. This is clear from the equation (69) p. 66, Vol. I. which may be written for the present case.

$$\iint (la + mb + nc)\, dS = -\iiint \left(\frac{da}{dx} + \frac{db}{dy} + \frac{dc}{dz}\right) dx\, dy\, dz \quad \ldots \quad (44)$$

since the quantity on the right is zero identically.

The truth of equation (44) may be seen from the following considerations. The expression $(da/dx + db/dy + dc/dz)\, dx\, dy\, dz$ represents the sum, for a small rectangular parallelepiped of the substance having its edges parallel to the axes, of the products of the average value of the component of induction, at each surface of the element into the area of the face. The integral on the right of (44) simply expresses the aggregate value of these sums for such elements making up the portion of the body considered. Now clearly if we imagine the body divided into small elements, then each face of these will be common to two elements, except those faces which abut on the surface. For every common face the products of induction into area for the two elements are equal and opposite, and cancel one another. We are left then with the aggregate of the products for the faces at the surface, and it is clear by projection that the sum of the products of induction for these faces is

$$\iint (la + mb + nc)dS.$$

Hence the theorem.

We may of course imagine a magnetic field divided up into unit tubes of induction, that is, tubular surfaces bounded by lines, and such that the magnetic induction over the cross section of each is everywhere unity. The magnetic induction over any surface is then measured by the number of unit tubes (or, as it is frequently put, by the number of "lines") of induction which pass through it.

It is clear from the result that the magnetic induction over any closed surface is zero, that the surface integral of magnetic induction over an unclosed surface depends only on the bounding curve. For consider the surface closed by a cap fitted to the boundary and not enclosing any part of the magnetic distribution, and let the integration be extended to the whole surface. The total integral is then zero, and therefore the integral taken over the cap is equal and opposite to that over the original surface. This holds if the cap close the surface, whatever be its form and position otherwise; hence the integral taken over the surface depends only on the form and position of the boundary.

Integral of B over an unclosed surface = a line integral round the bounding curve.

It follows that we can express the surface integral of magnetic induction over an unclosed surface by the integral of a certain quantity taken round the bounding curve. This quantity must be directed, since its sign must change with that of the magnetic induction. The sign of the integral will therefore depend on the direction of integration round the curve. Thus let F, G, H be functions of the co-ordinates of a point (x, y, z) on the curve, dx, dy, dz, the projections on the axes of an element ds of the curve we have

Vector Potential of Magnetic Induction.

$$\iint (la + mb + nc)dS = \int (Fdx + Gdy + Hdz) \quad . \quad . \quad . \quad (45)$$

F, G, H, have been called by Clerk Maxwell the components of the vector potential of magnetic induction. We shall now find the values of a, b, c, in terms of these quantities.

It is evidently possible to draw on the surface a series of curves cutting at right angles, so as to divide the surface into a series

Investigation of Line Integral. of rectangular areas (so small that each may be taken as plane) with incomplete rectangles round the bounding curve. The area of these incomplete elements is evidently vanishingly small in comparison with the sum of the areas of the complete elements, and therefore the induction over that portion of the area may be neglected. Now we can find the line integral of the vector potential round any element traced on the surface by calculating its average component along each side of the element, multiplying by the length of the side, and adding the results. Thus let du, dv be two adjacent sides of an elementary rectangle, and let U, V be the mean values of the components of vector potential along du and dv respectively, then for the integral round the element we have

$$Udu + Vdv + \frac{dV}{du}dudv - \left(Udu + \frac{dU}{dv}dvdu\right) - Vdv$$

$$= \left(\frac{dV}{du} - \frac{dU}{dv}\right)dudv \quad \ldots \ldots \quad (46)$$

Now writing dS for the area $dudv$ of the element and equating the magnetic induction over the element to the value just found, we get

$$\left(\frac{dV}{du} - \frac{dU}{dv}\right)dS = (la + mb + nc)dS \quad \ldots \quad (47)$$

if l, m, n, be the direction cosines of the normal to dS. Taking the line integral as above, and in the same direction, round all the elements of area into which the surface is divided, and adding the results together, we have plainly only the integral round the bounding curve, since each side which is common to two elements of surface contributes two equal and opposite elements to the sum, and it is easy to see that for each triangle left round the edge the line integral along the two rectangular sides can, in the limit, be replaced by the integral along the third side formed by the boundary, so that a complete series of elementary integrals, having the same direction round the boundary, is obtained. Hence integrating round the curve, and over the surface, we have finally

$$\int \mathbf{A}\cos\phi \, ds = \iint (la + mb + nc)dS \quad \ldots \ldots \quad (48)$$

where $\mathbf{A}$ is the vector potential, and ϕ the angle between its

VECTOR POTENTIAL OF MAGNETIC INDUCTION

direction and the element ds of the curve. Substituting the components of **A** parallel to the axes, we have

$$\int \left(F\frac{dx}{ds} + G\frac{dy}{ds} + H\frac{dz}{ds} \right) ds = \iint (la + mb + nc) dS. \quad . \quad (49)$$

If now our circuit be a small rectangle of sides $dydz$ at right angles to the axis of x, we get at once from (47)

and in the same way

$$\left. \begin{array}{l} a = \dfrac{dH}{dy} - \dfrac{dG}{dz} \\[4pt] b = \dfrac{dF}{dz} - \dfrac{dH}{dx} \\[4pt] c = \dfrac{dG}{dx} - \dfrac{dF}{dy} \end{array} \right\} \quad . \quad . \quad . \quad . \quad (50)$$

Components of **B** *expressed in terms of Components of Vector Potential.*

It is clear, as we have seen, that F, G, H, are directed quantities, and their signs must be reversed by reversing the signs of a, b, c. In what follows we shall take the positive direction of integration round any circuit as the direction in which a person must be imagined to go round the circuit so as to have the area always on his left, and the positive direction of the magnetic induction as across the element from the person's feet to his head.

The vector potential **A** may be specified as follows. Consider an element, volume δv, of the magnetized substance, at which the intensity of magnetization is **I**. The magnetic moment of the element is $\mathbf{I}\delta v$. Then (as will be seen below) the vector potential produced by this element at a point distant r from it is numerically $\mathbf{I}\delta v \cdot \sin\theta/r^2$, where θ is the angle between the positive direction of magnetization, and the radius vector r. The direction of the vector potential is at right angles to the plane passing through the directions of **I** and r; and by the convention stated above appears to an eye looking in the negative direction of **I** to be drawn in the counter-clockwise direction.

Specification of Vector Potential: (1) Elementary Magnet.

To verify this specification let λ, μ, ν, be the direction cosines of **I**, x, y, z, the co-ordinates of the magnetic element, ξ, η, ζ, those of the point considered, then we have

Verification.

$$\mathbf{I}\delta v \frac{\sin\theta}{r^2} = \frac{\mathbf{I}\delta v}{r^3} \left[\{\mu(\zeta - z) - \nu(\eta - y)\}^2 + \&c. \right]^{\frac{1}{2}} \quad . \quad (51)$$

from which the values of dF, dG, dH can be inferred by inspection.

VOL. II. D

Vector Potential (2) Finite Magnet.

Writing in (51) u for $1/r$, and for $I\lambda$, $I\mu$, $I\nu$, their values A, B, C, and integrating throughout the whole magnetized body, we get for a finite magnet

$$\left. \begin{aligned} F &= \iiint \left(B\frac{du}{dz} - C\frac{du}{dy} \right) dx\,dy\,dz \\ G &= \iiint \left(C\frac{du}{dx} - A\frac{du}{dz} \right) dx\,dy\,dz \\ H &= \iiint \left(A\frac{du}{dy} - B\frac{du}{dx} \right) dx\,dy\,dz \end{aligned} \right\} \quad \ldots \quad (52)$$

Verification.

From the equation $a = dH/d\eta - dG/d\zeta$ we get by (52), remembering that $du/d\xi = -du/dx$, &c.

$$a = -\frac{d}{d\xi}\iiint \left(A\frac{du}{dx} + B\frac{du}{dy} + C\frac{du}{dz} \right) dx\,dy\,dz$$

$$-\iiint A\nabla^2 u\,dx\,dy\,dz \quad \ldots \quad (53)$$

The first term of this expression is simply the force a at the point (ξ, η, ζ), since the first integral is the potential at the point (ξ, η, ζ). The second term of the expression is zero unless the point (ξ, η, ζ), fall within the limits of integration. In the latter case it is $-4\pi A'$ if A' be the value of A at the point (ξ, η, ζ), for evidently we may regard u as the potential at (x, y, z), due to a pole of strength A' at (ξ, η, ζ) and we know by Poisson's theorem that then the integral has the value stated. Hence in general we have by (52) $a = a + 4\pi A'$, where A' is the component of magnetization, and a the magnetic force, where a is taken. Similarly we could find from (52) $b = \beta + 4\pi B'$, $c = \gamma + 4\pi C'$, where β, γ, B', C', are the corresponding components of force and magnetization. Thus the general expressions (52) for the components of the vector potential are completely verified.

Energy of a Magnet.

Returning now to the determination of the energy of a magnet in a magnetic field, we have proved (p. 22 above) that

$$\mathbf{E} = -\iiint (A\alpha + B\beta + C\gamma)\,dx\,dy\,dz \quad \ldots \quad (54)$$

From the manner in which this expression has been found it is plain that it measures the increase of potential energy which takes place when the magnet is caused to take up the given

ENERGY OF MAGNETIC DISTRIBUTIONS

position against the action of **magnetic forces, that is,** it is equal to the work which must be done by **external forces in** bringing the **magnet into the field.** We shall now apply this result to the determination of the whole work which must be done to build up any two distributions (A) and (B) of magnetism. Plainly this may be regarded as consisting of three parts, E_1, the work done if (A) be supposed given in an infinite number of small parts at an infinite distance from one another, which are then put together to form the distribution, that is the work done in bringing these elements into the field simultaneously created by their aggregation to form the magnet:* E_2, the work done in similarly building up the other distribution; and E_3, the work done in carrying one magnetic distribution into the field of the other. Calling the components of force due to the distribution (A), a_1, β_1, γ_1, those due to the distribution (B), a_2, β_2, γ_2, and denoting by $A_1, B_1, C_1, A_2, B_2, C_2$, the corresponding **magnetization components, we have**

<small>Case of two Magnetic Distributions (A) and (B).</small>

$$\mathsf{E}_1 = -\tfrac{1}{2}\iiint (A_1 a_1 + B_1 \beta_1 + C_1 \gamma_1) dx\,dy\,dz \quad \cdots \quad (55)$$

<small>Energy of (A).</small>

$$\mathsf{E}_2 = -\tfrac{1}{2}\iiint (A_2 a_2 + B_2 \beta_2 + C_2 \gamma_2) dx\,dy\,dz \quad \cdots \quad (56)$$

<small>Energy of (B).</small>

Also we have

$$\mathsf{E}_3 = -\iiint (A_2 a_1 + B_2 \beta_1 + C_2 \gamma_1) dx\,dy\,dz \quad \cdots \quad (57)$$

in which the integration **is extended** throughout the volume of the magnet B. We **have** of course also by Green's **theorem,** or by the principle that the energy of (A) in the field **of** (B) must be equal to **the energy** of (B) in the field **of** (A),

<small>Mutual Energy of the Two Distributions.</small>

$$\mathsf{E}_3 = -\iiint (A_1 a_2 + B_1 \beta_2 + C_1 \gamma_2) dx\,dy\,dz \quad \cdots \quad (58)$$

---

* The manner in which **this is done is of importance.** From the fact that similar magnets **magnetized with equal intensity have similar magnetic** fields, so that **equal forces are produced at** points **similarly situated** at distances proportional **to the** linear dimensions **of the magnets, it** follows that **no work is done in** breaking a magnet up into **a number** of parts all similar, and removing them to an infinite distance from one another. **If on the other hand** the magnet **be** broken up into an infinite number of infinitely thin **filaments** (each very long in comparison with its thickness) taken along **the** lines of magnetization, and these be then separated **to** infinite distance from one another, the work done has the value given in the **text.** See Sir W. Thomson's *Electrostatics and Magnetism*, **p.** 441.

The coefficient $\frac{1}{2}$ in the two first expressions arises from the fact that with the annulment of the distribution its field disappears. Hence the total energy may be written

Total Energy.

$$\mathsf{E} = \mathsf{E}_1 + \mathsf{E}_2 + \mathsf{E}_3 = -\tfrac{1}{2}\iiint \{(A_1+A_2)(a_1+a_2) + (B_1+B_2)(\beta_1+\beta_2) + (C_1+C_2)(\gamma_1+\gamma_2)\} dx dy dz \quad . \quad (59)$$

$$\text{or } \mathsf{E} = -\tfrac{1}{2}\iiint (Aa + B\beta + C\gamma) dx dy dz$$

if $A, B, C, a, \beta, \gamma$, be put for A_1+A_2, &c., a_1+a_2 &c.

The integral may evidently be taken throughout all space, since at any point not within either of the distributions of magnetism, each of the quantities A, B, C, is identically zero.

Alternative form for Magnetic Energy.

We may put this expression into another form, thus: substituting for A, B, C, their values $(a-a)/4\pi$, $(b-\beta)/4\pi$, $(c-\gamma)/4\pi$, we find

$$\mathsf{E} = \frac{1}{8\pi}\int_{-\infty}^{+\infty}\!\!\iint (a^2+\beta^2+\gamma^2) dx dy dz$$

$$-\frac{1}{8\pi}\int_{-\infty}^{+\infty}\!\!\iint (aa+b\beta+c\gamma) dx dy dz \quad . \quad . \quad . \quad . \quad (60)$$

Now remembering that $a = -dV/dx$, $\beta = -dV/dy$, $\gamma = -dV/dz$, and integrating the second integral by parts we see that it vanishes, since Va, Vb, Vc, are each zero at an infinite distance, and a, b, c, fulfil the solenoidal condition (41) above. Hence we have

$$\mathsf{E} = \frac{1}{8\pi}\int_{-\infty}^{+\infty}\!\!\iint \mathbf{H}^2 dx dy dz \quad . \quad . \quad . \quad . \quad . \quad . \quad (61)$$

where $\mathbf{H}$ denotes the resultant magnetic force at the point x, y, z.

If $\mathbf{H}_1, \mathbf{H}_2$, denote the resultant forces produced at the point x, y, z, by the distributions (A) and (B) respectively, and θ the angle between $\mathbf{H}_1$ and $\mathbf{H}_2$, we have by elementary trigonometry.

$$\mathbf{H}^2 = \mathbf{H}_1^2 + \mathbf{H}_2^2 + 2\mathbf{H}_1\mathbf{H}_2\cos\theta.$$

Hence substituting in (61) we find by interpretation

$$E_1 = \frac{1}{8\pi} \int\int\int_{-\infty}^{+\infty} \mathbf{H}_1^2 dx dy dz$$

$$E_2 = \frac{1}{8\pi} \int\int\int_{-\infty}^{+\infty} \mathbf{H}_2^2 dx dy dz \quad \quad \dots \quad (62)$$

$$E_3 = \frac{1}{4\pi} \int\int\int_{-\infty}^{+\infty} \mathbf{H}_1 \mathbf{H}_2 \cos\theta\, dx dy dz$$

Expressions for Intrinsic and Mutual Energies.

which can be verified by (55), (56), (57), (58), it being remembered that by the definition of E_1, E_2, we must take $a_1 = a_1 + 4\pi A_1$, &c., $a_2 = a_2 + 4\pi A_2$, &c.

These expressions are obviously capable of generalization for any number of magnetic distributions, or a single distribution regarded as composed of any number of parts. They may be taken as expressing the fact that the energy may be regarded as residing in the medium in which the magnetized bodies are placed.*

We shall see later that magnetic force exists at every point in the space surrounding a conductor carrying an electric current, that in fact the molecular magnets composing any magnetized body are most probably produced by electric currents flowing in molecular circuits, which are devoid of resistance, so that the current continues to flow without diminution of strength from generation of heat. We shall then find that if a, b, c, be the components of magnetic induction $\mathbf{B}$, and α, β, γ, those of magnetic intensity $\mathbf{H}$, at any point in the field, the total magnetic energy $\mathbf{E}$ is given by the equation

General Expression for Magnetic Energy whether due to Currents or Magnets.

$$\mathbf{E} = \frac{1}{8\pi} \int\int\int_{-\infty}^{+\infty} (a\alpha + b\beta + c\gamma) dx dy dz$$

or

$$\mathbf{E} = \frac{1}{8\pi} \int\int\int_{-\infty}^{+\infty} \mathbf{BH}\, dx dy dz \cos\theta \quad \quad \dots \quad (63)$$

Assuming this we see that by drawing successive equipotential surfaces so that the difference of potential between each pair of consecutive surfaces is unity, and supposing these cut by unit tubes, we can divide the whole field up into cells, each of which may be regarded as containing $1/8\pi$ of a unit of magnetic energy.

---

* See Vol. I. p. 34.

Section III.

APPLICATIONS OF GENERAL THEORY.
MAGNETIC SHELLS. LAMELLAR DISTRIBUTION.
UNIFORMLY MAGNETIZED ELLIPSOID.

Magnetic Shell.   A most important form of magnetic distribution for consideration is that in which we have a thin sheet of matter magnetized normally to its surface. Such a sheet is called a *magnetic shell*. Its importance arises from the fact proved by Ampère that every linear circuit carrying a current is equivalent in magnetic action to a magnetic shell of a certain uniform intensity of magnetization, and having its bounding edge coincident with the circuit. A magnetic shell, it may be here stated, may be altered in position elsewhere than at its boundary, in any way whatever, without affecting its magnetic action at any given point, provided only the shell be not so changed in position as to cause the point to pass through it, and that its magnetic moment per unit of area be uniform, and kept constant throughout the changes of position. The chief properties of magnetic shells are investigated in what immediately follows, and the results will be directly available when we come to consider the magnetic action of electric currents.

If $d\nu$ be the thickness of the sheet at any element dS, the volume of the element is $d\nu \cdot dS$. If I then be the intensity of magnetization at the element, the magnetic moment of this portion is $I d\nu \cdot dS$. The product $I d\nu$ is called the *strength* of the shell, and is usually denoted by Φ. This may vary from point to point of the shell.

THEORY OF SIMPLE MAGNETIC SHELL

We shall consider first a simple shell, that is one for which Φ has the same value at every point. By (14) above, if we consider any element dS of the shell, and θ be the angle between the direction of magnetization of the shell, taken positive when drawn from the negative to the positive side, and a line drawn from the element to a point P at distance r, the potential at P due to the element is $\Phi dS \cos\theta/r^2$. But $dS \cos\theta$ is the projection of the element at right angles to r, and therefore $dS \cos\theta/r^2$ is the area $d\omega$, traced out on the surface of a sphere of unit radius, having its centre at P, by a line passing through P, and carried round the boundary of the element, that is, it is the solid angle subtended at P by the element. It follows therefore that the potential V at P produced by the whole shell is given by the equation

Simple Magnetic Shell.

Potential of Simple Magnetic Shell at any Point.

$$V = \Phi\omega \quad \ldots \ldots \ldots (64)$$

where ω is the total solid angle subtended by the shell at P.

This is also, of course, the potential energy of the shell in the field due to unit magnetic pole placed at P.

It is evident that the value of V depends only on the strength of the shell and its boundary, and hence we have the remarkable result, that any two shells of equal strength, which have the same boundary, produce equal potentials at the point P, provided P does not lie between them.

Magnetic Potential and Magnetic Force due to closed Shell.

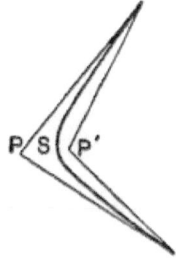

Fig. 7.

If the shell be closed its potential at any external point is zero, since the solid angle is then zero. Such a shell therefore produces no magnetic effect at any external point. At every internal point in such a shell however the potential is $-4\pi\Phi$ (if the positive side be outwards, or $+4\pi\Phi$ if the positive side be inwards) since the solid angle is then 4π. There is therefore no magnetic force at any internal point.

40   MAGNETIC THEORY

Reckoning of Solid Angles. In the reckoning of solid angles in this connection we shall adhere to the following convention. Let P, P' be adjacent points on opposite sides of a shell S, (Fig. 7), of which P is on the positive side. Then supposing the solid angle subtended at P by the shell to be ω, that subtended by the shell at P' is to be taken as $\omega - 4\pi$; for, plainly, if the generating lines of the cone which meet at P' were turned round the edge of the shell from meeting at P' to meeting at P the solid angle would change in the process by 4π, and we must take it as being increased by that amount.

Solid Angle at Opposite Sides of Shell. Or, the difference between the solid angles may be seen thus: consider the two simple shells A, B, of which a section by the plane of the paper is shown in Fig. 8, which have a common boundary b, b, and form a closed simple shell, the positive face of which is the outside. Let P, P' be infinitely near points, the

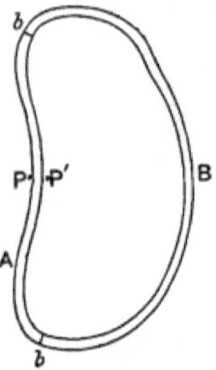

Fig. 8.

former on the outside, the latter on the inside of A. Let the potential due to A at P be V_1, and at P', V_2. The potentials at P and P' produced by B will be the same. V', say. But we have $V_1 = \Phi\omega$, $V_2 + V' = -4\pi\Phi$, and $V_1 + V' = 0$. Thus we get $V_2 = \Phi(\omega - 4\pi)$ as already stated.

Hence the potential of the shell V, varies, as the point at which it is measured changes in position from P to P' round the edge of the shell, from the value $\Phi\omega$ to the value $\Phi(\omega - 4\pi)$. If the point pass from a position infinitely near the negative side

POTENTIAL DUE TO LAMELLAR MAGNET

through the shell to an adjacent position on the positive side, the potential increases by the amount $4\pi\Phi$.

Lamellar Magnet.

In some cases of magnetization, as for example the induced magnetization of soft iron in certain circumstances, the body may be regarded as made up of simple magnetic shells, either closed or having their edges in the surface of the body; in such cases the magnetization is said to be lamellar. If we take ϕ to denote for such a body the sum of the strengths of the shells encountered by a point made to pass within the magnet from any given position to any other position (x, y, z), we easily see that

Condition fulfilled by.

$$A = \frac{d\phi}{dx}, B = \frac{d\phi}{dy}, C = \frac{d\phi}{dz} \quad \ldots \ldots \quad (65)$$

ϕ is called the *potential of magnetization*, since the quantities A, B, C, are derived from it by differentiation. When they can be so derived they are said to fulfil the lamellar condition. Now we have seen, (20) above, that the potential V at any point (ξ, η, ζ) due to a finite magnet is given by the equation

$$V = \iiint \left(A \frac{du}{dx} + B \frac{du}{dy} + C \frac{du}{dz} \right) dx\,dy\,dz$$

if u be written for the reciprocal of the distance r from (x, y, z) to $(\xi, \eta, \zeta.)$ Hence for a lamellar magnet this becomes

$$V = \iiint \left(\frac{d\phi}{dx}\frac{du}{dx} + \frac{d\phi}{dy}\frac{du}{dy} + \frac{d\phi}{dz}\frac{du}{dz} \right) dx\,dy\,dz \quad \ldots \quad (66)$$

Potential due to Lamellar Magnet.

Integrating this expression by parts, and putting l, m, n, for the direction cosines of the normal drawn outwards to an element dS of the surface we get

$$V = \iint \phi \left(l \frac{du}{dx} + m \frac{du}{dy} + n \frac{du}{dz} \right) dS$$

$$- \iiint \phi \nabla^2 u \, dx\,dy\,dz \quad \ldots \ldots \quad (67)$$

in which the first integral is taken over the surface of the magnet,

MAGNETIC THEORY

Case in which Point is within Magnet.

the second through its substance. Each element of the surface integral may be written in the form $\phi \cos\theta\, dS/r^2$, where θ is the angle between the normal and the direction of r. Each element of the second integral is zero unless the point (ξ, η, ζ) fall within the limits of integration. In the latter case the integral has the value (see above, p. 34) $-4\pi\phi'$ if ϕ' be the value of ϕ at (ξ, η, ζ). Hence in general we have for a lamellar magnet

$$V = \iint \frac{1}{r^2} \phi \cos\theta\, dS + 4\pi\phi'. \quad \ldots \quad (68)$$

Continuity of Potential at surface of Lamellar Magnet.

The value of V given in (68) is continuous at the surface of the magnet. For plainly we may regard the surface integral as the potential at P of a magnetic shell coinciding with the surface, and of strength ϕ, varying from point to point. The potentials of this shell at two adjacent points, one just outside, the other just inside, differ only by the potential due to the portion of the shell immediately between the points. Thus denoting the surface integral by Ω, if Ω_e, Ω_i denote the values of the surface integral at the external and internal points respectively we have

$$\Omega_e = \Omega_i + 4\pi\phi' \quad \ldots \quad \ldots \quad \ldots \quad (69)$$

and as the term $4\pi\phi'$ of (68) disappears in the passage from the inside to the outside of the surface, the potential is unchanged by the passage.

Definiteness of Value of V.

But the value of V whether at an internal or an external point at first sight seems indefinite, since the value of ϕ depends upon the zero of reckoning chosen for it. This is, however, not the case, for if any arbitrary value of ϕ be taken for a point in the surface, its value is thereby fixed for any other point, and it is clear that choosing any other value for that point would simply increase the strength of the shell by the same amount at every point, that is, would superimpose a simple closed shell of strength c, say, on the former. The value of ϕ at every internal point would also be increased by the amount c. Hence, for the potential V at an internal point we should have

$$V = \iint \frac{1}{r^2} \phi \cos\theta\, dS - 4\pi c + 4\pi(\phi' + c)$$

that is, its value would remain unaltered. At an external point the additional potential would be that of a simple closed shell of constant strength, which is zero.

MUTUAL ENERGY OF TWO SHELLS

The external and internal action of the lamellar magnet thus depends only on the variation of strength from point to point, and not on its actual value. For an external point therefore it depends only on the variation of ϕ from point to point along the surface. But by the values of $A, B, C,$ in (65) it is clear that the rate of variation of ϕ in any direction along the surface is the tangential component of magnetization in that direction. Hence the external action of the shell is given if the tangential component of magnetization is given for every point on the surface.

Action of Lamellar Magnet depends only on Variation of Strength from Point to Point.

Since in a lamellar distribution of magnetism we have

$$V = \Omega + 4\pi\phi$$

and $A, B, C = d\phi/dx$, &c., $\alpha, \beta, \gamma = -dV/dx$, &c., we have

$$a, b, c = -\frac{d\Omega}{dx}, -\frac{d\Omega}{dy}, -\frac{d\Omega}{dz} \quad \ldots \quad (70)$$

Potential of Magnetic Induction in Lamellar Magnet.

respectively. Ω is called the potential of magnetic induction.

It is plain that in a lamellar distribution the direction of magnetization is everywhere at right angles to the surfaces $\phi = c$, that is, the surfaces of equal potential of magnetization.

The potential energy of a simple magnetic shell in a magnetic field is given by equation (29) above, modified so as to suit the case of the shell. If dS be an element of area, l, m, n the direction cosines of the normal to the shell drawn from its negative to its positive side, Φ the (uniform) strength of the shell, and $\delta\nu$ its thickness, we have $A\delta\nu = l\Phi$, &c., and therefore

Energy of Simple Shell

$$\mathsf{E} = -\Phi \iint (l\alpha + m\beta + n\gamma)\,dS \quad \ldots \quad (71)$$

that is, it is the product of the strength of the shell into the surface integral of magnetic induction over the surface. Hence, by (45) above, the energy of the shell in the field may be expressed by a line integral taken round its boundary.

expressible as a Line Integral round Edge.

We have an interesting and extremely important case when the field is produced by another simple shell. In this case the mutual energy of the shells is expressible as a double line integral taken round their boundaries. Calling the energy in this case E_{ss}, we have at once by (45) and (71)

Mutual Energy of Two Shells.

$$\mathsf{E}_{ss} = -\Phi \int \left(F\frac{dx}{ds} + G\frac{dy}{ds} + H\frac{dz}{ds} \right) ds \quad \ldots \quad (72)$$

where ds is an element of the boundary of the shell, and F, G, H are given by (See equations (52), p. 34.)

Vector Potential of Magnetic Shell.

$$\left.\begin{aligned} F &= \Phi' \iint \left(m'\frac{du}{dz'} - n'\frac{du}{dy'}\right) dS' \\ G &= \Phi' \iint \left(n'\frac{du}{dx'} - l'\frac{du}{dz'}\right) dS' \\ H &= \Phi' \iint \left(l'\frac{du}{dy'} - m'\frac{du}{dx'}\right) dS' \end{aligned}\right\} \quad \ldots \ldots \quad (73)$$

in which the accented letters and the integrations refer to the shell producing the field, and u is the reciprocal of the distance between a point (x, y, z) in one shell and a point (x', y', z') in the other. Now by writing in the first of (73) $m'dS' = dz'dx'$, $n'dS' = dy'dx'$, it is easy to see that F is equal to the line integral of $udx'/ds'\cdot ds'$ taken round the boundary of the shell. The same thing may be proved by equations (49) and (50) of p. 33, by putting there $F = u$, $G = 0 = H$, and using accented variables. Similarly G and H in (73) may be dealt with. Hence we find for E_{ss} the equation

Mutual Energy expressed as a double Line Integral round Edges of Shells.

$$E_{ss} = -\Phi\Phi' \iint u\left(\frac{dx}{ds}\frac{dx'}{ds'} + \frac{dy}{ds}\frac{dy'}{ds'} + \frac{dz}{ds}\frac{dz'}{ds'}\right) ds\, ds'$$

$$= -\Phi\Phi' \iint \frac{1}{r}\cos\theta\, ds\, ds' \quad \ldots \ldots \quad (74)$$

where θ is the angle between ds and ds'.

Energy of Lamellar Distribution.

For a lamellar distribution of magnetism we have by (54) and (65)

$$\mathsf{E} = \iiint \left(\frac{d\phi}{dx}\frac{dV}{dx} + \frac{d\phi}{dy}\frac{dV}{dy} + \frac{d\phi}{dz}\frac{dV}{dz}\right) dx\,dy\,dz$$

which integrated by parts becomes, since $\nabla^2 V = 0$,

$$\mathsf{E} = \iint \phi \frac{dV}{d\nu}\, dS$$

where $dV/d\nu$ is the rate of variation of V along a normal to the shell drawn from the negative to the positive side.

Complex Shell.

Hitherto we have dealt only with simple shells, or with lamellar distributions built up of simple shells either closed or

POTENTIAL DUE TO CIRCULAR MAGNETIC SHELL.

having their edges in the surface of the magnet. A complex shell is a thin plate of substance normally magnetized, but varying in strength from point to point. It may be conceived as made up of overlapping simple shells. A magnet made up of complex shells fulfils the condition that the direction of magnetization at every point is normal to a family of surfaces; but the intensity is not derivable from a potential of magnetization. The condition just stated is obtained as follows. The equation of a line of magnetization is $A/dx = B/dy = C/dz$, and the condition that this line should be at right angles to a system of surfaces $\phi(x, y, z) = c$, where c is a variable parameter, is that

Condition fulfilled by Direction of Magnetization.

$$hA = \frac{d\phi}{dx}, \quad hB = \frac{d\phi}{dy}, \quad hC = \frac{d\phi}{dz}$$

where h is a factor, in this case a function of the co-ordinates. Hence we get

$$h(A\,dx + B\,dy + C\,dz) = d\phi$$

where $d\phi$ is a perfect differential. Applying the criterion of a perfect differential, viz. $d(hC)/dy = d(hB)/dz$, &c., we obtain the condition

$$A\left(\frac{dC}{dy} - \frac{dB}{dz}\right) + B\left(\frac{dA}{dz} - \frac{dC}{dx}\right) + C\left(\frac{dB}{dx} - \frac{dA}{dy}\right) = 0.$$

It will be convenient to develop here expressions for the potential at an external point due to a circular magnetic shell, and also for the mutual potential energy of two such shells. The results arrived at will be directly available in calculations of the electromagnetic constants of coils of circular section.

Potential due to a Circular Magnetic Shell.

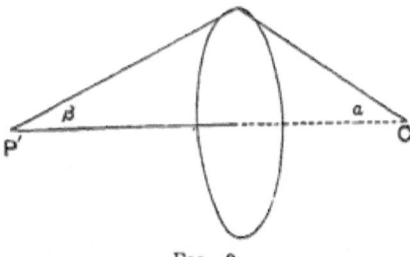

Fig. 9.

We take as the origin of co-ordinates a point C on the axis, distant c from any point on the circle, and consider first a point

P' on the **axis of the** shell, at a distance z from the origin. Now if β be the angle between the **axis** and a line drawn from P' to the boundary of the shell the solid angle subtended by the shell at P is $2\pi(1-\cos\beta)$ or $2\pi\{1-(z-c\cos a)/(z^2+c^2-2zc\cos a)^{-\frac{1}{2}}\}$. We may **prepare** this for expansion by writing it in the form

$$2\pi\left\{1-\frac{\dfrac{z}{c}-\cos a}{\left(1+\dfrac{z^2}{c^2}-\dfrac{2z}{c}\cos a\right)^{\frac{1}{2}}}\right\},$$

or,

$$2\pi\left\{1-\frac{1-\dfrac{c}{z}\cos a}{\left(1+\dfrac{c^2}{z^2}-\dfrac{2c}{z}\cos a\right)^{\frac{1}{2}}}\right\}$$

according as $c >$ or $< z$. Now we know by the definition of a zonal surface harmonic that we have

$$\left.\begin{aligned}\left(1+\frac{z^2}{c^2}-\frac{2z}{c}\cos a\right)^{-\frac{1}{2}} &= {}_aZ_0 + {}_aZ_1\frac{z}{c}+{}_aZ_2\left(\frac{z}{c}\right)^2+\cdots \\ &\qquad +{}_aZ_i\left(\frac{z}{c}\right)^i+\cdots \\ \text{or}\qquad\qquad\qquad & \\ \left(1+\frac{c^2}{z^2}-\frac{2c}{z}\cos a\right)^{-\frac{1}{2}} &= {}_aZ_0+{}_aZ_1\frac{c}{z}+{}_aZ_2\left(\frac{c}{z}\right)^2+\cdots \\ &\qquad +{}_aZ_i\left(\frac{c}{z}\right)^i+\cdots \end{aligned}\right\} \quad (75)$$

where ${}_aZ_i$ is written for the zonal harmonic of the ith order, taken for the angle a.

Spherical Harmonic Expansion for Potential at Point on axis. Substituting these expansions in the expressions given above, and writing ω for the solid angle we obtain

$$\omega = 2\pi\left\{1+\cos a+({}_aZ_1\cos a-{}_aZ_0)\frac{z}{c}+({}_aZ_2\cos a-{}_aZ_1)\left(\frac{z}{c}\right)^2\right.$$
$$\left.+\cdots+({}_aZ_i\cos a-{}_aZ_{i-1})\left(\frac{z}{c}\right)^i+\cdots\right\} \quad (76)$$

MUTUAL ENERGY OF TWO CIRCULAR SHELLS

or

$$\omega = 2\pi \left\{ (_aZ_0 \cos a - {_aZ_1}) \frac{c}{z} + (_aZ_1 \cos a - {_aZ_2})\left(\frac{c}{z}\right)^2 + \ldots \right.$$
$$\left. + (_aZ_{i-1} \cos a - {_aZ_i})\left(\frac{c}{z}\right)^i + \ldots \right\} \quad (77)$$

where it is to be remembered $_aZ_0 = 1$, and $_aZ_1 = \cos a$, so that $_aZ_0 \cos a - {_aZ_1}$, the first term in the second series, is zero.

The solid angle ω at a point P not on the axis can now be obtained at once. If θ be the angle between CP and the axis, and we denote CP by r, and the ith zonal harmonic with respect to θ by $_\theta Z_i$, we have, taking the strength of the shell as unity,

Potential at any Point.

$$\omega = 2\pi \left\{ 1 + \cos a + (_aZ_1 \cos a - {_aZ_0})\, _\theta Z_1 \frac{r}{c} \right.$$
$$\left. + (_aZ_2 \cos a - {_aZ_1})\, _\theta Z_2 \left(\frac{r}{c}\right)^2 + \ldots \right\} \quad (78)$$

or

$$\omega = 2\pi \left\{ (_aZ_1 \cos a - {_aZ_2})\, _\theta Z_1 \left(\frac{c}{r}\right)^2 + (_aZ_2 \cos a - {_aZ_3})\, _\theta Z_2 \left(\frac{c}{r}\right)^3 + \ldots \right\} \ldots (79)$$

according as $c >$ or $< r$. But by the fundamental relations of spherical harmonics we have, writing μ for $\cos\theta$ and Z'_i for $dZ_i/d\mu$,

$$\left. \begin{array}{l} \mu Z_i - Z_{i-1} = -\dfrac{1}{i}(1-\mu^2)Z'_i \\[4pt] Z_i - \mu Z_{i-1} = -\dfrac{1}{i}(1-\mu^2)Z'_{i-1} \end{array} \right\} \quad \ldots \ldots (80)$$

Hence the series for ω may be written in the more compact forms

$$\omega = 2\pi \left\{ 1 + \cos a - \sin^2 a \sum \frac{1}{i}\, _aZ'_i \cdot {_\theta Z_i}\left(\frac{r}{c}\right)^i \right\} \quad \ldots (81)$$

or

$$\omega = 2\pi \sin^2 a \sum \frac{1}{i+1}\, _aZ'_i \cdot {_\theta Z_i}\left(\frac{c}{r}\right)^{i+1} \quad \ldots \ldots (82)$$

We can from these expressions find the potential energy of two circular magnetic shells in any relative positions. We may without loss of generality suppose for simplicity that the shells are segments of spherical surfaces having a common centre C,

Mutual Potential Energy of Two

Circular Shells.

1. When the Shells have a Common Axis.

and radii c_1 and c_2, and assume first that they are so placed as to have a common axis, and have like faces turned towards C. If $c_1 > c_2$ the solid angle subtended by the second shell at any point P of the first is

$$\omega = 2\pi \sin^2 a \sum \frac{1}{i+1} \, _aZ_i \cdot {}_0Z_i \left(\frac{c_2}{c_1}\right)^{i+1} \quad \ldots \quad (83)$$

Now if dS be the area of an element of the first shell, Φ_1, Φ_2, the strengths of the first and second shells respectively, the potential energy of the element is $-\Phi_1\Phi_2 \, d\omega/dc_1 \cdot dS$. We may take as dS a ring of breadth $c_1 d\theta$ so that its area is $2\pi c_1^2 \sin\theta d\theta$. Hence by (83) we have

$$-\Phi_1\Phi_2 \frac{d\omega}{dc_1} dS = 4\pi^2 \Phi_1\Phi_2 \sin^2 a \sum {}_aZ_i \cdot {}_0Z_i \left(\frac{c_2}{c_1}\right)^i \sin\theta \, d\theta;$$

and denoting by μ_1 the cosine of the angle which a line drawn from C to the edge of the first shell makes with the axis, we obtain for the total potential energy, E, the equation

$$\mathsf{E} = 4\pi^2 \Phi_1\Phi_2 c_2 \sin^2 a \sum {}_aZ_i \left(\frac{c_2}{c_1}\right)^i \int_{\mu_1}^{1} Z_i d\mu. \quad \ldots \quad (84)$$

But by the differential equation satisfied by a zonal harmonic, viz.

$$i(i+1)Z_i + \frac{d}{d\mu}\left\{(1-\mu^2)\frac{dZ_i}{d\mu}\right\} = 0$$

$$i(i+1)\int_{\mu}^{1} Z_i d\mu = (1-\mu^2)Z'_i \quad \ldots \quad (85)$$

Hence writing in (84) a_2 for a, and a_1 for $\cos^{-1}\mu_1$, we get

$$\mathsf{E} = 4\pi^2 \Phi\Phi_2 c_2 \sin^2 a_1 \sin^2 a_2 \sum \frac{1}{i(i+1)} \, _{a_1}Z'_i \cdot {}_{a_2}Z'_i \left(\frac{c_2}{c_1}\right)^i. \quad (86)$$

2. When the Shells are in any Relative Positions.

Lastly, if (Fig. 10) the axis of either shell be turned round through an angle θ about C, the potential energy is got by

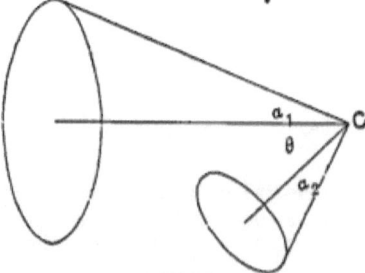

Fig. 10.

MUTUAL ENERGY OF TWO CIRCULAR SHELLS

multiplying the general term of the series just found by the ith zonal harmonic for θ. Thus

$$E = 4\pi^2 \Phi_1 \Phi_2 \sin^2 a_1 \sin^2 a_2 c_2 \sum \frac{1}{i(i+1)} \, {}_{a_1}Z'_i \cdot {}_{a_2}Z'_i \cdot {}_{\theta}Z_i \left(\frac{c_2}{c_1}\right)^i.$$

From this result the moment of the forces tending to increase θ is obtainable by differentiation with respect to θ. Thus

$$\frac{dE}{d\theta} = -4\pi^2 \Phi_1 \Phi_2 \sin^2 a_1 \sin^2 a_2 \sin\theta c_2$$

$$\sum \frac{1}{i(i+1)} \, {}_{a_1}Z'_i \cdot {}_{a_2}Z'_i \cdot {}_{\theta}Z'_i \left(\frac{c_2}{c_1}\right)^i \quad \ldots \quad (87)$$

The value of Z'_i can be calculated for values of i from 1 to 3, from the values of Z_i given at p. 23. By the second of the relations of (95) the value of Z_i can be found when that of Z_{i-1} is known, hence values of Z_i beyond those given can be found if required.

When c_2/c_1 is nearly equal to unity the series for E does not converge rapidly. To meet this difficulty, however, in the very important case of two parallel coaxial circular shells, we can find E by transforming the double integral round the circular boundaries which we get from (74) into an expression involving elliptic integrals.

We shall call the radii of the circles a_1, a_2, b the distance between their planes, and ϕ_1, ϕ_2 the angles which radii from the centres to elements ds_1, ds_2 respectively, make with a fixed plane, passing through the centres of the circles. Then since the co-ordinates of the centres of ds_1, ds_2, reckoned from the centre of the first circle as origin, are $a_1 \cos\phi_1$, $a_1 \sin\phi_1$, 0, and $a_2 \cos\phi_2$, $a_2 \sin\phi_2$, b,

Mutual Potential Energy of Two Parallel Coaxial Circular Shells.

$$r^2 = a_1^2 + a_2^2 + b^2 - 2a_1 a_2 \cos(\phi_1 - \phi_2).$$

Also $a_1 d\phi_1 = ds_1$, $a_2 d\phi_2 = ds_2$, and since the radii are at right angles to ds_1, ds_2, $\theta = \phi_1 - \phi_2$. Hence

$$E = -\Phi_1 \Phi_2 \int_0^{2\pi} \int_0^{2\pi} \frac{a_1 a_2 \cos(\phi_1 - \phi_2)}{\{a_1^2 + a_2^2 + b^2 - 2a_1 a_2 \cos(\phi_1 - \phi_2)\}^{\frac{1}{2}}} d\phi_1 d\phi_2 \quad (88)$$

50   MAGNETIC THEORY

in terms of Elliptic Integrals.   To transform this expression to elliptic integrals we write the integral thus

$$\int_0^{2\pi} d\phi_2 \int_0^{2\pi} \frac{a_1 a_2 \cos(\phi_1 - \phi_2)\, d\phi_1}{\{a_1^2 + a_2^2 + b^2 - 2a_1 a_2 \cos(\phi_1 - \phi_2)\}^{\frac{1}{2}}}.$$

Now it is clear that it is immaterial which element of either of the shells we take as the starting point for the integration round the other. Suppose then we begin from that element of the second shell the angular coordinate of which is ϕ_2, and integrate from this round the first circle, we have $ds_1 = a_1 d(\phi_1 - \phi_2)$, ϕ_2 being regarded as a constant; and we obtain the same result as if ϕ_2 had not been here introduced. Hence in the last expression we may change the variable of the integral taken first to $\phi_1 - \phi_2$ and integrate with respect to this from 0 to 2π, and finally integrate the result with respect to ϕ_2 from 0 to 2π.

$$\mathsf{E} = -\Phi_1 \Phi_2 \int_0^{2\pi} d\phi_2 \int_0^{2\pi} \frac{a_1 a_2 \cos(\phi_1 - \phi_2)\, d(\phi_1 - \phi_2)}{\{a_1^2 + a_2^2 + b^2 - 2a_1 a_2 \cos(\phi_1 - \phi_2)\}^{\frac{1}{2}}}$$

$$= -2\pi \Phi_1 \Phi_2 \int_0^{2\pi} \frac{a_1 a_2 \cos\psi\, d\psi}{\{a_1^2 + a_2^2 + b^2 - 2a_1 a_2 \cos\psi\}^{\frac{1}{2}}}$$

$$= 4\pi \Phi_1 \Phi_2 \sqrt{a_1 a_2} \left\{ \left(k - \frac{2}{k}\right) \int_0^{\frac{\pi}{2}} \frac{d\frac{\psi}{2}}{\{1 - k^2 \sin^2 \frac{\psi}{2}\}^{\frac{1}{2}}} \right. $$

$$\left. + \frac{2}{k} \int_0^{\frac{\pi}{2}} \{1 - k^2 \sin^2 \frac{\psi}{2}\}^{\frac{1}{2}} \frac{d\psi}{2} \right\}$$

$$= 4\pi \Phi_1 \Phi_2 \sqrt{a_1 a_2} \left\{ \left(k - \frac{2}{k}\right) F + \frac{2}{k} E \right\} \quad \ldots \quad (89)$$

where $k = 2\sqrt{a_1 a_2}/\sqrt{(a_1 + a_2)^2 + b^2}$, and F and E are the complete elliptic integrals of the first and second kind to modulus k. We shall find this expression for E extremely useful when we come to treat of electro-magnetic induction.

SYNTHESIS OF UNIFORM MAGNETIZATION

We shall now consider the potential and force at external and internal points in one or two important cases of magnetization. We shall deal first with the magnetization of a body of uniform susceptibility when placed in a uniform field. The magnetization of the body will also be uniform, and we shall suppose it known in amount. We shall deal with its relation to the magnetizing force later when we consider determinations of susceptibility.

Potential and Force in Particular Cases.

Any case of uniform magnetization may be regarded as produced by supposing two uniform volume distributions of magnetism, equal in density but opposite in sign, to be made coincident with the body and the negative distribution to be then displaced through a small distance in the direction opposite to that of magnetization.* The (finite) product of the volume density ρ, supposed infinitely great, into this displacement, supposed infinitely small, is the intensity I of magnetization. Now if ρU be the potential at the point P produced by the positive distribution, the potential at the same point produced by the negative distribution displaced relatively through a distance $-\delta s$, will be equal in amount and opposite in sign to that which the positive distribution would produce at P, if P were displaced through an equal and opposite distance $+\delta s$, that is, $-\rho(U + dU/ds \cdot \delta s)$. Hence if V denote the potential at P due to the magnet, we have

Synthesis of Uniform Magnetization by Couches de Glissement

$$V = -\frac{dU}{ds} \cdot \rho \cdot \delta s = -I\frac{dU}{ds} \quad \ldots \ldots \quad (90)$$

If λ, μ, ν be the direction cosines of I we have $A, B, C = \lambda I, \mu I, \nu I$, as before, and therefore (90) may be written

Potential at Internal Point

$$V = -\left(A\frac{dU}{dx} + B\frac{dU}{dy} + C\frac{dU}{dz}\right) = AX + BY + CZ \quad . \quad (91)$$

if X, Y, Z, be the components of force at P due to the positive distribution. From this expression the components of magnetic force can be obtained by differentiation.

We might also obtain equation (91) by remembering that any element of volume, δv, distant r from P, has the magnetic moment $I\delta v$, and produces therefore a magnetic potential at P, of amount $I\delta v \cos\theta/r^2$, where θ is the angle between I and r. But this is the component force at P (on unit mass) in the direction of I due to a material element of volume δv, and

represented by a Gravitational Attraction.

---

* See Vol. I. p. 122, and above, p. 27.

density **I**. Hence the whole magnetic potential at P is numerically equal to the resultant force at P due to a uniform distribution of matter, coinciding with the body, and of density **I**; which is what (91) expresses.

Uniformly Magnetized Ellipsoid. We shall now consider the case of a uniformly magnetized ellipsoid. Let the axes be a, b, c, and the intensity of its magnetization **I**, it is required to find the magnetic potential of the ellipsoid at an external point $P(\xi, \eta, \zeta)$. By the last paragraph this problem will be solved if we find the axial components of force at P due to a uniform ellipsoid of any density ρ.

Attraction of Elliptic Homœoid on External Particle at Surface Now we know that the force at the surface of a thin elliptic homœoid* is at right angles to the surface, and equal to $4\pi\tau\rho$, where τ is the thickness of the homœoid at the point. Now by Maclaurin's theorem of attractions† (extended to confocal homœoids) the attraction of an elliptic homœoid at a point $P(\xi, \eta, \zeta)$ is equal to the attraction at P due to a confocal homœoid the external surface of which passes through P. The equation of the surface of the given ellipsoid is

$$\frac{x^2}{a^2} + \frac{y^2}{b^2} + \frac{z^2}{c^2} = 1 \quad \ldots \ldots \quad (92)$$

and the equation of a similar and similarly situated surface within and concentric with it is

$$\frac{x^2}{\theta^2 a^2} + \frac{y^2}{\theta^2 b^2} + \frac{z^2}{\theta^2 c^2} = 1 \quad \ldots \ldots \quad (93)$$

where $1 > \theta > 0$. If $\delta\theta$ be small the equation of the inner surface of a thin homœoid for which (93) is the equation of the outer surface is got from (93) by multiplying θ^2 by $1 - 2\delta\theta/\theta$. Hence the thickness of the homœoid ‡ at any point is $p\delta\theta/\theta$ where p is the perpendicular let fall from the centre on the tangent plane at the point.

applied to find Attraction of Homogeneous Ellipsoid at External Point. Again the equation of an ellipsoid confocal with the outer surface, and passing through P is

$$\frac{\xi^2}{a'^2} + \frac{\eta^2}{b'^2} + \frac{\zeta^2}{c'^2} = 1 \quad \ldots \ldots \quad (94)$$

where $a'^2 = \theta^2(a^2 + \phi^2)$, $b'^2 = \theta^2(b^2 + \phi^2)$, $c'^2 = \theta^2(c^2 + \phi^2)$, so

---

* A shell bounded by two similar and similarly situated concentric ellipsoidal surfaces. Vol. I. p. 47 et seq.
† Vol. I. p. 50.
‡ See Vol. I. p. 49.

ATTRACTION OF HOMOGENEOUS ELLIPSOID

that ϕ is a function of θ given by (94). Let p' be the perpendicular let fall from the centre of this ellipsoid to a tangent plane touching at the point (ξ, η, ζ), then the thickness of a thin elliptic homœoid having (ξ, η, ζ), on its outer surface is $\frac{1}{2}p'\nu$ where ν is a constant small quantity. Now the mass of this homœoid is to be the same as that of thickness $p\delta\theta/\theta$, so that we are to have, if dS', dS be elements of the areas of the surfaces,

$$\tfrac{1}{2}\nu \iint p' dS' = \frac{\delta\theta}{\theta} \iint p\, dS$$

or

$$\nu = 2\frac{abc}{a'b'c'}\theta^2 . \delta\theta.$$

Hence the attraction is $2\pi\rho p'\nu$ or $4\pi\rho p'\theta^2\delta\theta . abc/a'b'c'$, and its direction is along p'. But the direction cosines of p' are $p'\xi/a'^2$, $p'\eta/b'^2$, $p'\zeta/c'^2$, and therefore the x-component of attraction is $4\pi\rho\xi p'^2\theta^2\delta\theta\, abc/a'^3 b'c'$. Hence we get for the x-component of attraction due to the whole ellipsoid

$$X = 4\pi\rho\, abc\, \xi \int_0^1 \frac{p'^2\theta^2 d\theta}{a'^3 b' c'} \quad \ldots \ldots \quad (95)$$

We may simplify this equation by substituting for $p'^2\theta^2 d\theta$ its value $-\theta^5 \phi d\phi$ obtained by differentiating (94). Equation (95) becomes then

Attraction of Homogeneous Ellipsoid at External Point

$$X = 2\pi\rho\, abc\, \xi \int_{\phi_1^2}^{\infty} \frac{d(\phi^2)}{\{(a^2+\phi^2)^3(b^2+\phi^2)(c^2+\phi^2)\}^{\frac{1}{2}}} \quad . \quad (96)$$

where ϕ_1^2 is the positive root of the cubic for ϕ^2 given by (94) when $\theta^2 = 1$. The corresponding expressions for Y and Z may at once be written down by symmetry. It may be noticed that if (ξ, η, ζ) be on the surface of the given ellipsoid the limits of ϕ^2 are 0 and ∞.

Writing now $X = L\xi$, $Y = M\eta$, $Z = N\zeta$, we have by (91)

applied to find Magnetic Potential and Force due to Ellipsoid uniformly Magnetized.

$$V = AL\xi + BM\eta + CN\zeta.$$

Hence the components of magnetic force a, β, γ, at the point (ξ, η, ζ) are

$$a = -A\left(L + \xi\frac{dL}{d\xi}\right), \quad \beta = -B\left(M + \eta\frac{dM}{d\eta}\right), \quad \gamma = -C\left(N + \zeta\frac{dN}{d\zeta}\right) \quad (97)$$

At an internal point the force due to the elliptic homœoid external to the point is zero, and we have only to calculate the force due to a uniform ellipsoid similar and similarly situated to the given ellipsoid. Let the semi-axes be θa, θb, θc respectively, and substitute in (96), observing that the limits of integration are now 0 and ∞.

Then

$$X = 2\pi\rho\theta^3 abc\xi \int_0^\infty \frac{d(\phi^2)}{\{(\theta^2 a^2 + \phi^2)^3 (\theta^2 b^2 + \phi^2)(\theta^2 c^2 + \phi^2)\}^{\frac{1}{2}}}$$

or writing χ^2 for ϕ^2/θ^2, we get

$$X = 2\pi\rho abc\xi \int_0^\infty \frac{d(\chi^2)}{\{(a^2 + \chi^2)^3 (b^2 + \chi^2)(c^2 + \chi^2)\}^{\frac{1}{2}}} \quad . \quad . \quad (98)$$

and similarly for Y and Z. It is to be observed that the integrals are now independent of (ξ, η, ζ). Hence we have, writing L, M, N for the multipliers of ξ, η, ζ, in these expressions,

$$V = AL\xi + BM\eta + CN\zeta \quad . \quad . \quad . \quad . \quad (99)$$

and

$$\alpha = -AL, \ \beta = -BM, \ \gamma = -CN \quad . \quad . \quad . \quad (100)$$

that is, the magnetic force is uniform in value within the ellipsoid, and is the same for the same intensity of magnetization within similar ellipsoids. The direction of the force is not however that of magnetization unless the latter coincide with the direction of one of the axes of the surface; then the force acts in the opposite direction.

Results in Particular Cases. The integral can be easily evaluated in finite terms when the ellipsoid is one of revolution. Thus to find L, we write $(a^2 + \chi^2)^{\frac{1}{2}} = 1/v$, and the integral then reduces to a form at once integrable. Similarly M and N may be dealt with. The results are

(1) Prolate Ellipsoid. (1) For a prolate ellipsoid of eccentricity e ($b = c = a\sqrt{1-e^2}$)

$$\left. \begin{array}{l} L = 4\pi\rho \dfrac{1-e^2}{e^2}\left(\dfrac{1}{2e}\log\dfrac{1+e}{1-e} - 1\right) \\[2mm] M = N = 2\pi\rho \dfrac{1}{e^2}\left(1 - \dfrac{1-e^2}{2e}\log\dfrac{1+e}{1-e}\right) \end{array} \right\} \quad . \quad . \quad (101)$$

(2) For an oblate ellipsoid of eccentricity e, ($b = c = a/\sqrt{1-e^2}$),

$$L = \frac{4\pi\rho}{e^2}\left(1 - \frac{\sqrt{1-e^2}}{e}\sin^{-1} e\right)$$
$$M = N = 2\pi\rho\frac{\sqrt{1-e^2}}{e^2}\left(\frac{1}{e}\sin^{-1} e - \sqrt{1-e^2}\right) \quad \ldots \quad (102)$$

(2) Oblate Ellipsoid.

From these results (writing 1 for ρ) we can easily find formulæ for special cases. Thus if the ellipsoid be infinitely long, (1) gives $L=0$, $M=N=2\pi$.

Thus the magnetic force within an infinitely long uniformly magnetized cylinder is zero if the magnetization is parallel to the axis, and is perpendicular to the axis and equal to $-2\pi\mathbf{I}$ if the cylinder is magnetized transversely.

Again let the ellipsoid be spherical, that is, let $e=0$, and let the direction of magnetization be parallel to the axis of x. Then the force is

$$-\mathbf{I}L = -\tfrac{4}{3}\pi\mathbf{I} \quad \ldots \quad \ldots \quad (103)$$

since $4\pi/3$ is the value of the vanishing fraction which L is in this case.

These two results might have been inferred from the investigation on p. 27 above, of the force within a spherical or cylindrical hollow cut within a magnet.

Lastly, let the ellipsoid be very oblate, a disc in fact, then $M = N = \pi^2 a/c$, and the force at right angles to it is

$$-\mathbf{I}L = -4\pi\mathbf{I} \quad \ldots \quad \ldots \quad (104)$$

These forces are all in the opposite direction to that of the magnetization, and therefore act as demagnetizing forces. We shall consider them fully when we deal with induced magnetization and measurements connected therewith.

SECTION IV.

INDUCED MAGNETIZATION.

We now pass to the consideration of induced magnetization, and shall consider here only the problem of the magnetization produced in a homogeneous body when placed in a uniform magnetic field. The essential

MAGNETIC THEORY

Weber's Theory of Magnetic Induction.
nature of the magnetization of the body is not known to us; but probably Weber's theory is substantially true, viz. that the body consists of particles already magnetized, but so arranged in the unmagnetized mass as to give no resultant external magnetic effect. It has been conceived that the magnetization of each of these small particles may consist in rotatory motion of the ether; and if this be true the direction of rotation is what corresponds to the notion of polarization.

According to Weber's theory, when a body, unmagnetized in mass, is submitted to the action of a magnetic field, the molecular magnets undergo an alignment, so that like extremities are turned preponderatingly in the same direction. Each particle experiences a couple tending to turn its axis into coincidence with the direction of the magnetic force, and, unless prevented from turning by frictional or other resistance, it moves towards that position until brought to rest by an equilibrating couple due to mutual action between the molecule and the surrounding particles. Thus the molecular magnets are in general prevented from coming every one into coincidence with the direction of the magnetic force, in which case no further magnetization would be possible, and we know that by increasing the magnetic force we can increase the magnetization, although not in an unlimited degree.

Again, when the magnetizing force is removed the substance does not in general return to its former unmagnetized state, but does so only to a certain extent, retaining under some circumstances a very considerable amount of magnetization.

INDUCED MAGNETIZATION

This property of resisting magnetizing action, and of retaining residual magnetization, is sometimes called *coercive force*. It has been attributed to something analogous to frictional resistance, which prevents the magnetic particles from moving freely in obedience to the magnetizing force and from returning freely when it is removed. A theory in which the mutual action of the molecular magnets plays the chief part has been put forward by Ewing, and will be considered in Chapter XIII., but it may be stated here that mechanical agitation, such as jarring or tapping an iron wire or bar, in general increases the magnetization while the body is under the influence of magnetic force, and diminishes the magnetization when the magnetizing force has been removed. The mechanical disturbance enables the particles to obey more completely the magnetizing or demagnetizing action, as the case may be.

Coercive Force.

If a piece of iron be subjected to a gradually increasing magnetic force, and then to a gradually decreasing one, the two magnetizations for the same magnetic force, are, in consequence of residual magnetism, not identical. This phenomenon we shall see indicates dissipation of energy in the magnetized iron. We shall deal with this, and with other phenomena when treating of the experimental work on this subject.

We shall consider first the case of a spherical portion of an æolotropic body placed in a uniform magnetic field, and examine the magnetization which it receives on the following supposition:

The total magnetization which the magnet receives is the resultant of the magnetizations which the several parts of the magnetizing system would produce if each acted alone.

This implies, first, that if the intensity of the field at each point is altered in any ratio, the magnetization is simply altered

Æolotropic Sphere in Uniform Field.

58   MAGNETIC THEORY

Assumption as to Superposition of Magnetizations. in intensity in this ratio at each point without change of direction; second, that magnetizations in different directions are produced in the substance, and are superposed as if no other magnetization were present. In point of fact these conditions do not hold, and their assumption gives only an approximation to the result in any actual case. The magnetic susceptibility is, as we shall see, a function of the magnetizing force; and the magnetic behaviour of the material is further complicated in a great many ways not here taken into account.* The investigation now to be given however yields results of great theoretical interest which are of much importance in the theory of diamagnetism and magnecrystallic action.

The supposition made above gives for A, B, C expressions of the form

Magnetization Intensities parallel to the Axes.
$$\left. \begin{array}{l} A = p\alpha + u\beta + t\gamma \\ B = u'\alpha + q\beta + s\gamma \\ C = t'\alpha + s'\beta + r\gamma \end{array} \right\} \quad \ldots \quad (105)$$

where α, β, γ are as usual the components of the resultant magnetic force **H**, and $p, q, r, s, s', t, t', u, u'$, coefficients. We shall see directly that $s = s'$, $t = t'$, $u = u'$, that is, that a magnetic force of a certain intensity acting in the direction of the axis of y produces the same intensity of magnetization parallel to the axis of x, as is produced parallel to the axis of y, by an equal force acting parallel to the axis of x, and so for any other two axes.

Taking the sphere as of unit volume the component magnetic moments due to the magnetization are simply A, B, C. Hence **Couples acting on Sphere.** the sphere in the field is acted on by a couple round each of the axes, the moment of which is, for the z axis, $\beta A - \alpha B \, (= \mathbf{N}, \text{say})$. Now let the axes be fixed in the sphere, and let it then be turned round the axis of z. If θ be the angle which the direction of **H** makes with the axis of z, ϕ the angle which the projection of **H** on the plane of xy makes with the axis of x, we have

$$\mathbf{H} \sin \theta \cos \phi = \alpha, \quad \mathbf{H} \sin \theta \sin \phi = \beta, \quad \mathbf{H} \cos \theta = \gamma.$$

Hence the work done in increasing ϕ by the small angle $d\phi$ is

$$\mathbf{N} \, d\phi = - \mathbf{H}^2[\{u' \cos^2\phi - u \sin^2\phi + (q-p) \sin \phi \cos \phi\} \sin^2\theta \\ + (s \cos \phi - t \sin \phi) \sin \theta \cos \theta] \, d\phi \quad . \quad . \quad (106)$$

---

* For a detailed examination of the theory of inductive magnetization the reader may refer to Duhem's *L'Aimantation par Influence*. Paris, 1888.

The work done in turning the body through a complete revolution is therefore

$$\int_0^{2\pi} N d\phi = \pi \mathbf{H}^2 \sin^2 \theta (u - u') \quad \ldots \quad (107)$$

Now since the body has come back to its former position its magnetization is by hypothesis the same as before, and no work can, on the whole, have been spent or gained in the revolution, otherwise the body would be either a continual source of energy, that is, a perpetual motion, or a place where energy is continually dissipated. We shall see later that there can be no such dissipation of energy on the supposition of constant magnetic susceptibility. Assuming then that the work is zero we have $u' = u$, and in the same way we could show that $s = s'$, $t = t'$. The equations (105) above can therefore be written

Reciprocal Relation between Coefficients.

$$\left. \begin{array}{l} A = p\alpha + u\beta + t\gamma \\ B = u\alpha + q\beta + s\gamma \\ C = t\alpha + s\beta + r\gamma \end{array} \right\} \quad \ldots \quad (108)$$

The magnetization in any direction the cosines of which are $l, m, n,$ is

Magnetization in any direction

$$lA + mB + nC$$
$$= pl\alpha + qm\beta + rn\gamma + u(l\beta + m\alpha) + s(m\gamma + n\beta) + t(n\alpha + l\gamma) \quad (109)$$

If l, m, n be the direction cosines of $\mathbf{H}$ this equation becomes

$$lA + mB + nC = \mathbf{H}(pl^2 + qm^2 + rn^2 + 2ulm + 2smn + 2tnl) \quad (110)$$

Now if we consider the quadric surface of which the equation is

determined by reference to a certain Quadric surface.

$$r^2(pl^2 + qm^2 + rn^2 + 2ulm + 2smn + 2tnl) = 1 \quad (111)$$

we see that the quantity within the brackets in (110) is inversely proportional to the square of the radius of this surface drawn in the direction of l, m, n. Hence for different directions of $\mathbf{H}$ the magnetic moment of the sphere in the direction of $\mathbf{H}$ is $\mathbf{H}/r^2$, where r is the radius in that direction of the quadric surface of which (111) is the equation.

Further since by (108) $A = \mathbf{H}(pl + um + tn)$, &c., we see that the resultant magnetization is in the direction of the normal drawn to the quadric at the point at which the radius in the direction of $\mathbf{H}$ cuts the surface, or, in other words, is at right angles to the diametral plane of all radii in the direction of l, m, n.

Relation to Direction of Resultant Magnetization.

Principal Axes of Magnetization.
It follows from this that along the principal axes of the quadric surface represented by (111) the magnetization coincides in direction with the magnetizing force. Now the directions of the axes of this quadric are given by the equations

$$\left.\begin{array}{l}pl + um + tn = kl \\ ul + qm + sn = km \\ tl + sm + rn = kn\end{array}\right\} \quad \ldots \quad (112)$$

where k is a constant. Eliminating l, m, n, we get

$$\begin{vmatrix} p-k, & u, & t, \\ u, & q-k, & s, \\ t, & s, & r-k \end{vmatrix} = 0 \quad \ldots \quad (113)$$

which is a cubic from which k can be found. The three roots k_1, k_2, k_3, of this equation successively substituted in (112) enable l, m, n to be calculated for each of the three axes.

Now let **H** be in the direction of one of the axes (l, m, n) thus found. The magnetization in that direction is $lA + mB + nC$.

Principal Magnetic Susceptibilities.
Substituting the values of A, B, C given by (108) and having regard to (112), we see that the magnetization has the value $k\mathbf{H}$. Hence k_1, k_2, k_3 are the magnetic susceptibilities in the direction of the axes just found. They are called the principal magnetic susceptibilities of the substance. If the axes just found be chosen for reference the coefficients s, t, u of (108) vanish, and we have $p = k_1, q = k_2, r = k_3$. We have thus also three principal magnetic inductive capacities, viz.

$$\mu_1 = 1 + 4\pi k_1, \quad \mu_2 = 1 + 4\pi k_2, \quad \mu_3 = 1 + 4\pi k_3. \quad (114)$$

The physical meaning of the principal susceptibilities is apparent from their mode of derivation; that of a principal magnetic inductive capacity μ can at once be inferred. Cut a crevasse at right angles to the axis in question, and suppose the magnetization unaffected in the remainder of the body. Then if **H** be the magnetizing force in the direction of the axis and **B** the induction across the crevasse we have $\mathbf{B} = \mu\mathbf{H}$.

Principal Magnetic Inductive Capacities.

Magnetization in Direction of Principal Axes.
Choosing now the principal axes as axes of reference, and putting a, β, γ for the components of the intensity of the externally produced field, we get by (103) for the components of magnetic force within the sphere the values $a - 4/3 . \pi A$, $\beta - 4/3 . \pi B$, $\gamma - 4/3 . \pi C$. Hence $A = k_1(a - 4/3 . \pi A)$, $B = \&c.$, and therefore

$$A = \frac{k_1}{1 + \tfrac{4}{3}\pi k_1}a, \quad B = \frac{k_2}{1 + \tfrac{4}{3}\pi k_2}\beta, \quad C = \frac{k_3}{1 + \tfrac{4}{3}\pi k_3}\gamma \quad (115)$$

The sphere is acted on as we have seen by three component couples, of values $\gamma B - \beta C$, $aC - \gamma A$, $\beta A - aB$, round the axes of x, y, z respectively, in the positive direction, viz. from x to y, y to z, z to x. Denoting these by **L, M, N**, we get by (115)

Couples on Sphere in Terms of Principal Susceptibilities.

$$\begin{aligned}
\mathbf{L} &= \frac{k_2 - k_3}{(1 + \tfrac{4}{3}\pi k_2)(1 + \tfrac{4}{3}\pi k_3)}\beta\gamma \\
\mathbf{M} &= \frac{k_3 - k_1}{(1 + \tfrac{4}{3}\pi k_3)(1 + \tfrac{4}{3}\pi k_1)}\gamma a \\
\mathbf{N} &= \frac{k_1 - k_2}{(1 + \tfrac{4}{3}\pi k_1)(1 + \tfrac{4}{3}\pi k_2)}a\beta
\end{aligned} \right\} \quad \ldots \quad (116)$$

The sphere thus tends to turn so as to bring the axis of greatest magnetic susceptibility into coincidence with the direction of the magnetizing force. The sphere is therefore in stable or in unstable equilibrium according as the axis of greatest or the axis of least susceptibility is in this direction.

Positions of Stable and Unstable Equilibrium.

The total magnetic induction through the sphere across a central section at right angles to **B** is $\pi r^2 B$, where r is the radius of the sphere. Now the components of **B** are $\mu_1 a'$, $\mu_2 \beta'$, $\mu_3 \gamma'$, where a', β', γ' are the magnetic forces within the sphere in the directions of the principal axes. Let for simplicity the sphere be so placed that the field of force is at right angles to the axis of z. Then $\mathbf{B} = (\mu_1^2 a'^2 + \mu_2^2 \beta'^2)^{\frac{1}{2}}$. But

$$a' = a - 4\pi A/3 = a/(1 + 4\pi k_1/3) = 3a/(\mu_1 + 2);$$

and similarly $\beta' = 3\beta/(\mu_2 + 2)$. Hence substituting, and putting ϕ for the angle between the direction of **H** and the axis of x, and S for the surface integral $\pi r^2 B$, we find

Magnetic Induction through Sphere.

$$S = 3\pi r^2 \mathbf{H}\left\{\left(\frac{\mu_1}{\mu_1 + 2}\right)^2 \cos^2\phi + \left(\frac{\mu_2}{\mu_2 + 2}\right)^2 \sin^2\phi\right\}^{\frac{1}{2}} \quad (117)$$

If μ_1 be the greatest magnetic inductive capacity, and μ_2 the least, the greatest number of unit tubes of induction (or, as they are commonly called, "lines of magnetic force") which can pass through the sphere per unit of the impressed magnetizing force is $3\pi r^2 \mu_1/(\mu_1 + 2)$, and the least number $3\pi r^2 \mu_2/(\mu_2 + 2)$. The sphere therefore tends to set itself so that the magnetic induction through it is a maximum.

Couple on Sphere in terms of Permeabilities.

We may express the resultant couple in terms of S and the magnetic inductive capacities. Clearly if $\gamma = 0$, its value is **N** of (116), and this may be written

$$\mathbf{N} = \frac{9}{4\pi} \frac{\mu_1 - \mu_2}{(\mu_1 + 2)(\mu_2 + 2)} \mathbf{H}^2 \sin\phi \cos\phi$$

$$= -\frac{1}{24\pi^2 r} \frac{(\mu_1 + 2)(\mu_2 + 2)}{\mu_1 + \mu_2 + \mu_1\mu_2} \frac{d(S^2)}{d\phi} \quad \ldots \quad (118)$$

by (117) above. Here πr^3 has been put equal to its value $3/4$, so that the formula is suited to a sphere of any radius.

Case of Isotropic Sphere.

In the particular case of an isotropic sphere $k_1 = k_2 = k_3$, $= k$ say, and the susceptibility and magnetic inductive capacity are the same in all directions. Thus the coefficients of α, β, γ, in (115) are the same, and the common value is $k/(1 + 4/3 \cdot \pi k)$. If k be great this is approximately $3/4\pi$. Hence the magnetization intensity of a highly susceptible sphere is always less than, but nearly equal to $3/4\pi \cdot \mathbf{H}$. It is thus useless to attempt to determine the susceptibility of a highly susceptible substance by experiments on a portion of it of a spherical shape. In the comparison of different specimens, the influence of slight differences of form would completely mask differences of susceptibility.

The couples calculated above vanish for an isotropic sphere, and the sphere is in equilibrium in all positions. The magnetic induction through a central section at right angles to the field

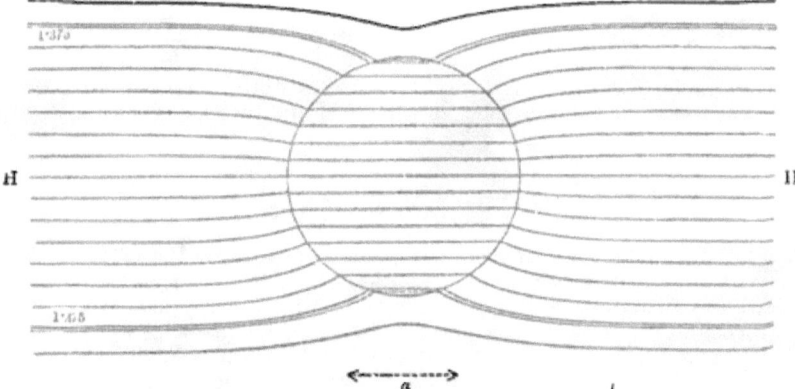

Fig. 11.

ÆOLOTROPIC ELLIPSOID IN UNIFORM FIELD

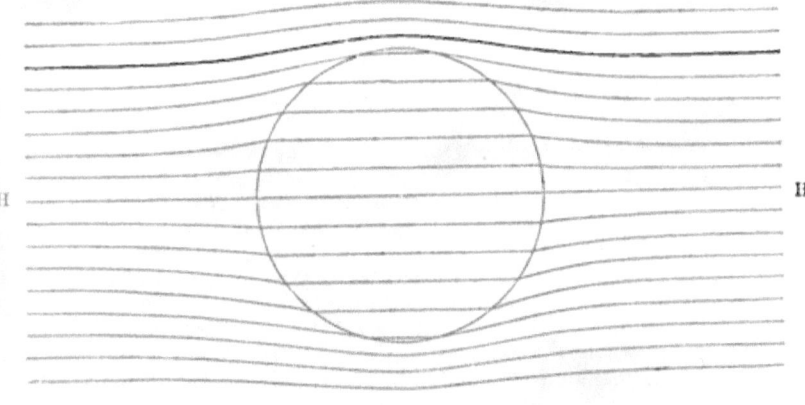

FIG. 12.

is now $3\mu/(\mu + 2) \cdot \pi r^2 \mathbf{H}$. This is greater or less than $\pi r^2 \mathbf{H}$ according as $\mu >$ or < 1. In the former case the body is said to be paramagnetic, in the latter diamagnetic. Thus the number of tubes of induction through the central section is increased by the presence of the substance if paramagnetic, diminished if it is diamagnetic. The field outside and inside in both cases is shown in Figs. (11) and (12), which are Figs. 30, 31, now repeated from Vol. I., pp. 125, 126.

We shall here consider, very briefly, the problem of an æolotropic ellipsoid in a uniform field. We shall suppose the ellipsoid cut with its axes of figure coincident with the principal axes of susceptibility. The force within the body has now the values in the direction of the axes

Æolotropic Ellipsoid in Uniform Field.

$$a' = a - AL, \quad \beta' = \beta - BM, \quad \gamma' = \gamma - CN \ . \ (119)$$

where L, M, N have the values given in (101) above. The second term in each expression is the component force due to the magnetization. Its effect is to oppose the magnetizing component; that is, AL, BM, CN are components of a demagnetizing force. Therefore if k_1, k_2, k_3 be the three principal susceptibilities the values of A, B, C are given by the equations

Components of Magnetizing Force.

$$A = \frac{k_1 a}{1 + k_1 L}, \quad B = \frac{k_2 \beta}{1 + k_2 M}, \quad C = \frac{k_3 \gamma}{1 + k_3 N} \ . \ (120)$$

Hence if the field be at right angles to the axis of z, $\gamma = \gamma' = 0$, and the couple on the ellipsoid is by (120) (if f, g, h be the semi-axes

Couple on Ellipsoid.

$$\mathbf{N} = \tfrac{4}{3}\pi fgh(\beta A - \alpha B) = \tfrac{4}{3}\pi fgh \frac{k_1 - k_2 - k_1 k_2(L - M)}{(1 + k_1 L)(1 + k_2 M)} \alpha\beta \quad (121)$$

Particular Cases:

1. Weakly Magnetic Bodies.

From (120) it follows that if k_1, k_2, k_3 be so small that their second powers may be neglected, $A = k_1 \alpha$, $B = k_2 \beta$, $C = k_3 \gamma$; that is, the internal demagnetizing forces AL, BM, CN, are without sensible effect. These forces depend (see p. 54) upon the form of the body; hence in weakly magnetic substances it is of little consequence whether the body be of elongated shape or not. In fact for such bodies the shape of the specimen experimented on is without influence on the magnetization.

2. Strongly Magnetic Bodies.

If however the values of k_1, k_2, k_3 be very great the magnetization of the body depends almost entirely on the shape of the body, since then the values of A, B, C depend mainly on L, M, N. Thus in highly magnetic bodies such as iron, the magnetization is principally affected by the shape of the specimen. For example, the magnetization in the direction of the axis (say that of x) of a very elongated ellipsoid is practically independent of L, since by (120) we have $A = k_1 \alpha$ simply. On the other hand for a very short ellipsoid (or disc) we have, since $L = 4\pi$, $A = k_1 \alpha/(1 + k_1 L) = (\mu_1 - 1)\alpha/4\pi\mu_1$. In the case of a diamagnetic body (see p. 65 below) k_1 is negative, and hence if the body were shaped so as to give $L = -1/k_1$, a finite magnetizing force would give infinite diamagnetization.

Limiting Values of Couple for Weakly Magnetic and Strongly Magnetic Bodies.

The couple acting on the body has two different limiting values according as the susceptibilities are very small or very great. If the former the couple is

$$\mathbf{N} = \tfrac{4}{3}\pi fgh \frac{(k_1 - k_2)\alpha\beta}{1 + k_1 L + k_2 M} \quad \cdots \quad (122)$$

and the ellipsoid tends to turn its axis of greatest susceptibility parallel to the direction of the field.

If the susceptibilities be very great the couple is approximately $\tfrac{4}{3}\pi fgh(M - L)\alpha\beta/LM$, and this is positive or negative according as $M >$ or $< L$. Hence by the values of M and L (see above, p. 54) the ellipsoid will set itself with its longest dimension parallel to the lines of force.

The influences of form and æolotropy may be made to counteract one another, and, under certain circumstances, by properly shaping the body it may be made to remain in neutral equilibrium when movable, as supposed above, about an axis in the magnetic field.

DIAMAGNETISM

In the case of a homogeneous isotropic ellipsoid the values of A, B, C and of the turning couple are obtained by putting $k_1 = k_2 = k_3 = k$ say, in (120), (121) above. It is obvious at once that the magnetization is not parallel to the resultant magnetic force, but makes with it the angle

$$\cos^{-1}(aA + \beta B + \gamma C)/\{(a^2 + \beta^2 + \gamma^2)(A^2 + B^2 + C^2)\}^{\frac{1}{2}}$$

which vanishes for a sphere. Further we also see that the ellipsoid will turn itself with its longest axis parallel to the lines of force, and this whether k be positive or negative.

Regarding k as a constant, we have, if $a, b, c, a, \beta, \gamma$, be the components of magnetic induction and magnetic force at any point of the medium, the equations

$$a = a(1 + 4\pi k), \quad b = \beta(1 + 4\pi k), \quad c = \gamma(1 + 4\pi k) \quad (123)$$

and therefore since the solenoidal condition holds for the magnetic induction (a, b, c), it also holds for the magnetic force (a, β, γ). Hence also it holds for the induced magnetization (A, B, C), that is, this magnetization is solenoidal.

But since $a = -dV/dx$, &c., and $A = ka$, &c., where k is a constant, we see that the magnetization is also lamellar (p. 41).

It is to be very carefully observed that these results follow from the fact that k is a constant, and do not hold in the general case (unless the magnetization be uniform) in which k is a function of the magnetization, and which is the case of actual practice. The *total* magnetization, made up of the pre-existent magnetization, if any, and the induced magnetization, is not solenoidal, unless the former is itself solenoidal.

We have considered incidentally above the consequences of a negative value of k, and have stated that in that case the substance is said to be diamagnetic. The phenomena of diamagnetism are to some extent explicable on a theory of negative or differential susceptibility. The substance placed in the field behaves as if its polarity were opposite to that which an ordinarily magnetic, or *paramagnetic*, body would receive in the same circumstances.

Hitherto we have been supposing that the medium in which the magnetized substance is placed is of zero susceptibility, that is, possesses unit magnetic inductive capacity. We shall now show that a paramagnetic body placed in a medium of greater magnetic inductive capacity than its own will behave diamagnetically. The medium, being in the field, will be magnetized; and if μ, μ' be the magnetic inductive capacities of the medium

Homogeneous Isotropic Ellipsoid.

If Susceptibility be Uniform,

Induced Magnetization is Solenoidal,

and also Lamellar.

Explanation of Diamagnetism as Differential Effect between Medium and Body

and the substance imbedded in it, respectively, we have, from the continuity of the normal component of magnetic induction at every point of the separating surface, the equation

$$\mu \frac{dV}{dr} + \mu' \frac{dV'}{dr'} = 0 \quad \ldots \ldots \quad (124)$$

in which r, r' denote normals drawn from the surface into the respective media. But we have also the characteristic equation of the surface (Vol. I., p. 28),

$$\frac{dV}{dr} + \frac{dV'}{dr'} + 4\pi\sigma = 0 \quad \ldots \ldots \quad (125)$$

where σ is the surface density of free magnetism on the separating surface at the point where the normals are drawn. These two equations give

$$\sigma = \frac{\mu' - \mu}{4\pi\mu} \frac{dV'}{dr'} = \frac{\mu - \mu'}{4\pi\mu'} \frac{dV}{dr} \quad \ldots \quad (126)$$

One Medium within another showing Diamagnetism. The first multiplier is positive and the second negative if $\mu' > \mu$, that is if the substance be of higher susceptibility than the medium. Hence the surface density of magnetization where the lines of force pass from the medium to the body is positive or negative, and where they pass from the body to the medium negative or positive, according as $\mu' <$ or $> \mu$. In the latter case the body behaves as a paramagnetic, in the former as a diamagnetic substance. If μ be put $= 1$ in the above results, we have the case of a paramagnetic or diamagnetic body in a medium of zero susceptibility. We shall see later, however, that this explanation does not account for all the facts of diamagnetism.

These results are in accordance with and explain the behaviour of a solution of a magnetic salt of iron suspended in a tube within another solution of greater or less strength, and the whole placed in a magnetic field. In the former case the suspended salt behaves as a diamagnetic, in the latter as a paramagnetic.

Magnetic Poles. A great deal of research and theoretical discussion has been spent on the question of the distribution of magnetism in straight bar magnets of rectangular or circular section, and it may be well, before leaving the subject of magnetic theory, to refer to the widely prevalent idea of the existence of poles in ordinary magnets. It has been tacitly supposed by many persons that there are two definite points or poles, one near each end of a regularly* magnetized bar at which the whole of the free magnetism of the bar may be supposed concentrated, the negative at one, the positive at the other, and much time and labour have been spent in determining the positions of these poles. Now certainly, according to the theory given above, there is a certain amount of free positive and an equal amount of free negative magnetism in every magnet, but so far as the action of the magnet at external or internal points is concerned, there are no such definite points, except in the theoretical case of an infinitely thin and uniformly magnetized filament, in which case the poles are at its extremities. *(non-existent in sense of 'Centres of Gravity' of Magnetism;)*

In an accurate sense the magnet can be said to have poles or points at which the free magnetism may be supposed concentrated, when the couple it experiences when placed in a uniform field is considered. If its axis is at right angles to the lines of force, the couple experienced is equal to the magnetic moment of the magnet, and this may be regarded as due to two forces, each of which is the resultant of the parallel forces on the elements of free magnetism. The centres of these two systems of parallel forces, or, which is the same thing, the "centres of mass" of the two distributions of free magnetism are the poles in this connection. The idea of pole is not here of any utility, as the poles could not be determined; what we are concerned with *(existent, but useless, in sense of 'Centres of Mass' of Magnetism.)*

---

* That is, without "consequent points," not necessarily uniformly.

is the magnetic moment only, which is the resultant of the couples exerted on the molecular magnets composing the body.

Case in which Existence of Poles may be Assumed.   But as a matter of approximation the existence of poles in the sense of points at which, if the free magnetisms were concentrated, the action of the magnet at an external point would be the same as it actually is, can be assumed in certain cases, and their position assigned. For example, when we consider the mutual forces between two magnets, each symmetrical about its axis and about a plane at right angles to the axis, and at a distance apart which is great in comparison with any dimension of either, such positions of the poles can be found, and the distance between them used as the virtual length of the magnet.

CHAPTER II

DETERMINATION OF THE HORIZONTAL COMPONENT OF THE EARTH'S MAGNETIC FIELD

ALL the methods by which galvanometers may be graduated so as to measure currents and potentials in absolute units, involve, directly or indirectly, a comparison of the indications of the instrument to be graduated with those of a standard instrument, of which the constants are fully known for the place at which the comparison is made. There are various forms of such standard instruments, as, for example, the tangent galvanometer which Joule made, consisting of a single coil of large radius and a small needle hung at its centre, or the Helmholtz modification of the same instrument with two large equal coils placed side by side at a distance apart equal to the radius of either; or some form of "dynamometer," or instrument which instead of the needle of the galvanometer has a movable coil, in which the whole or a known fraction of the current in the fixed coil flows. The measurement consists essentially in determining the couple which must be exerted by the earth's magnetic force on the needle or suspended coil, in order to equilibrate that exerted by the current. But the former depends on the value, usually denoted by H,[*] of the horizontal

[*] We use, it is to be noted, the symbol H to denote the resultant magnetic force at any point of a magnetic field.

component of the earth's magnetic force, and it is necessary therefore, except when some such method as that of Kohlrausch, referred to below, is used, to know the value of that quantity in absolute units.

Determination of H. The value of H may be determined in various ways, but we shall here describe only one or two methods which are most convenient in practice. We shall take first that due to Gauss,* which consists in finding (1) the angle through which the needle of a magnetometer is deflected by a magnet placed in a given position at a given distance, (2) the period of vibration of the magnet when suspended horizontally in the earth's field, so as to be free to turn round a vertical axis. The first operation gives an equation involving the ratio of the magnetic moment of the magnet to the horizontal component H of the terrestrial magnetic force, the second an equation involving the product of the same two quantities. We shall describe this method somewhat in detail, but with direct reference to physical laboratory work only, and shall therefore not enter into a discussion of the instruments and methods peculiar to magnetic observatories, which would occupy more space than can here be spared.

Magnetometer. A very convenient form of magnetometer is that devised by Mr. J. T. Bottomley, and made by hanging within a closed chamber, by a silk fibre from 6 to 10 cms. long, one of the little mirrors with attached magnets used in Thomson's reflecting galvanometers. The fibre is carefully attached to the back of the mirror, so that the magnets hang horizontally and the front of the mirror is vertical. The closed chamber for the fibre and mirror is very readily made by cutting a narrow groove to within

---

* "Intensitas vis magneticae ad mensuram absolutam revocata."— *Comment. Soc. Reg. Gotting.* 1833.

a short distance of each end, along a piece of mahogany about 10 cms. long. This groove is widened at one end to a circular space a little greater in diameter than the diameter of the mirror. The piece of wood is then fixed with that end down in a horizontal base-piece of wood furnished with three levelling screws. The groove is thus placed vertical; and the fibre carrying the mirror is suspended within it by passing the free end of the fibre through a small hole at the upper end of the groove, adjusting the length so that the mirror hangs within the circular space at the bottom, and fixing the fibre at the top with wax. When this has been done, the chamber is closed by covering the face of the piece of wood with a strip of glass, which may be kept in its place either by cement, or by proper fastenings which hold it tightly against the wood. By making the distance between the back and front of the circular space small, and its diameter very little greater than that of the mirror, the instrument can be made very nearly "dead beat,"—that is to say, the needle when deflected through any angle comes to rest at once, almost without oscillation about its position of equilibrium. A magnetometer can be thus constructed at a trifling cost, and it is much more accurate and convenient than the magnetometers furnished with long magnets frequently used for the determination of H; and as the poles of the needle may always in practice be taken at the centre of the mirror, the calculations of results are much simplified.

The instrument is set up with its glass front in the magnetic meridian, and levelled so that the mirror hangs freely inside its chamber. The foot of one of the levelling screws should rest in a small conical hollow cut in the table or platform, of another in a V-groove the axis of which is in line with the hollow, and the third on the plane surface of the table or platform. When thus set up the instrument is perfectly steady, and if disturbed can in an instant be replaced in exactly the same position. A beam of light passes through a slit, in which a thin vertical cross-wire is fixed, from a lamp placed in front of the magnetometer, and is reflected, as in Thomson's reflecting galvanometer, from the mirror to a scale attached to the lamp-stand, and facing the mirror. The lamp and scale are moved nearer to or farther from the mirror, until the position at which the image of the cross-wire of the slit is most distinct is obtained. It is convenient to make the horizontal distance of the mirror from the scale for this position if possible one metre. The lamp-stand should also have three levelling screws, for which the arrangement of conical hollow, V-groove, and plane should be adopted. The scale should be straight, and placed with its length in the magnetic

Adjustment of Magnetometer.

72   THE EARTH'S MAGNETIC FIELD

north and south line; and the lamp should be so placed that the incident and reflected rays of light are in an east and west vertical plane, and that the spot of light falls near the middle of the scale. To avoid errors due to variations of length, the scale, if of paper, should be glued to the wooden backing which carries it, not simply fastened with drawing pins as is often the case.

Preparation of Deflecting Magnets.

The magnetometer having been thus set up, four or five magnets, each about 10 cms. long and ·1 cm. thick, and tempered glass-hard, are made from steel wire. This is best done as follows. From ten to twenty pieces of steel wire, each perfectly straight and having its ends carefully filed so that they are at right angles to its length, are prepared. These are tied tightly into a bundle with a binding of iron wire and heated to redness in a bright fire. The bundle is then quickly removed from the fire, and plunged with its length vertical into cold water. The wires are thus tempered glass-hard without being seriously warped. They are then magnetized to saturation in a helix by a strong current of electricity.

Deflection Experiments

A horizontal east and west line passing through the mirror is now laid down on a convenient platform (made of wood put together without iron and extending on both sides of the magnetometer) by drawing a line through that point at right angles to the direction in which a long thin magnet hung by a single silk fibre there places itself (see also p. 80).

FIG. 13.

One of the magnets is placed, as shown in Fig. 12, with its length in that line, and at such a distance that a convenient deflection of the needle is produced. This deflection is noted and the deflecting magnet turned end for end, and the deflection again noted. In the same way a pair of observations are made with the magnet at the same distance on the opposite side of the magnetometer; and the mean of all the observations is taken. These deflections from zero ought to be as nearly as may be the same, and if the magnet is properly placed, they will exactly agree; but the effect of a slight error in placing the magnet will

'End-on' Position.

be nearly eliminated by taking the mean deflection. The distance in cms. between the two positions of the centre of the magnet is also noted, and is taken as twice the distance of the centre of the magnet from that of the needle. The same operation is gone through for each of the magnets, which are carefully kept apart from one another during the experiments. The results of each of these experiments give an equation involving the ratio of the magnetic moment of the magnet to the value of H. Thus if M denote the magnetic moment of the magnet, M' the magnetic moment of the needle, r the distance of the centre of the magnet from the centre of the needle, 2λ the distance between the "poles" (p. 68) of the magnet which, for a nearly uniformly magnetized magnet of the dimensions stated above, is nearly equal to its length, and $2\lambda'$ the distance between the poles of the needle, r, λ, and λ' being all measured in cms., we have for the repulsive force (denoted by F in Fig. 13) exerted on the blue* pole of the needle by the blue pole of the magnet, supposed nearest to the needle, as in Fig. 13, the value $M/2\lambda \cdot M'/2\lambda' \cdot 1/(r-\lambda)^2$, since the value of λ' is small compared with λ. Similarly for the attraction exerted on the same pole of the needle by the red pole of the magnet, we have the expression $M/2\lambda \cdot M'/2\lambda' \cdot 1/(r+\lambda)^2$. Hence the total repulsive force exerted by the magnet on the blue pole of the needle is

$$\frac{MM'}{4\lambda\lambda'}\left\{\frac{1}{(r-\lambda)^2}-\frac{1}{(r+\lambda)^2}\right\} \text{ or } M\frac{M'}{\lambda'}\frac{r}{(r^2-\lambda^2)^2}.$$

Proceeding in a precisely similar manner, we find that the magnet of moment M exerts an attractive force equal to $MM'r/\lambda'(r^2-\lambda^2)^2$ on the red pole of the magnet. The needle is therefore acted on by a "couple" which tends to turn it round the suspending fibre as an axis, and the amount of this couple, when the angle of deflection is θ, is plainly equal to $2MM'r\cos\theta/(r^2-\lambda^2)^2$. But for equilibrium this couple must be balanced by $M'H\sin\theta$: hence we have the equation—

$$\frac{M}{H}=\frac{(r^2-\lambda^2)^2}{2r}\tan\theta \quad . \quad . \quad . \quad . \quad (1)$$

If the arrangement of magnetometer and straight scale described above is adopted, the value of $\tan\theta$ is easily obtained, for the number of divisions of the scale which measures the

---

* The convention according to which the end of the needle which has magnetic polarity of the same kind as that of the earth's northern regions is coloured blue, and the other red is here adopted. The letters B, R, b, r, in the diagrams denote blue and red.

74 THE EARTH'S MAGNETIC FIELD

'Side-on' Position. deflection, divided by the number of such divisions in the distance of the scale from the mirror, is then equal to tan 2θ.

Instead of in the east and west horizontal line through the centre of the needle, the magnet may be placed, as represented in Fig. 13', with its length east and west, and its centre in the horizontal north and south line through the centre of the needle. If we take M, M', λ, λ', and r to have the same meaning as before, we have, for the distance of either pole of the magnet from the needle, the expression $\sqrt{r^2 + \lambda^2}$. Let us consider the force acting on one pole, say the red pole of the needle. The

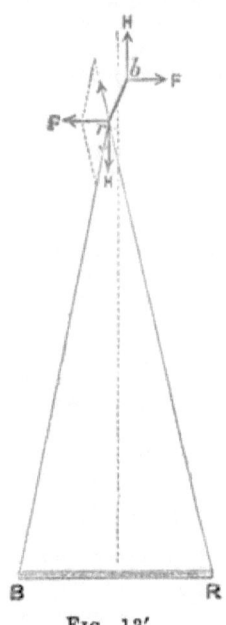

FIG. 13'.

red pole of the magnet exerts on it a repulsive force, and the blue pole an attractive force. Each of these forces has the value $M/2\lambda \cdot M'/2\lambda' \cdot 1/(r^2 + \lambda^2)$. But the diagram shows that they are equivalent to a single force, F, in a line parallel to the magnet, tending to pull the red pole of the needle towards the left. The magnitude of this resultant force is plainly $2M/2\lambda \cdot M'/2\lambda' \cdot \lambda/(r^2 + \lambda^2)^{\frac{3}{2}}$ or $MM'/2\lambda'(r^2 + \lambda^2)^{\frac{3}{2}}$. In the same way i can be shown that the action of the magnet on the red pole of the needle is a force of the same amount tending to pull the blue pole of the needle towards the right. The needle is, therefore, subject to no force tending to produce motion of translation, but simply to a "couple" tending to produce rotation. The magnitude of this couple when the needle has been turned through an angle θ, is $MM'/2\lambda' \cdot 2\lambda' \cos \theta/(r^2 + \lambda^2)^{\frac{3}{2}}$, or $MM' \cos \theta/(r^2 + \lambda^2)^{\frac{3}{2}}$. If there be equilibrium for the deflection θ, this couple must be balanced by that due to the earth's horizontal force, which, as before, has the value $M'H \sin \theta$. Hence equating these two couples we have—

$$\frac{M}{H} = (r^2 + \lambda^2)^{\frac{3}{2}} \tan \theta \quad . \quad . \quad . \quad . \quad . \quad (2)$$

Still another position of the deflecting magnet relatively to the needle may be found a convenient one to adopt. The magnet may be placed still in the east and west line, but with its centre vertically above the centre of the needle. The couple in this case also is given by the formula just found, in which the symbols have the same meaning as before.

The greatest care should be taken in all these experiments, as well as in those which follow, to make sure that there is no movable iron in the vicinity; and the instruments and magnets should be kept at a distance from any iron nails or bolts there may be in the tables on which they are placed.

We come now to the second operation, the determination of the period of oscillation of the deflecting magnet when under the influence of the earth's horizontal force alone. The magnet is hung in a horizontal position in a double loop formed at the lower end of a single fibre of unspun silk, attached by its upper end to the roof of a closed chamber. A box about 30 cms. high and 15 cms. wide, having one pair of opposite sides, the bottom, and the roof made of wood, and the remaining two sides made of plates of glass, one of which can be slid out to give access to the inside of the chamber, answers very well. The fibre may be attached at the top to a horizontal axis which can be turned round from the outside so as to wind up or let down the fibre when necessary. The suspension-fibre is so placed that two vertical scratches, made along the glass sides of the box, are in the same plane with it when the magnet is placed in its sling, and the box is turned round until the magnet is at right angles to the glass sides. A paper screen with a small hole in it is then set up at a little distance in such a position that the hole is in line with the magnet, and therefore in the same plane as the scratches. The magnetometer should be removed from its stand

Determination of Period of Oscillation of Deflecting Magnet.

and this box and suspended needle put in its place. If the magnet be now deflected from its position of equilibrium and then allowed to vibrate round a vertical axis, it will be seen through the small hole to pass and re-pass the nearer scratch, and an observer keeping his eye in the same plane as the scratches can easily tell without sensible error the instant when the magnet passes through the position of equilibrium. Or, a line may be drawn across the bottom of the box so as to join the two scratches, and the observer keeping his eye above the magnet and in the plane of the scratches may note the instant when the magnet, going in the proper direction, is just parallel to the horizontal line. The operator should deflect the magnet by bringing a small magnet near to it, taking care to keep this small deflecting magnet always as nearly as may be with its length in an east and west line passing through the centre of the suspended magnet. If this precaution be neglected the magnet may acquire a pendulum motion about the point of suspension, which will interfere with the vibratory motion in the horizontal plane. When the magnet has been properly deflected and left to itself, its range of motion should be allowed to diminish to about 3° on either side of the position of equilibrium before observation of its period is begun. When the amplitude has become sufficiently small, the person observing the magnet says sharply the word "Now," when the nearer pole of the magnet is seen to pass the plane of the scratches in either direction, and another observer notes the time on a watch having a seconds hand. With a good watch having a centre seconds hand moving round a dial divided into quarters or fifths of a second, the instant of time can be determined with greater accuracy in this way than by means of any of the usual appliances for starting and stopping watches, or for registering on a dial the position of a seconds hand when a spring is pressed by the observer. The person observing the magnet again calls out "Now" when the magnet has just made ten complete to and fro vibrations, again after twenty complete vibrations, and, if the amplitude of vibration has not become too small, again after thirty; and the other observer at each instant notes the time by the watch. By a complete vibration is here meant the motion of the magnet from the instant when it passes through the position of equilibrium in either direction, until it next passes through the position of equilibrium going in the same direction. The observers then change places and repeat the same operations. In this way a very near approach to the true period is obtained by taking the mean of the results of a sufficient number of observations, and from this the value of the product of M and H can be calculated.

For a small angular deflection θ of the vibrating magnet from the position of equilibrium the equation of unresisted motion is

$$\frac{d^2\theta}{dt^2} + \frac{MH}{\mu}\theta = 0,$$

where μ is the moment of inertia of the vibrating magnet round an axis through its centre at right angles to its length. The solution of this equation is

$$\theta = A \sin\left\{\sqrt{\frac{MH}{\mu}}\, t - B\right\}$$

and therefore for the period of oscillation T we have

$$T = 2\pi \sqrt{\frac{\mu}{MH}}.$$

Hence we have

$$MH = \frac{4\pi^2\mu}{T^2}.$$

Now, since the thickness of the magnet is small compared with its length, if W be the mass of the magnet and $2l$ its actual length, μ is $Wl^2/3$, and therefore

$$MH = \frac{4\pi^2 l^2 W}{3 T^2} \quad \ldots \ldots \quad (3)$$

Combining this with the equation (1) already found we get for the arrangement shown in Fig. 12,

$$M^2 = \frac{2}{3} \cdot \frac{\pi^2 (r^2 - \lambda^2)^2 l^2 W \tan\theta}{T^2 r} \quad \ldots \quad (4)$$

and

$$H^2 = \frac{8}{3} \cdot \frac{\pi^2 l^2 r W}{T^2 (r^2 - \lambda^2)^2 \tan\theta} \quad \ldots \quad (5)$$

If either of the other two arrangements be chosen we have from equations (2) and (3)

$$M^2 = \frac{4}{3} \cdot \frac{\pi^2 l^2}{T^2} (r^2 + \lambda^2)^{\frac{3}{2}} W \tan\theta \quad \ldots \quad (6)$$

and

$$H^2 = \frac{4}{3} \cdot \frac{\pi^2 l^2 W}{(r^2 + \lambda^2)^{\frac{3}{2}} T^2 \tan\theta}. \quad \ldots \quad (7)$$

78 THE EARTH'S MAGNETIC FIELD

Corrections. Various corrections which are not here made are of course necessary in a very exact determination of H. The magnetic distribution must be taken account of (see pp. 85 and 92 below). Allowances should be made for the magnitude of the arc of vibration; the torsional rigidity if sensible (see below, Chapter XIV.) of the suspension fibre of the magnetometer in the deflection experiments and of the suspension fibre of the magnet in the oscillation experiments; the frictional resistance of the air to the motion of the magnet; the virtual increase of inertia of the magnet due to motion of the air in the chamber; and the effect of induction and, if necessary, of changes of temperature in producing temporary changes in the moment of the magnet. The correction for an arc of oscillation of 6° is a diminution of the observed value of T of only $\frac{1}{80}$ per cent., and for an arc of 10° of $\frac{1}{20}$ per cent. Of the other corrections

Correction for Induction Error. that for induction is no doubt the most important; but its amount for a magnet of glass-hard steel, nearly saturated with magnetism, and in a field so feeble as that of the earth, may, if only a roughly accurate result is required, be neglected.

This correction arises from the fact that the magnet in the deflection experiments is placed in the magnetic east and west line, whereas in the oscillation experiments it is placed north and south, and is therefore subject in the latter case to an increase of longitudinal magnetization from the action of terrestrial magnetic force. The increase of magnetic moment may be determined by the following method, which is due to Prof. Thomas Gray. Place the magnet within, and near the centre of, a helix, considerably longer than the magnet and made of insulated copper wire. Place the helix and magnet in position either as shown in Fig. 13, or as in Fig. 13', for giving a deflection of the magnetometer needle, and read the deflection. Then pass such a current through the wire of the helix as will give by electromagnetic induction a magnetic field within the helix nearly equal to the horizontal component of the earth's field, and again observe the deflection of the magnetometer needle. The field-intensity within the helix at points not near the ends is given, as will be seen in Chap. VI., in C.G.S. units by the formula $4\pi nC$, where n is the number of turns per centimetre of length of the helix, and C is the strength of the current in C.G.S. units. Experiments may be made with different strengths of current, and the results put down in a short curve, from which the correction can be at once read off when the approximate field has been determined by the method of deflection and oscillation described above. Care must of course be taken in experimenting to eliminate the deflection of the magnetometer needle caused by the current in the coil.

DETERMINATION OF H

This is easily done by observing the deflection produced by the current when the magnet is not inside the coil and subtracting this from the previous deflection, or by arranging a compensating coil through which the same current passes. This plan, as will be seen below, has several advantages. The change of magnetic moment produced in hard steel bars, the length of which is 12 cms. and diameter ·2 cm., and previously magnetized to saturation, is, according to Prof. Thomas Gray's experiments, about $\frac{1}{20}$ per cent. Particulars of actual experiments are given below.

The deflection experiments are, as stated above, to be performed with several magnets, and when the period of oscillation of each of these has been determined, the magnetometer should be replaced on its stand, and the deflection experiments repeated, to make sure that the magnets have not changed in strength in the meantime. The length of each magnet is then to be accurately determined in centimetres, and its weight in grammes; and from these data and the results of the experiments the values of M and of H can be found for each magnet by the formulas investigated above. Equation (5) is to be used in the calculation of H when the arrangement of magnetometer and deflecting magnet, shown in Fig. 13, is adopted, equation (7) when that shown in Fig. 13' is adopted.

The object of performing the experiments with several magnets, is to eliminate as far as possible errors in the determination of weight and length. The mean of the values of H, found for the several magnets, is to be taken as the value of H at the place of the magnetometer.

The following is an account of a determination of H made by this method, with several improvements in the practical carrying of it out, by Mr. Thomas Gray in the Physical Laboratory of the University of Glasgow, during the summer of 1885. The apparatus and its arrangement is shown in Fig. 14. T represents a table which supports the magnetometer M, two stands A and B for the deflecting magnets, and a lamp and scale S. The magnetometer consisted, as described above, of a light mirror about ·8 cm. in diameter, suspended by a single silk fibre within a recess in a block of wood, and carrying on its back two magnets each 1 cm. long and ·08 cm. in diameter. Two holes cut in the wood at right angles to one another (and plugged when not in use) permitted the position of the mirror and magnets to be seen and adjusted. [A preferable form of magnetometer since adopted consists of a mirror and attached magnet suspended within a glass tube from a brass mounting at the upper end which allows the fibre to be wound up or down. For definiteness

Account of Actual Experiments.

80   THE EARTH'S MAGNETIC FIELD

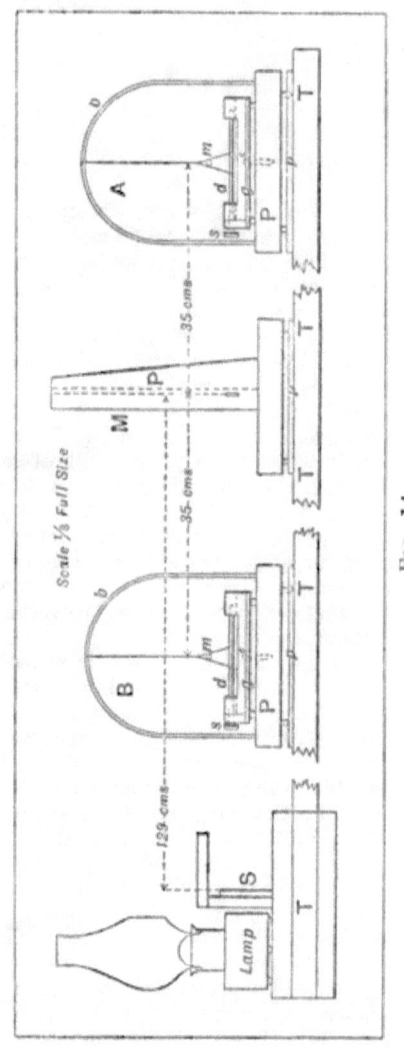

Fig. 14.

DETERMINATION OF H.

and ease of determination of the magnetic centre of the needle, a single small cylindrical magnet is used, carried at the lower end by a short strip of aluminium to which the mirror is attached. At the lower end half the tube is cut away over a length of two or three cms., and the part remaining closes the back of the chamber in which the mirror hangs. The sides of the chamber are of wood attached to the base piece, and the front, or side toward the lamp, is closed by a panel of glass. The vibration of the needle can be checked by a small coil, placed near the needle and in circuit with a cell and reversing key. By depressing the key for an instant in the direction to oppose the motion of the needle when it is passing through the central part of its swing, the needle can be quickly stopped in its position of equilibrium.] The sole plate P, made of mahogany, is supported on three brass feet, which rest in a hole, slot, and plane arrangement cut, as described above, in a horizontal plate of glass cemented to the table.

The deflector stands, A, B, rest each on a base plate P, of mahogany, supported, according to the hole, slot, and plane device, in precisely the same way as the magnetometer, on plates of glass p, p, cemented to the table T. Each stand consists of a horizontal carriage for the deflector magnet, and is constructed as follows: A strip of hard wood, about 13 cms. long and 4 cms. broad, has a V-shaped groove run along its whole length in the middle of one side. One end is faced with a plate of brass in which a brass screw works, and the piece is cemented with the groove upwards to a plate of glass g. This plate is supported on three feet of hard wood, resting on the mahogany sole plate P, and is free to turn in azimuth round a closely fitting centre pivot c fixed in the sole plate. The apparatus is so adjusted that the bottom of the V-groove is just over the pivot c. The magnet when placed in the carriage lies along the groove, and the screw s serves to give a fine adjustment of the position of one end which abuts against it. Over each carriage a wire of brass or copper bent into a semi-circle serves as a support for a suspension fibre with double loop, by which the deflector can be suspended for purposes of adjustment or for the oscillation experiments. A glass shade can be placed on the plate P to prevent currents of air from disturbing the magnet in the oscillation experiments.

In Fig. 14 the deflecting magnets d, d, are shown in positions at equal distances east and west of the magnetometer, at a distance of 70 cms. between their centres. Four plates of glass are fixed to the table in two end-on positions and in two side-on positions, each pair of positions being at equal distances from the magnetometer needle, and on opposite sides of it. The scale

Arrangement of Deflector.

S, shown at a distance from the mirror of 129 cms., is a millimetre scale carefully divided on transparent glass so that the spot of light may be observed either from the front or the back.

The first adjustment, made in setting up the apparatus, was to place the table so that the line joining the centres of A and B should be exactly at right angles to the magnetic meridian. This was done by one or other of the following two methods according as (*a*) the end-on, or (*b*) the side-on position was required. (*a*) After the adjustment had been first roughly made, a plane circuit was formed by stretching a thin wire along the line joining the centres of A, B under the magnetometer needle, and then carrying the wire back, either above the magnetometer, or below it, at a greater distance, in a vertical plane. An electric current was then sent through the wire, and the table T, with the apparatus, turned until the current produced no deflection of the needle. (*b*) One of the deflecting magnets was placed in its carriage, either south or north of the needle, and lifted out of the V-groove by the suspension fibre, and the table turned until the suspended magnet produced no deflection of the magnetometer needle. The magnet and needle were then in one line, and if the needle was in its proper position this line produced through the centre of the needle passed through the position of the deflector on the other side. The deflector was placed on the opposite side of the needle, and the table T, turned until no deflection was obtained. The position of the needle was then altered, if necessary, by the levelling screws until the positions of the table for no deflection, with the magnet first on one side then on the other of the magnetometer, were coincident. If this could not be done the plates p were not placed with sufficient accuracy, and their position had to be changed. This process gave the direction of the magnetic meridian with accuracy and ensured that the plates p in the north and south line were properly placed on the table. The two methods taken together ensured that all four plates p were properly placed.

Observation of Deflections.
Deflectors of different relative lengths and thicknesses, and of different degrees of hardness, were used. These were originally magnetized by placing them between the poles of a large Ruhmkorff magnet excited by a considerable current, and afterwards by the same magnet excited by a much stronger current. The relative strengths of the magnets were unchanged by the second magnetization, and their absolute strengths only very slightly. The dimensions are given in the table of results, p. 90 below. The method of observing the deflections was as follows: According to a suggestion of Sir William Thomson two

deflectors were used at the same time, one on each side of the magnetometer. This arrangement was more symmetrical than that of a single deflector, and, what was of very great importance, it enabled a readable deflection to be obtained with the magnets at a much greater distance from the needle, thus diminishing error due to uncertainty as to the actual magnetic distribution. As each magnet was transferred on its carriage from one glass plate to another the magnets were not handled during the experiments. One deflector A was placed east, another B west of the magnetometer, and the plate g turned for each until their lengths were accurately in the east and west line, with their poles so pointing that each magnet gave a deflection of the needle to the same side of zero ; and the deflection was then noted. The plates g were then turned through 180°, and the deflection on the opposite side of zero read off. The carriages were then turned back to the first position and the deflection again read. The difference between the mean of the first and third readings and the second reading gave twice the deflection for the position of the magnets. The same operation was then repeated with the deflectors in interchanged positions. Two similar series of observations were next made with the magnets in the north and south line through the magnetometer and at equal distances on opposite sides of the needle. The mean deflection for the east and west positions, and that for the north and south positions, were calculated, and the results were used in the calculation of H in the manner described below.

After the deflection observations for a particular magnet had been completed, the magnetometer was removed and the deflector stand put in its place. The magnet was suspended from the brass bow b over its carriage by a length of single cocoon fibre, in a double stirrup formed by twice doubling the lower end of the fibre and knotting. The suspension thus obtained was sufficiently fine to be practically devoid of inertia, and long enough to give a negligible moment of torsion. The magnet was deflected in the manner already described (p. 76 above), and then left to oscillate. The period was observed in some cases by noting the times of the successive transits of the needle across the vertical cross wire of an observation telescope ; but the method finally adopted was to attach to the stirrup as shown in Fig. 15 a light silvered mirror m (·3 cm. in diameter and ·01 gramme in mass), and to use the same lamp and scale as in the deflection experiments. This latter arrangement enabled the amplitude of oscillation to be reduced to less than a degree and so reduced to zero the correction necessary for arc. The moment

Observation of Oscillations.

of inertia of the mirror was only about $\frac{1}{10000}$ of that of the deflector, and its neglect therefore introduced an error of only $\frac{1}{100}$ per cent.

Time was observed in these experiments by means of a very accurate watch provided with a centre seconds hand moving round a dial divided into quarter seconds. When two observers were available, one counted the oscillations and called sharply "Now" at the end of every four or five periods, while the other observed the time at each call. When only one observer counted the oscillations he used a chronometer beating half seconds. Having read time, he counted the beats until he could observe a transit. He then counted the beats until he observed another transit. From the result he estimated the number of periods in

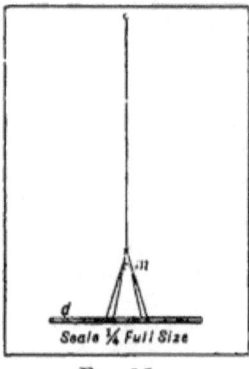

FIG. 15.

one minute, and therefore observed the time of the first transit after each minute so long as there was sufficient amplitude. The fractions of half seconds were estimated from the positions of the magnet at the beat next before and the beat next after the transit. With the mirror and scale arrangement these observations could be made with great accuracy.

Reduction of Observations.

The observations were combined in the following manner* so as to give the most probable value of the period. Supposing the number of observations to have been even, $2n$ say. The interval between the nth observation and the $(n + 1)$th, three times that between the $(n - 1)$th and the $(n + 2)$th, five times that between

---

* See Appendix: **On Errors of Observation** and the Combination of Experimental **Results**.

the $(n-2)$th and the $(n+3)$th, and so on to that between the 1st and the $2n$th were added together, the sum divided by the sum of the series $1^2 + 3^2 + 5^2 + \ldots + (2n-1)^2$, and the result by the number of periods (which was the same in each case) between each successive pair of observations. This gave the average period to a high degree of approximation. If an odd number of observations $(2n+1)$ was taken, the interval between the nth and the $(n+2)$th, twice that between the $(n-1)$th and the $(n+3)$th, three times that between the $(n-2)$th and the $(n+4)$th, and so on to the 1st and $(2n+1)$th, were added together, and the sum divided by twice the sum of the series $1^2 + 2^2 + 3^2 + \ldots + n^2$. The result divided by the number of periods in each interval gave the average period. The period adopted was always the mean of those given by two closely agreeing sets of observations.

Assuming that the magnet has two definite poles, that is (in this connection) points at which the whole of the free magnetism in each half of the magnet may be supposed concentrated in considering the external action of the magnet (an assumption not seriously erroneous in the case of the thin magnets and the distances used); the distance between them can be calculated from the results of deflection experiments in the side-on and end-on positions obtained as described above, since the effect of the distribution is opposite in the two cases. For if r be the distance, θ the deflection, for the end-on position, and r', θ' the distance and deflection for the side-on position, we have by equating the values of M/H given by equations (1) and (2): {Correction for distribution.}

$$\frac{(r^2 - \lambda^2)^2}{2r(r'^2 + \lambda^2)^{\frac{3}{2}}} = \frac{\tan \theta'}{\tan \theta} \quad \ldots \ldots \quad (8)$$

Expanding the numerator and denominator of each side and neglecting terms smaller than those of the second order we get:

$$\lambda^2 = \frac{r^3\theta - 2r'^3\theta'}{2r\theta + 3r'\theta'} \quad \ldots \ldots \quad (9)$$

By this equation the value of λ used in the calculation of H and M was found. The results for magnets of different lengths and diameters are interesting in themselves.

The moment of inertia of the bar was found by weighing the bar and carefully measuring its length and cross-section, and calculating for a vertical axis through the centre of the magnet supposed hung horizontally. The axis of suspension of the magnet in any case was not, however, that vertical, but another near it owing to the compensation for the tendency of the magnet to dip in the earth's field. The distance between these two axes can be found approximately for each magnet from the magnetic moment, mass, and length as given in the table below, and is so small that any error caused by supposing the magnet simply to vibrate round the former vertical is well within the possible limit of accuracy.

Theoretical Results.

For a cylindrical magnet of mass W, actual length $2l$ and diameter d, the moment of inertia is $W(l^2/3 + d^2/16)$. Hence (3) becomes :

$$MH = \frac{4}{3} \frac{\pi^2 (l^2 + \tfrac{3}{16}d^2) W}{T^2} \quad \ldots \ldots \quad (10)$$

Hence for a single deflector we get instead of equations (4), (5), (6), (7) equations obtained from these by substituting instead of l^2, $l^2 + 3d^2/16$.

If two deflectors be used, each of the actual length $2l$, and diameter d, but of masses W_1, W_2, periods T_1, T_2, and nearly equal effective lengths which give a mean, λ, we get from (1) and (2) instead of (5) and (6) for the end-on and side-on positions respectively :

$$H^2 = \frac{8}{3} \frac{\pi^2 r (l^2 + \tfrac{3}{16}d^2)(T_1^2 W_2 + T_2^2 W_1)}{(r^2 - \lambda^2)^2 T_1^2 T_2^2 \tan \theta} \quad \ldots \quad (11)$$

$$H^2 = \frac{4}{3} \frac{\pi^2 (l^2 + \tfrac{3}{16}d^2)(T_1^2 W_2 + T_2^2 W_1)}{(r'^2 + \lambda^2)^{\frac{3}{2}} T_1^2 T_2^2 \tan \theta'} \quad \ldots \quad (12)$$

In these formulas θ and θ' are the angular deflections found from the mean readings taken as described above (p. 83).

Corrections for alteration of Moment.

There are two corrections for alteration of moment of the magnet, produced (1) by variation of temperature, (2) by induction when the magnet is in or near the magnetic meridian when oscillating. The first correction was found by placing the magnet within a bath, in one of two principal positions at such a distance from the magnetometer needle that a deflection of 1,000 divisions was obtained, and then raising the temperature through about 40° C. It was found that such a rise of temperature produced a change of deflection of only about two divisions. Thus the magnets changed in magnetic moment by only $\frac{1}{200}$ per cent. for a change of temperature of 1° C. Hence as the variation of temperature in the experiments never exceeded 2° C. or 3° C. this correction was neglected.

Correction for Induction.

The correction for induction was found by immersing the deflecting magnet in an artificially produced magnetic field of known strength, and ascertaining the alteration of magnetic moment which resulted. The field was produced by surrounding

DETERMINATION OF H

the magnet with a magnetizing coil, and its intensity calculated from the number of turns of wire per unit of length of the coil and the current-strength, which was measured. The coil was sufficiently long to project beyond the magnet at each end some distance, so that the magnetic field was uniform, and equal to $4\pi nC$, where n is the number of turns per cm. of length, and C the current strength in C.G.S. units. Fig. 16 shows the arrangement

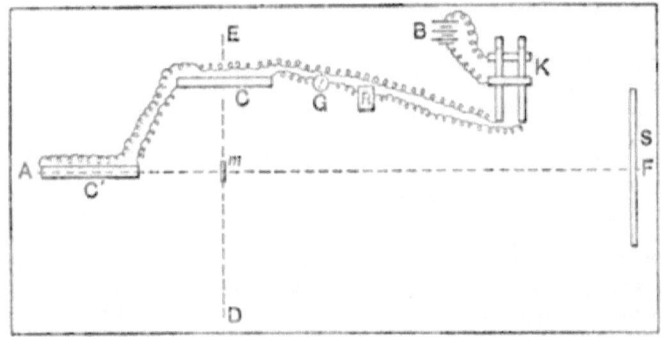

FIG. 16.

of apparatus for these experiments; m is the magnetometer needle, C, C' are coils each consisting of silk-covered copper wire wound on glass tubes 5 cms. in external diameter, S is the lamp scale, R a box of resistance coils, G the current galvanometer, K a reversing key, and B a battery. DE represents a horizontal line through the needle and in the magnetic meridian, and AF a horizontal line at right angles to DE, and also passing through the centre of the needle. As shown in the diagram the coil C was placed with its axis parallel to AF and its centre on the line DE. C' had its axis in the line AF, and the relative distances of the coils from the magnetometer needle were so adjusted that the magnetic effect of the current passing through the coils was zero at the needle, although the current flowing was made many times greater than that used in the experiments.

The magnet for which the induction correction was to be determined was then placed in one of the coils and the deflection read while as yet no current flowed. A field of about $\frac{1}{10}$ of a C.G.S. unit was then produced by passing a current, and the deflection was once more read. The current was then reversed, and the deflection again noted. The same operations were then repeated with greater and greater currents until a field of from

1 to 2 units had been reached. The magnet was then transferred to the other coil, and a similar series of observations made. It was found that a field of considerably greater intensity than the highest thus used is required to produce any permanent change of the magnetic moment of hard-tempered magnets. Each increase of magnetic moment being plotted as an ordinate of a curve, with the field-intensity for the corresponding abscissa, enabled the change produced by the earth's field to be obtained by interpolation as described above (p. 78).

A comparison of the results obtained with the two coils showed that the percentage change of deflection produced by the field was smaller for the coil C than for the coil C'. This was undoubtedly due to change of magnetic distribution, the effect of which on the deflection is opposite in the two cases. Assuming that the magnet has an effective half-length λ, the deflection in the first case is given by (1) and in the other by (2). Thus by using the coils in the two positions as described, the change of distribution as well as the change of moment can be approximately estimated. The plan of having two coils has also the advantage of allowing the change of magnetic moment to be obtained free from any error caused by want of exact compensation between the two coils of their direct effect upon the needle.

The results of the experiment showed that to make the effect of induction small the magnet should be hard tempered, and its length should be at least 40 times its diameter. The results are shown in the table on p. 91 below.

Effects of Variations of Earth's Field. The effects of variations in the intensity and direction of the earth's magnetic field were quite marked. The latter showed itself by changes of the magnetometer zero, which were eliminated by reading the zero before and after each deflection, and by reversing the magnets. The effect of change of intensity was allowed for by observing the period of a permanent magnet kept suspended for the purpose. This period was observed at the beginning of the experiment, after the deflection experiment, and again after the oscillation experiment. The necessary correction was estimated from the results and applied. It will be observed that the effect of diurnal variation is quite perceptible. The results in the table on p. 90 are tabulated in the order in which they were obtained, and it will be noticed that the earlier results of each day are generally the smaller. On some occasions on account of magnetic storms it was found impossible to obtain results at all. This was notably the case on Sept. 1, 1885.

Effect in Inductive Correction. The results of this determination are shown in the following two tables. The variation of the effect of induction on the magnetic moment with different ratios of the length of the

DETERMINATION OF H

deflecting magnet to its diameter is shown in the curve of Fig. 17.

of Varying Thickness of Magnet.

Curve illustrating the effect of Ratio of Length to Diameter on the Inductive Coefficient.

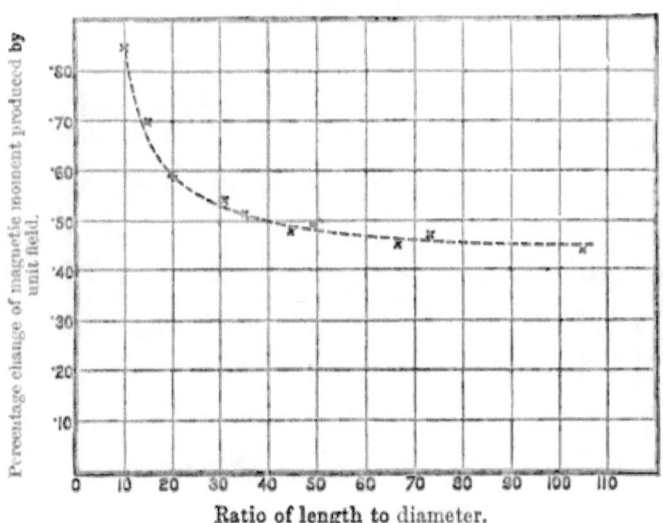

Ratio of length to diameter.

FIG. 17.

It will be observed that the effect of induction diminishes, rapidly at first, then more and more slowly, towards a constant value of about ·4 per cent. for unit field for glass-hard magnets of the kind of steel experimented on.

TABLE I.

Date 1885	Number of deflector	Length of deflector, in centimetres	Diameter of deflector, in centimetres	Weight of deflector, in grammes	Distance of centre of deflector from the magnetometer-needle, in cms. (East and West positions)	Distance of centre of deflector from the magnetometer-needle, in cms. (North and South positions)	Distance of the scale from the magnetometer mirror	Effective length of the deflector	Magnetic moment per gramme of the deflector	Horizontal Intensity in C.G.S. units	Mean of each set of results	Remarks
May 27	1	8.03	0.25	3.054	32.06	28.75	108.7	6.91	44.9	.1520	......	
,, 27	2	8.05		3.063	,,	,,	,,	7.10	58.5	.1524	......	
,, 28	3	8.05		3.075	,,	,,	,,	6.31	54.1	.1521	......	
,, 29	4	8.05		3.067	,,	,,	,,	7.11	52.3	.1522	.1522	
June 5	5	4.00	0.25	1.526	30.00	,,	,,	3.12	35.2	.1524	......	
,, 5	6	3.00		1.522	,,	7.25	,,	3.72?	33.7	.1524	......	
,, 5	7	4.00		1.525	,,	,,	,,	2.33	31.7	.1527	.1524	
June 10	8	14.933	0.25	5.646	51.90	38.85	,,	13.22	55.8	.1527	......	
,, 11	9	15.030		5.727	,,	,,	,,	12.82	64.3	.1525	......	
,, 11	10	15.021		5.666	,,	,,	,,	13.58	54.5	.1527	.1526	
Aug. 21	11	10.01	0.2	2.318	35.00	30.00	128.9	9.14	71.0	.1526	......	Double deflector method.
,, 21	12	10.01		2.336	,,	,,	,,		62.7	.1527		
,, 26	11	10.01		2.318	35.00	30.00	128.9	8.98	71.0			
,, 26	12	10.01		2.336	,,	,,	,,		62.7			
,, 31	13	10.00		2.318	35.00	30.00	128.9	9.14	70.0	.1527		
,, 31	14	10.005		2.336	,,	,,	,,		61.8	.1526	.1526*	

* Corrected to noon for diurnal variation.

TABLE II.—Showing the effect of Length and of Hardness on the Induction-Coefficient of Magnets.

Length of bar in centimetres.	Ratio of length to diameter.	Unit field.		Mean of numbers in columns 3 and 4.	Magnetic moment per gramme.	Remarks.
		Apparent percentage increase of moment for unit field: side-on position.	Apparent percentage increase of moment for unit field: end-on position.			
3	10	0·80	0·90	0·85	27	Glass hard.
4	16	0·67	0·73	0·70	32	,,
4	16	0·67	0·70	0·69	35	,,
6	20	0·51	0·67	0·59	36	,,
7	31	0·51	0·58	0·54	39	,,
8	32	0·51	0·58	0·54	54	,,
8	32	0·51	0·58	0·54	52	,,
10	34	0·46	0·56	0·51	40	,,
10	44	0·40	0·56	0·48	43	,,
7	47	0·46	0·51	0·49	57	,,
10	50	0·44	0·58	0·51	67	,,
10	50	0·48	0·54	0·51	60	,,
10	50	0·46	0·55	0·51	53	,,
10	50	0·46	0·52	0·49	71	,,
10	50	0·46	0·56	0·51	60	,,
10	67	0·41	0·51	0·46	65	,,
7	73	0·41	0·50	0·46	64	,,
10	105	0·42	0·45	0·44	66	,,
10	34	0·47	0·53	0·50	41·5	Glass hard.
10	34	0·63	0·67	0·65	44·5	Yellow.
10	34	0·84	0·98	0·91	54·1	Blue.
10	48	0·32	0·40	0·36	45	Glass hard.
10	43	0·43	0·55	0·49	46	Yellow.
10	48	0·53	0·67	0·60	71	Blue.

The method given above for the determination of the correction for the non-uniform magnetization of the deflecting magnet, gives of course only a first approximation to the true correction, but under the condition that the length of the bar is sufficiently small in comparison with the distance r, say from $\frac{1}{8}$ to $\frac{1}{16}$ of r, and on the supposition that the magnet is reversed at the position on either side of the needle, it is generally sufficient.

Elimination of Effect of Magnetic Distribution.
The following method eliminates to a high degree of accuracy the effect of the magnetic distribution. Let two deflections be taken by reversing the deflecting magnet at a distance r_1 on the west side of the needle, and similarly two deflections at the same distance on the east side, and let D_1 be the mean of the tangents of these four deflections. Let this process be repeated for a second distance r_2, and let D_2 be the mean tangent for this distance. It is easy to prove that, approximately

$$\frac{2M}{H} = \frac{r_1^5 D_1 - r_2^5 D_2}{r_1^2 - r_2^2} \quad \ldots \ldots \quad (13)$$

For if we make no particular supposition as to the distribution we may write instead of equation (1)

$$\frac{Hr^3}{2M} \tan \theta = 1 + \frac{A}{r} + \frac{B}{r^2} + \&c. \quad \ldots \ldots \quad (14)$$

the series on the right converging. Therefore denoting by θ_1, θ_1', the first two deflections obtained as described above, we have

$$\tfrac{1}{2}\frac{Hr_1^3}{M} \tan \theta_1 = 1 + \frac{A}{r_1} + \frac{B}{r_1^2} + \frac{C}{r_1^3} + \&c. \quad \ldots \quad (15)$$

Now reversing the magnet without altering its distance is obviously equivalent to shifting it to the same distance on the other side of the magnetometer without reversing, that is to altering the sign of r_1. Hence, by (15),

$$\tfrac{1}{2}\frac{Hr_1^3}{M} \tan \theta_1' = 1 - \frac{A}{r_1} + \frac{B}{r_1^2} - \frac{C}{r_1^3} + \&c. \quad \ldots \quad (16)$$

Thus four values of $\tfrac{1}{2}Hr_1^3 \tan \theta/M$ are obtained which give

$$\tfrac{1}{2}\frac{Hr_1^3}{M} D_1 = 1 + \frac{B}{r_1^2} + \frac{D}{r_1^4} + \&c. \quad \ldots \ldots \quad (17)$$

… Similarly from the other two pairs of deflections at the distance r_2 we get

$$\tfrac{1}{2}\frac{Hr_2^3}{M} D_2 = 1 + \frac{B}{r_2^2} + \frac{D}{r_2^4} + \&c. \quad \quad \quad (18)$$

Multiplying (17) by r_1^2 and (18) by r_2^2, and subtracting, we have finally, neglecting all terms beyond the second in each equation,

$$\frac{H}{2M}(r_1^5 D_1 - r_2^5 D_2) = r_1^2 - r_2^2,$$

the relation expressed in (13).

It will be shown in the appendix on Reduction of Observations that if approximately $r_1 = 1\cdot 32 r_2$, the effect of errors in the observed deflections on the value of M/H will be a minimum for these distances.

If long thin bars are used in the determination of H, their magnetic distribution could be accurately found by Rowland's method (Chap. XIV. below) and the proper corrections applied. On the other hand, short thick bars of hard steel have the advantage of giving greater magnetic moment for a given length, and they can therefore be placed at a comparatively greater distance from the needle, so that the correction for the distribution becomes of less importance. So far, then, as the deflection experiments are concerned, it is better to use thick strong magnets of the hardest steel, and to place them at such a distance from the needle that the error, caused by neglecting the distribution, becomes vanishingly small. On the other hand, the magnets must be sufficiently long and thin to render it possible to determine with accuracy their moments of inertia, and therefore to reduce correctly the results of the vibration experiments. When the distance is so great that the effect of distribution is negligible, we may use the approximate formula

Elimination of Effect of Magnetic Distribution.

$$M = \frac{r^3}{2} H \tan \theta \quad \quad \quad \quad \quad (19)$$

for the position shown in Fig. 13, or

$$M = r^3 H \tan \theta \quad \quad \quad \quad \quad (20)$$

for the position shown in Fig. 13'.

A magnetic survey of horizontal force, in the neighbourhood of a place for which H has been determined, may very readily be made with one of the magnets used in the deflection experiments, by simply observing its period of vibration at the various

Magnetic Survey.

places for which a knowledge of H is desired. The magnetic moment M of the magnet being of course known from the previous experiments, H can be found by equation (5) or (7) above.

By keeping a magnetometer set up with lamp and scale in readiness, the magnetic moments of large magnets can be found with considerable accuracy by placing them in a marked position, at a considerable distance from the needle, and observing the deflection produced. By having a graduated series of distances for each of which the constant $\frac{1}{2}r^3H$, or r^3H, as the case may be, by which tan θ must be multiplied to give M, has been calculated, the magnetic moments can be very quickly read off.

Comparison of Moments of Large Magnets.

The magnetic moments of large magnets of hard steel, well magnetized, can be compared very conveniently with considerable accuracy by hanging them horizontally in the earth's field, and determining the period of a small oscillation about the equilibrium position. They should be hung by a bundle of as few fibres of unspun silk as possible, at least six feet long, so that the effect of torsion may be neglected. The suspension thread should carry a small cradle or double loop of copper wire, on which the magnet may be laid to give it stability, and to allow of its being readily placed in position or removed. Two vertical marks are fixed in the meridian plane containing the suspension thread, and the observer placing his eye in their plane, can easily tell very exactly when the magnet is passing through the equilibrium position, and so determine the period. Or, a north and south line may be drawn on the floor or table under the magnet, and the instant at which the magnet is parallel to this line observed by the experimenter, standing opposite one end of the magnet and looking from above. The value of M is given in terms of H by equation (3) above.

Care must of course be taken to avoid undue disturbance from currents of air, and to prevent the magnet, when being deflected from the meridian, from acquiring any pendulum swing under the action of gravity. The deflection from the meridian should be made with another magnet, brought with its length along the east and west line through the centre of the suspended magnet, near enough to produce the requisite deflection, and then withdrawn in the same manner.

Stroud's Magnetometer for Complete Determination of H.

A new form of magnetometer by which the determination of H is at once effected by direct observation of angular deflections, has been invented by Prof. W. Stroud of the Yorkshire College, Leeds. A steel ring (M of Fig. 18) is made by bending a piece of thin ribbon steel about 1 metre long, $\frac{1}{10}$ millimetre in thickness, and 3 millimetres broad, into a circle, and soldering the

DETERMINATION OF H

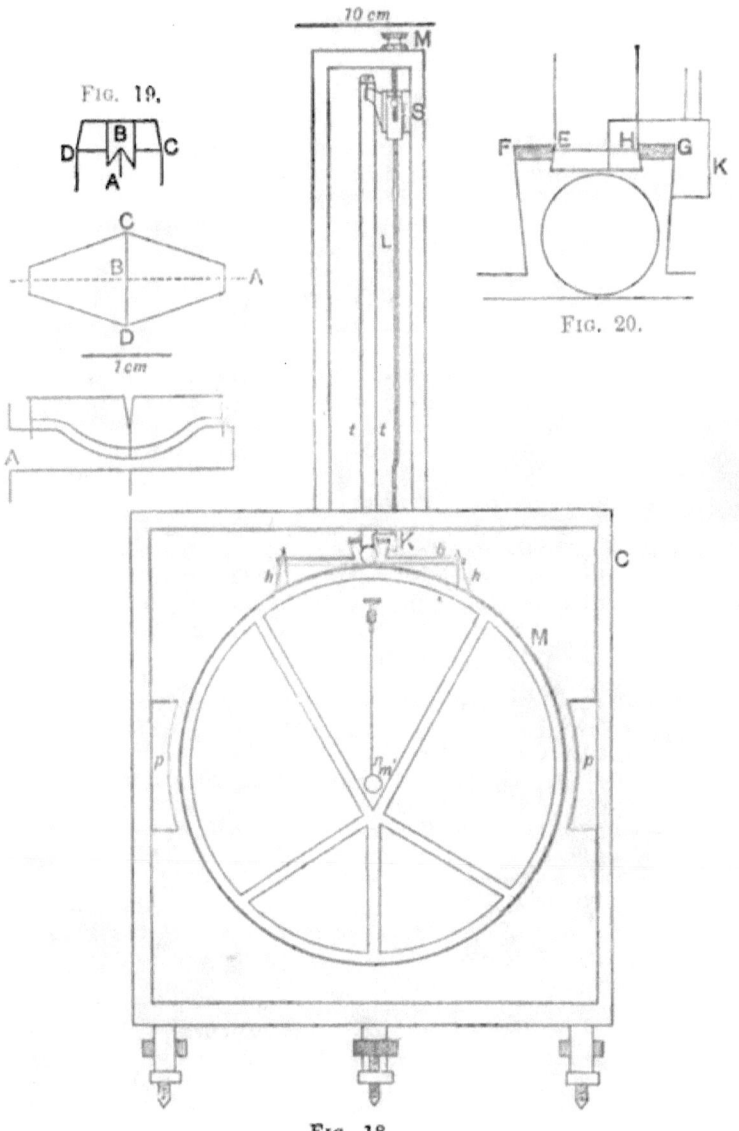

Fig. 19.

Fig. 20.

Fig. 18.

ends together with the overlap at the top or bottom of the ring. The shape is maintained as nearly as possible a perfect circle by means of a ring of tissue paper, or, better, aluminium with connecting arms as shown in Fig. 18.

When the bifilar is placed in an east and west (magnetic) vertical plane, it gives a means of measuring the couple exerted by the earth's horizontal field. That couple is proportional to MH, if M be the moment of the ring-magnet, that is, the couple tending to turn the magnet in a field of unit intensity and of direction at right angles to the plane of the ring.

This ring-magnet is hung within a case C, C, supported on levelling screws. The case is made partly of glass, so that the apparatus can be seen from the outside. The ring is hung by hooks h, h, from a brass crossbar b, by means of which it is attached to the bifilars t, t. The upper side of this bar is a knife-edge furnished with a V-notch near the end to receive one of the hooks h, and thus allow the wire to be removed, and replaced accurately reversed in position on the bar. A small plane mirror is carried above the centre of this bar, and serves to determine the position of the ring.

The details of the suspension are shown in Fig. 19. A is a piece of brass fixed to the wall of the instrument case. A knife-edge is worked on its upper side and on this rests a piece of aluminium of the shape shown in the lowest diagram of Fig. 19. To this piece is attached the bifilars, and the distance CD between them is about 1 cm.

The knife-edge bisects the distance to at least $\frac{1}{10}$ mm. The thread rests in a groove in the aluminium piece, so that the whole upper suspension arrangement is the equivalent of a pulley mounted on a knife-edge.

Stroud's Magnetometer.

The lower suspension is shown in Fig. 20, and consists of an aluminium piece to which the fibres are attached. One fibre comes from above to E, passes from E to F, thence round by G to H, and then up. The distance EH is, like CD, about 1 cm.

It is to be noticed that at the top the fibres lie outside the space CD, at the bottom inside EH, so that the product of the distances of the fibres apart at the top and bottom is accurately $CD \times EH$. CD and EH are measured by means of a micrometer gauge easily to $\frac{1}{100}$ mm. Error from effect of the pressure of the gauge does not enter, as CD is measured directly, then EF, HG, and FG, giving EH by difference; so that EH is as much too great in consequence of compression produced by the gauge as CD is too small. This arrangement also eliminates error arising from the thickness and flexural rigidity of the suspending fibres.

DETERMINATION OF H

The length of the fibres is **determined** as follows. A mirror K (Fig. 18) with a horizontal line on it is attached by a brass arm to a slider L, worked by a screw with milled head M at the top of the instrument. The screw is worked until the horizontal line on the mirror, the horizontal line given by the top of the piece GH, and the image of the latter in the mirror K behind it are in one line. By the motion of the screw, a mark on the nut at the top of the slider L is brought to some position on a brass scale S attached by brass connecting pieces to the piece A shown in Fig. 19. The length of the fibres is equal to the reading on the scale S increased by a constant quantity.

Any alteration in the length of the scale due to temperature, &c., is thus given by measurement in terms of divisions of a brass scale, so that the length can always be obtained with almost perfect accuracy. The residual temperature correction is indeed quite negligible for even large differences of temperature.

A small needle n is hung from an arm of brass which is attached to one side of the box, so that the needle, when in position, can hang with its centre as nearly as may be at that of the ring-magnet. A small mirror m' fixed at right angles to the axis of the needle is carried below it.

A forked piece of wood prevents the needle from turning round, and enables it to be placed at once very near the centre of the ring, while copper pieces p, p, on the sides of the case, damp the motion of the ring-magnet and limit the free space in which it swings to about 1 millimetre of clearance on each side.

Changes of the positions of the ring-magnet and of the small needle are read by means of a lamp and scale, or a telescope and scale, in the ordinary manner. (Of course a telescope and scale free from iron must be used.) By properly arranging the positions of the two mirrors a single telescope, with, if necessary, two scales, can be used to determine the deflections of both magnets.

The method of using the instrument and its theory are as follows. The bifilars are adjusted so that their plane is approximately east and west, then the ring-magnet is placed in position, and the deflections of the needle and of the bar carrying the ring read off by their mirrors. If θ be the angle which the plane of the ring makes with a vertical east and west (magnetic) plane, the magnetic couple on the ring due to H is $MH \cos \theta$. The total magnetic couple on the ring is thus $MH \cos \theta - L$, where L is a couple in the opposite direction due to the small needle at the centre of the ring. Since, if necessary, all the suspension threads may be single fibres of silk, or still better thin threads of quartz, the torsion of the bifilars may be neglected. Hence if a be the angle which the plane of the ring

Stroud's Magnetometer. Use and Theory of Instrument.

makes with a vertical east and west (magnetic) plane when the bifilar plane is vertical, the angle through which the bifilar has been turned is $\theta - a$, and if d, d' be the distances between the threads at top and bottom, l their length, and W the mass supported, the couple given by the bifilar is (Vol. I. p. 244) $Wdd' \sin(\theta - a)/4l$.

Hence we have

$$MH \cos\theta = \frac{Wdd'}{4l} \sin(\theta - a) + L \quad . \quad . \quad . \quad (21)$$

The small needle is likewise deflected through an angle ϕ. This can be measured by observing the positions of the needle with and without the ring-magnet in the instrument.

The component of the moment M' of the small needle at right angles to the plane of the ring is $M' \cos(\theta - \phi)$. Now if we suppose a small quantity of magnetism δm of the ring to be situated at a point the radius to which makes an angle χ with the horizontal diameter through the centre, the horizontal component force due to δm will be $\delta m \cos\chi/r^2$, or $\delta m/r^3 \cdot r \cos\chi$. It follows, if the length of the needle be taken as very small, and the breadth of the ribbon be neglected, that the moment of the couple deflecting the needle is $M'/r^3 \cdot \cos(\theta - \phi) \Sigma \delta m \, r \cos\chi$, where the summation is extended throughout the whole distribution of the ring-magnet. But $\Sigma \delta m \, r \cos\chi$ is evidently the magnetic moment M of the ring-magnet. The couple exerted by the ring on the needle is thus $MM' \cos(\theta - \phi)/r^3$, and this is equal and opposite to the couple L exerted on the ring by the magnet.

Hence for the equilibrium of the small needle we have neglecting the torsion of the thread

$$\frac{M}{r^3} \cos(\theta - \phi) = H \sin\phi \quad . \quad . \quad . \quad . \quad (22)$$

or

$$\frac{M}{H} = \frac{r^3 \sin\phi}{\cos(\theta - \phi)} \quad . \quad . \quad . \quad . \quad (23)$$

But we have also

$$MH \cos\theta = \frac{Wdd'}{4l} \sin(\theta - a) + L \quad . \quad . \quad . \quad (24)$$

If $L (= M'H \sin\phi)$ be small in comparison with $MH \cos\theta$, that is if $M' \sin\phi/M \cos\theta$ be a small quantity, L may be neglected in (24), and we get

DETERMINATION OF H

$$H^2 = \frac{Wdd'}{4lr^3} \frac{\sin(\theta - a)\cos(\theta - \phi)}{\cos\theta \sin\phi} \quad \ldots \quad (25)$$

If two experiments be made with the same weight on the bifilar, but with the ring-magnet reversed, we get if $\theta' + a$, ϕ', be the angular deflections of the ring and needle, respectively,

$$H^2 = \frac{Wdd'}{4lr^3} \frac{\sin(\theta' + a)\cos(\theta' - \phi')}{\cos\theta' \sin\phi'} \quad \ldots \quad (26)$$

Hence

$$\frac{\sin(\theta - a)}{\sin(\theta' + a)} = \frac{\cos(\theta' - \phi')\cos\theta \sin\phi}{\cos(\theta - \phi)\cos\theta' \sin\phi'} \quad \ldots \quad (27)$$

from which a can be found, so that H can be calculated from (25) or (26).

If the angles are all so small that they may be replaced by their sines, and the cosines may be put each equal to 1, we have

$$H^2 = \frac{Wdd'}{4lr^3} \frac{\theta - a}{\phi} = \frac{Wdd'}{4lr^3} \frac{\theta' + a}{\phi'} \quad \ldots \quad (28)$$

or

$$H^2 = \frac{Wdd'}{4lr^3} \frac{\theta + \theta'}{\phi + \phi'} \quad \ldots \quad (29)$$

Hence all that is necessary is to take the angular readings before and after the reversal of the ring. The differences of the readings in the two cases are $\theta + \theta'$ and $\phi + \phi'$.

The errors due to neglect of the couples, due to torsion of the fibres, the couple exerted on the ring by the small needle, and the error due to uncertainty of magnetic distribution in the thickness of the wire of the ring are not all of the same order of magnitude. The first may be made quite negligible even with silk fibres; the couple due to the small needle produced in Prof. Stroud's first instrument, which had a ring of pianoforte steel wire, gave an effect of about 1 in 700, and the thickness of the wire gave a possible extreme error of about 1 in 300. The two latter couples are made negligibly small by increasing M sufficiently, and making the ring of thin steel strip instead of wire. Of course the couple due to the small needle can always be approximately determined and allowed for.

Order of Magnitude of Errors.

Results obtained with Trial Instrument.

The following table contains examples of determinations of H made by Prof. Stroud with his first trial instrument.

	I. May 17, 1890.	II. May 29, 1890.	III. May 29, 1890.
Length of scale	100 cms.	200 cms.	} Same constants as in II.
Distance from centre of inst.	97 "	200 "	
Length of bifilars	30·07 "	27·81 "	
Distance apart above	1·293 "	1·293 "	
Distance apart below	1·314 "	1·302 "	
Diameter of ring	27·50 "	27·50 "	
Reading for ring before reversal	10·33 "	153·66 "	153·69 cms.
Reading for ring after reversal	52·59 "	77·34 "	77·72 "
Difference of readings	42·26 "	76·32 "	75·97 "
Reading for needle before reversal	6·21 "	191·76 "	192·51 "
Reading for needle after reversal	88·60 "	25·40 "	26·67 "
Difference of readings	82·48 "	166·36 "	165·84 "
Mass suspended from bifilars	11·37 grammes	11·69 grammes	11·69 grammes
Value of H in C.G.S. units, from these data	·1803	·1805	·1803

Other methods of determining H which depend on current induction will be explained in a later chapter.

CHAPTER III.

THEORY OF ELECTROMAGNETIC ACTION.

SECTION I.

ACTIONS BETWEEN CURRENTS AND MAGNETS.

THE action of a current on a magnet, discovered by Örsted in 1820, is the foundation of the modern science of electromagnetism, for from it has come by a steady process of discovery, at once inductive and deductive, the whole theory of the mutual action of magnets and currents, and of currents on one another, of the induction of currents by the motion of conductors in a magnetic field, and the great modern applications of electricity to telegraphy and telephony, lighting and transmission of power, and electric traction. We shall follow to a certain extent the historical order of development of this part of the subject, making use freely, however, for brevity and clearness, of the theorems contained in the digest of magnetic theory already given, and of the ideas and methods suggested by later writers, such as Thomson and Maxwell. Some account of speculations as to the nature of currents and the *rationale* of electromagnetic action generally will be given in Chapter V. *Örsted's Experiment.*

In Örsted's experiment, as commonly performed, a magnet is suspended horizontally in the magnetic meridian, and a conductor carrying a current is stretched parallel to the needle, above it or below it. The magnet is acted on by a couple which turns it round towards the position at right angles to the conductor, and it finally rests in equilibrium in a position in which this deflecting couple is balanced by the return couple due to the terrestrial magnetic field. The deflecting couple is reversed in direction by turning round through 180°, or " end for end," the conductor carrying the current, so that, for example, the current flows from south to north instead of from north to south; and it is likewise reversed when the conductor is transferred from a position above the needle to a position below the needle, and *vice versâ*. Thus the direction of the deflecting couple is not reversed when the conductor is both turned end for end and transferred from above to below, or from below to above; and we see therefore that if the current flow, say from north to south above the magnet and back from south to north below the magnet, the deflecting couples due to both currents are in the same direction. By multiplying the number of conductors or portions of one conductor thus carrying currents the effect on the needle is also enhanced. Hence by winding the conductor into a coil of a large number of turns, one part of each of which is above the other below the magnet, the actions of the various turns on the magnet are given all the same direction, and the magnet is acted on by a resultant couple round a vertical axis, made up of the component couples round such an axis furnished by the turns of wire in the coil. This is the construction and mode of action of the old form of "galvanic multiplier," and of the modern galvanometer.

EQUIVALENCE OF A CURRENT AND A MAGNET 103

Since the needle is deflected by the current just as it would be by bringing another magnet into its neighbourhood, we are led to regard the current as producing a magnetic field, which is superimposed on the terrestrial magnetic field so as to give a resultant field, parallel to a line of force of which the needle, if short, places its magnetic axis. In fact, the current produces the same effect as would a certain distribution of magnetism, and we have to inquire what is the nature of this distribution. This is set forth in the following general theorem given by Ampère:—*Every linear conductor carrying a current is equivalent to a simple magnetic shell, the bounding edge of which coincides with the conductor, and the moment of which per unit of area, that is, the strength of the shell, is proportional to the strength of the current.* The direction of magnetization of the shell is reversed when the current is reversed, and may be found in any given case as follows. Supposing an observer to be standing on the edge of the shell with its surface on his left hand, and to be looking in the direction in which the current is flowing,* the side of the shell towards the observer will be covered with northern magnetism. This may also be remembered by the rule, that the magnetism of the earth coincides in direction with that of a needle placed within it, and turned into position by currents circulating round the earth in the direction of the sun's apparent motion.

A Circuit carrying a Current produces a Magnetic Field.

Equivalence of a Current and a Distribution of Magnetism.

---

* From copper to zinc in the external part of the circuit of a voltaic cell, according to the ordinary convention.

Equivalence at Distant Points of Magnetic Actions of a Plane Circuit with Current and a Magnet.

The theorem of Ampère just stated depends on another theorem which we shall consider first. *The magnetic field produced by the current in a plane closed circuit is the same at all points, the distances of which from every part of the conductor are great in comparison with every dimension of the circuit, as that produced by a small magnet placed anywhere within the circuit, with its axis at right angles to the plane of the current, and having a magnetic moment proportional to the current flowing, and to the area of the circuit.*

Experimental Proof.

The truth of this theorem may be demonstrated by a simple experiment which has become a common laboratory exercise. A plane circuit of convenient form, for example circular, is arranged in a vertical position parallel to the magnetic meridian, by connecting to a circular coil, of one turn or more, the terminals of a battery placed at a considerable distance from every part of the apparatus being used in the experiment. It is easy to prove by separate experiments that the current in the part of the circuit consisting of the battery itself, and the wires connecting it to the circular conductor, produce no appreciable effect if the wires are twisted together, and are both joined as nearly as may be at the same point to the coil. The effective part of the circuit is then only the coil, and it is this only we mean when we refer in what follows to the "circuit." A magnetometer is placed with the centre of its needle on a horizontal magnetic east and west line passing through the centre of the circular conductor, which is so arranged that the distance of its centre from the magnetometer needle can be altered at pleasure. It is found by observing the deflections on the magnetometer that the magnetic forces produced at the centre of the needle are very nearly in the inverse ratio of the cubes of the distances of the centre of the needle from the centre of the coil, when these distances are great in comparison with the dimensions of the circular conductor. The same result may be obtained for a plane conductor of any other form by so placing it that the east and west line through the centre of the needle passes through the plane of the conductor within or near the circuit, and taking the distance as that between the plane and the needle's centre. Now, by equations (9) p. 9 above, this is precisely the result that we should have obtained for a small magnet placed as specified above with

EQUIVALENCE OF A CURRENT AND A MAGNET

regard to the circuit; and it is possible so to adjust the moment of the magnet that its action and that of the current may be identical.

It is further found experimentally that if we have a magnet and a current which produce the same magnetic force at distant* points upon an east and west line passing through the circuit, the magnet and the current produce the same magnetic effect at all other distant points. Finally, by altering the area of the circuit in any ratio, we find the magnetic force at every point altered in the same ratio. Hence the equivalence is completely proved.

We shall define the current strength in a given circuit as proportional to the intensity of the magnetic field which the current produces at a given point; and hence it is not necessary to prove that the moment of the equivalent magnet must be proportional to the current, since we know that the magnetic field due to a magnet at a given point so distant that the effect of distribution of magnetism does not enter into account, is proportional to the magnetic moment of the magnet. We shall find that this mode of measuring current-strength gives results consistent with those obtained from the definition based on the electrostatic system of units, viz. the quantity of electricity which passes across an equipotential surface in the circuit per unit of time. {Definition of Current-Strength.}

We here define unit current as that current which flowing in a circuit of unit area can be replaced by a magnet of unit magnetic moment. This definition depends on the unit of magnetism already defined, and, when the latter unit is 1 C.G.S. unit of magnetism, we have by the definition 1 C.G.S. unit of current. We shall find other, but equivalent, definitions of unit current. {Definition of Unit Current.}

The magnet equivalent at distant points to the plane circuit may be supposed broken up into an infinite number of equal short magnets uniformly distributed over the circuit with their centres in and their lengths at right angles to its plane. If the aggregate magnetic moment be the same as before, the same effect will be produced, since the position of the equivalent magnet within the circuit and its form do not affect the force which it produces at distant points. But this converts the equivalent magnet into a uniform magnetic shell the strength of which is, by the definition of unit current just given, simply the strength of the current. {Equivalent Magnet may be a Plane Magnetic Shell bounded by the Circuit.}

---

* "Distant" here, as elsewhere in a similar connection, means that the points are at distances from the circuit great in comparison with any of its dimensions.

Proof of General Theorem of Equivalence of a Linear Current and a Magnetic Shell.

Ampère's further proposition that any finite linear circuit carrying a current is equivalent to a magnetic shell, can now be proved at once. For let ABC be the circuit, in which we shall suppose a current of

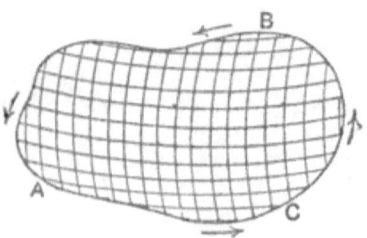

FIG. 21.

strength γ to be flowing. We may construct, as indicated in the figure, a network of conductors of which the circuit is the bounding edge, having each mesh so small that it may be considered plane. Round each of these meshes a current may be supposed to flow in the same direction as that of the current in the boundary. It is clear that this will give two equal and opposite currents in every conductor which is common to two meshes, and thus the system reduces simply to the current in the original conductor which forms the boundary. Each of these small circuits may, however, by the proposition just proved, be replaced by a small magnet, or by an infinite number of equal infinitely small magnets uniformly distributed over it, and the aggregate of these small magnets gives a magnetic shell bounded by the circuit.

WORK IN CARRYING POLE ROUND CURRENT

It is important to notice that the meshes may have any positions provided the boundary be undisturbed. Thus the shell is geometrically defined only by its boundary the conductor. It should also be observed that there is not here any restriction of the equivalence to the action at distant points; only, since the conductor must always in practice be a wire of finite thickness, the points at which the action is considered must be at a distance of several diameters of the wire from the boundary. *Equivalent Shell Geometrically Defined only by the Conductor.*

We can now at once show that the work done in carrying a unit pole from any point P in the field of a current round a closed path to the point P again is zero, if the path do not embrace the circuit, and is $4\pi\gamma$ if the path embrace the circuit once. For, let a position of the equivalent shell be chosen which does not intersect the closed path, if the latter does not embrace the circuit, and one close to the point P, if the closed path does pass round the circuit. In both cases the work done is equal to the total change of potential in passing round the path. In the former case this is zero. In the latter case let the pole be carried first from the point P to a point Q infinitely near to P on the opposite side of the shell. The change of solid angle in passing from P to Q is, as proved in p. 40 above, 4π, and therefore by the definition of current strength the work done is $4\pi\gamma$. Now although the shell was fixed in position in estimating the work done in carrying the unit pole from P to Q, it is not necessary to suppose it fixed in the same position in finding the work done in carrying the pole along the infinitely small part of the closed path which lies between Q and P. We may therefore suppose the shell in any other position clear of the element QP of path. The work done in carrying the pole from Q to P is therefore infinitely nearly zero, that is, the work done in carrying the pole round the closed path is $4\pi\gamma$. Another proof of this theorem is given below on p. 113. *Theorem of Work done in Carrying a Pole in Closed Path round a Current: First Proof.*

If the path be laced round the circuit any number, n, of times, the whole work done in carrying the pole round the path will be $4\pi n\gamma$. To see this we have only to join P to the points R, T, &c. (Fig. 22). The work done in carrying the pole round the path $PQRS\ldots P$, is equal to the work done in carrying the pole round the n closed paths $PQRP$, $RSTR$, $\ldots$, $VWPRTV$, since *Case in which the Path and Circuit Interlace any number of times.*

the portions PR, RT, &c. are each traversed twice but in opposite directions, so that the work done in traversing them in one direction cancels the work done in traversing them in the other.

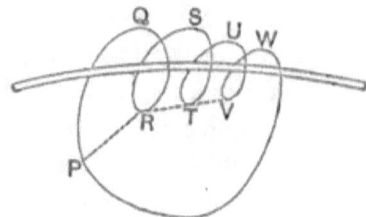

Fig. 22.

In the same way, we can prove that if the circuit pass n times through the path, the work done in carrying the pole round the path is $4\pi n\gamma$. For, consider the case represented in Fig. 23, in

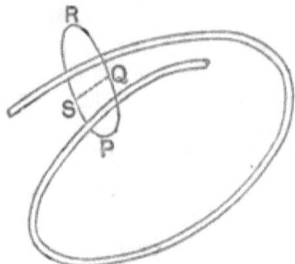

Fig. 23.

which the circuit passes twice through the path, and join the two points Q, S of the path by the line QS passing between the two portions of the circuit. The work done in carrying a unit pole round the path is plainly equal to the work done in carrying it round the two closed paths $PQSP$, $QRSQ$, since SQ is traversed in opposite directions in the two cases. But in each case the work done is $4\pi\gamma$, and hence the whole work done is $2 \times 4\pi\gamma$. Hence, proceeding in the same way for further interlacing of the circuit with the path, we obtain the general result stated above.

Any combination of the two kinds of interlacing will give a result which can be calculated according to the circumstances of the case by combining the two results just found.

Ampère's theorem is confirmed by quantitative experiments on the magnetic effects of a long straight conductor carrying a current. With the arrangement of horizontal conductor and horizontal needle, it is found as has been stated (p. 102) that, according as the conductor is above or below the level of the needle, the latter is deflected in one direction or the other, and hence when the conductor is in the same horizontal plane with the needle, no deflection is produced. If the free period of oscillation of the needle be observed when the conductor is present in the same horizontal plane it is found to be the same as when no current is flowing. *Magnetic Field of a Long Straight Conductor carrying a Current.*

These results show that the current produces then no component force in the horizontal direction on a magnetic pole, and it follows that the resultant force is in the vertical direction. The force on a magnetic pole is therefore at right angles to the plane through the conductor and the pole. *Direction of Force.*

The same thing is shown by the fact observed by Ampère, that the position of the needle is at right angles to the conductor in a plane parallel to it when there is no force acting on the needle except that due to the current; for this proves that there is no component in the plane through the current and a magnetic pole on which the current acts.

It is found that the magnitude of the magnetic force due to the current in a straight conductor, at points not opposite the ends, and at distances from the conductor small in comparison with its length, varies inversely as the distance of the point considered from the conductor. Its direction is, as we have seen, at right angles to the plane through the conductor and the point considered. A magnetic pole free to move in a circular groove with the conductor for its axis would move round the groove in the same direction and would be acted on by the same force, which would be everywhere tangential to the groove. In fact the lines of magnetic force round the conductor except near its ends, are circles having the conductor for their common axis. *Lines of Force Circles round Conductor.*

These results for a straight conductor are proved by a number of simple experiments. That the intensity of the magnetic field varies inversely as the distance from a thin conductor was shown by Biot and Savart,* who placed a horizontal conductor at right angles to the magnetic meridian, and at different distances above and below the centre of a horizontally suspended needle, and observed the periods of oscillation when the needle was under the influence of the earth's force alone, and again when a current was made to flow in the conductor. If T, T' be the periods in

---

* *Ann. de Chim. et de Phys.* t. xv. 1820.

the two cases, M the magnetic moment, and μ the moment of inertia of the magnet, the intensity of the field due to the current is given by the expression $4\pi^2\mu/M \cdot (1/T''^2 - 1/T^2)$.

Law of Force found Experimentally: Method of Maxwell.

The law of variation of force with distance is also shown by the following elegant experiment apparently suggested by Maxwell.* A conductor is placed in a vertical position and a light carriage of non-magnetic material is suspended so as to be free to turn round the conductor as an axis. It is found that when a magnet is fixed on this carriage there is no couple tending to turn the carriage round the conductor. Consider a thin uniformly magnetized bar magnet attached to the carriage. It may be regarded as composed of two equal and opposite magnetic poles at its extremities. The moment round the axis on one pole must be equal and opposite to the moment exerted on the other, whatever the position of the magnet on the carriage may be. Let F_1, F_2 be the forces on the poles at right angles to the planes through them and the conductor, r_1, r_2 the distances of the poles from the conductor supposed to be a thin wire. The moments round the conductor give

$$F_1 r_1 + F_2 r_2 = 0$$

and therefore

$$-\frac{F_1}{F_2} = \frac{r_2}{r_1} \quad \ldots \ldots \ldots (1)$$

or the forces have opposite moments and are inversely as the distances from the axis.

Deduction of Law of Force from Equivalent Magnetic Shell.

We may deduce the results stated above for a long straight conductor from Ampère's theorem of the equivalence of a current and a magnetic shell. We have seen that the shell is defined only by its bounding edge and the strength of the current. If we consider an infinitely long straight conductor carrying a current γ, the equivalent shell is geometrically defined only by its edge, and we may take the shell as a plane surface, otherwise in any position we please. Let the shell be at right angles to the plane of the paper, A (Fig. 24) the projection of the conductor, AB of the shell, P the position of the magnetic pole, PC ($= a$) its distance from the plane of the shell, and CA ($= b$) the distance of C from A. Let E be the projection of an element of the shell, the distance $CE = y$, the distance of the element from the plane of the paper z, and its area $dydz$. The radius vector from P to the element has for length $(y^2 + z^2 + a^2)^{\frac{1}{2}}$, and the projection of the

---

* *El. and Mag.* vol. ii. p. 130 (2nd ed.).

FIELD OF INFINITELY LONG STRAIGHT CONDUCTOR

element at right angles to the radius vector is $adydz/(y^2+z^2+a^2)^{\frac{1}{2}}$. Hence the solid angle subtended by the element at P is $adydz/(y^2+z^2+a^2)^{\frac{3}{2}}$. The total solid angle ω subtended at P

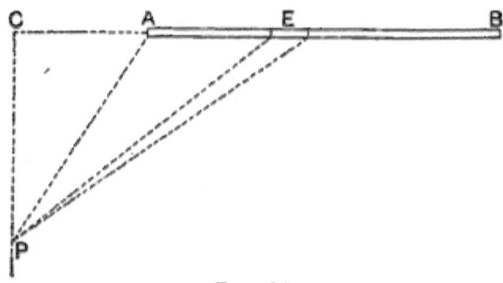

Fig. 24.

by the shell, supposing the positive side turned towards P, is given by the equation

$$\omega = \int_b^\infty \int_{-\infty}^\infty \frac{adydz}{(y^2+z^2+a^2)^{\frac{3}{2}}} = \pi - 2\tan^{-1}\frac{b}{a} \quad . \quad . \quad (2)$$

Hence for the potential V at P of the magnetic shell we have

$$V = \gamma\left(\pi - 2\tan^{-1}\frac{b}{a}\right) \quad . \quad . \quad . \quad . \quad (3)$$

where for $\tan^{-1} b/a$ is to be taken the angle between 0 and $\pi/2$ which has b/a for its tangent.

The same result may be obtained geometrically with great ease thus:—The solid angle subtended at P by a plane rectangle, of finite breadth and infinite length, is the area of the lune cut out of the unit sphere (centre P) by planes drawn through P and the edges of the rectangle. If θ be the angle between these planes the area is $4\pi \times \theta/2\pi = 2\theta$. Thus, if by the addition of a rectangular strip the edge of the shell were brought to C_1 the solid angle would be $2 \times \pi/2$ or π. But for this strip, $\theta = \tan^{-1} b/a$. Hence the actual solid angle is $\pi - 2\tan^{-1} b/a$.

The components of the magnetic force at P are $-dV/da$, $-dV/db$ respectively. Hence the resultant magnetic force at P is $\{(dV/da)^2 + (dV/db)^2\}^{\frac{1}{2}} = 2\gamma/(a^2+b^2)^{\frac{1}{2}}$, or if r be the distance of P from A it is $2\gamma/r$. The direction of the force is therefore in the plane of the paper, and at right angles to PA, and *from* that side of the plane through P and the conductor on which C lies, for we have

$$-dV/da = -2\gamma b/(a^2 + b^2), \quad -dV/db = 2\gamma a/(a^2 + b^2),$$

and the equation of the plane the projection of which is PA, is $bx - ay = 0$, if x be taken from P in the direction PC. The x and y direction cosines of a normal to this plane are respectively proportional to $-b$ and a, as are also the component forces parallel to x and y. By experiment it is found that the direction which the current must have in order that a positive or north-seeking pole should move as here specified is from below upwards through the paper. This agrees with the rule at the foot of p. 103. P is thus on the positive side of the shell.

Expression for Potential found from Law of Force.
We may proceed from the experimental fact, that the intensity of the magnetic field at any point is inversely as the distance, r, of the point from the straight conductor, to determine whether the current has a magnetic potential or not. First defining the unit of current so that the magnetic force is $2\gamma/r$, taking the origin at A, and the axes of x and y along AB and parallel to CP respectively, and putting x, y, z for the coordinates of the point P, we have for X, Y, Z, the components of force at P, the values $X = -2\gamma y/r^2$, $Y = 2\gamma x/r^2$, $Z = 0$, and hence

$$Xdx + Ydy + Zdz = -2\gamma \frac{d(y/x)}{1 + y^2/x^2}$$

that is, the expression in the left is a perfect differential of the function $-\gamma \tan^{-1} y/x + C$, which is therefore the potential at P. This is a many valued function of x, y, z; but since we have to deal only with the difference of potential between two points, that is with the work done in carrying a unit pole from one to the other, the expression will lead to no ambiguity.

We have here to take into account, as pointed out above, the difference in the work done in any closed path according as the path does or does not pass round the conductor. The work done in any closed path is zero, if the path can be supposed shrunk in upon any point within it without cutting the conductor, for, clearly, the work done in carrying a unit pole from any point P to another point Q is equal and opposite to the work done in carrying the pole from Q to P along the remaining part of the path.

WORK IN CARRYING POLE ROUND CURRENT

On the other hand if the path embrace the conductor this reasoning does not hold. It is clear that the work done in carrying a unit pole once round in a circle of which the conductor is the axis, say $TUVWT$ in Fig. 25, is $4\pi\gamma$. For the force at each point is tangential to the circle, and has the value $2\gamma/r$, while the length of path is $2\pi r$, and these give the product $4\pi\gamma$. Let now the given closed path, which may or may not be in a plane, for example $PQRSP$ in Fig. 25, be connected with the circle by the

Theorem of Work done in Carrying Unit Pole round Current Second Proof.

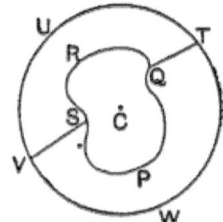

Fig. 25.

lines QT and SV. The work done in each of the closed paths $SRQTUVS$, $SVWTQPS$, is zero since neither embraces the conductor. Hence the whole work done in these two paths is zero. But if the pole be carried round these paths in the order stated above, the whole work done is the sum of the work done in carrying the pole round the circle $TUVWT$, and round the given closed path in the direction $PQRS$, since the work done in the paths SV and QT is zero, these being traversed twice in opposite directions. Hence the work done in the path $PQRS$ embracing the conductor is also numerically $4\pi\gamma$.

This method of proof leads also to the result, already proved above (p. 107), that the work done in carrying a pole round a conductor whether straight or not is $4\pi\gamma$. For if we suppose the conductor infinitely thin and to have finite curvature, and take a closed circular path infinitely near it, the pole will be acted on only by the portion of the conductor which is near it as compared with the rest of the circuit, and this may be considered as a long straight conductor. The work done in carrying a unit pole round the circular path is $4\pi\gamma$. Then by connecting with the circular path any other path embracing the conductor, the work done in carrying a pole round it may be proved, as above, to be $4\pi\gamma$.

If the circuit be not infinitely thin the actual conductor may be supposed made up of an infinite number of filamental conductors

VOL. II.   I coinciding with the lines of flow, and for each of these the work done in carrying a unit pole round a path embracing it is $4\pi \times$ the current in the filament. Hence in a closed path embracing the whole current γ, the work done upon a unit pole traversing it is $4\pi\gamma$. Thus the theorem is extended to non-linear conductors. The case of interlacing of the path and the conductor may be dealt with as before (see p. 108).

Relation of Current to Line Integral of Magnetic Force round Conductor.
If the current strength per unit area at right angles to the direction of flow at any point be denoted by q, and l, m, n, be the direction cosines of that direction, then we may call lq, mq, nq the components of the current along the axes. Denoting these by u, v, w, we have for the component of flow in any direction of which the cosines are λ, μ, ν, the expression $\lambda u + \mu v + \nu w$.

If now we take any closed path round a conductor, or portion of a conductor carrying currents, and take the line-integral of the magnetic force round the path, and the surface integral of the current across the surface, the theorem just discussed may be thus expressed

$$4\pi \int (\lambda u + \mu v + \nu w) dS = \int \left(a \frac{dx}{ds} + \beta \frac{dy}{ds} + \gamma \frac{dz}{ds} \right) ds \ . \quad (4)$$

The second integral may be transformed by the following process, which may also be employed to transform the expression on the right of (49), p. 33, and so give the values of a, b, c, in terms of the components of vector potential. Let ABC, Fig. 26,

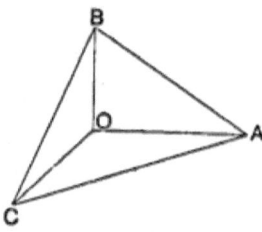

FIG. 26.

be one face of a tetrahedron the other three faces of which are OAB, OBC, OCA, and have edges OA, OB, OC in the direction of the axes of x, y, z.

Taking $\int \left(a \frac{dx}{ds} + \beta \frac{dy}{ds} + \gamma \frac{dz}{ds} \right) ds$ round the closed path AB, we see at once that it can be converted into the corresponding integrals round the three paths $OABO$, $OBCO$, $OCAO$, since the integral along each of the lines OA, OB, OC is thus taken twice in opposite directions. Thus we obtain

$$\int_{ABC}\left(\alpha\frac{dx}{ds} + \beta\frac{dy}{ds} + \gamma\frac{dz}{ds}\right)ds = \int_{OAB}\left(\alpha\frac{dx}{ds} + \beta\frac{dy}{ds}\right)ds + \int_{OBC}\left(\beta\frac{dy}{ds} + \gamma\frac{dz}{ds}\right)ds$$
$$+ \int_{OCA}\left(\gamma\frac{dz}{ds} + \alpha\frac{dx}{ds}\right)ds.$$

Now consider any one of these three integrals, say the first taken round $OABO$. Let α, β, γ be the components of the magnetic force at O. Then the values of the first two components at any point distant δx, δy from O in the plane xy are

$$\alpha + \frac{d\alpha}{dx}\delta x + \frac{d\alpha}{dy}\delta y, \qquad \beta + \frac{d\beta}{dx}\delta x + \frac{d\beta}{dy}\delta y.$$

If we take the tetrahedron so small that its edges OA, OB, OC are δx, δy, δz, the integral round $OABO$ may be found by taking the values of the components at the middle points of OA, AB, BO as the mean values over these distances. Thus we get for the integral

$$\left(\alpha + \tfrac{1}{2}\frac{d\alpha}{dx}\delta x\right)\delta x - \left(\alpha + \tfrac{1}{2}\frac{d\alpha}{dx}\delta x + \tfrac{1}{2}\frac{d\alpha}{dy}\delta y\right)\delta x$$
$$+ \left(\beta + \tfrac{1}{2}\frac{d\beta}{dx}\delta x + \tfrac{1}{2}\frac{d\beta}{dy}\delta y\right)\delta y - \left(\beta + \tfrac{1}{2}\frac{d\beta}{dy}\delta y\right)\delta y,$$

which reduces to

$$\tfrac{1}{2}\left(\frac{d\beta}{dx} - \frac{d\alpha}{dy}\right)\delta x \delta y = \left(\frac{d\beta}{dx} - \frac{d\alpha}{dy}\right) \times \text{area } AOB.$$

In the same way we obtain corresponding results for $OBCO$ and $OCAO$. But if ABC be taken as dS we have area $OBC = \lambda dS$, $OCA = \mu dS$, $OAB = \nu dS$.

Hence

$$4\pi(\lambda u + \mu v + \nu w) = \lambda\left(\frac{d\gamma}{dy} - \frac{d\beta}{dz}\right) + \mu\left(\frac{d\alpha}{dz} - \frac{d\gamma}{dx}\right) + \nu\left(\frac{d\beta}{dx} - \frac{d\alpha}{dy}\right)$$

Equations of Currents.

and therefore

$$u = \frac{1}{4\pi}\left(\frac{d\gamma}{dy} - \frac{d\beta}{dz}\right)$$
$$v = \frac{1}{4\pi}\left(\frac{d\alpha}{dz} - \frac{d\gamma}{dx}\right) \quad \ldots \ldots \quad (5)$$
$$w = \frac{1}{4\pi}\left(\frac{d\beta}{dx} - \frac{d\alpha}{dy}\right)$$

A Circuit and Magnet Equivalent in one Medium not necessarily so in another.

equations which will be found of great importance in the sequel. There is one remark on the equivalence of a current and a magnetic distribution which ought to be made here, though we have not space to deal fully with the matter. The mutual action between a current flowing in a conductor and a distribution of magnetism is independent of the nature of the medium in which they are placed, if that medium be the same throughout, but this does not hold for the mutual action between two distributions of magnetism.*

Effect of the Nature of the Medium.

The value of the magnetic force at any point, by definition, does not depend on the nature of the medium *at* the point in question, but only on the magnetization elsewhere. In a uniform medium, which has imbedded in it a conductor carrying a current, the potential at any point may be taken as made up of two parts, that which would be produced by the circuit alone, in a medium of unit inductive capacity, and that due to the magnetization which the medium receives in consequence of its specific inductive capacity differing from unity. Now the second part of the potential is single valued, and hence the line-integral of its variation round a closed curve is zero. If the induced magnetization of the medium is solenoidal (as it always is when κ is uniform) and the medium extends indefinitely in all directions, no force due to the magnetization of the medium is experienced by a magnetic pole placed anywhere; but the action is precisely the same as if the circuit and pole were situated in air. Of course if the medium is different in different parts, as for example when it consists partly of iron, partly of air, the magnetization of the different parts must be taken into account in assigning the value of the magnetic force at any point. In the case of solenoidal distribution this is effected by taking into account the virtual surface distribution, resulting from the discontinuity of the magnetization at the separating surfaces.

---

* Neglect of this difference in the two cases has led to the assignment of wrong dimensions to unit quantity of magnetism in electrostatic units. See a discussion in the *Phil. Mag.*, 1882.

TOTAL MAGNETIC INDUCTION THROUGH CIRCUIT

We come now to the action of a magnetic system upon a current. The theorem of the equivalence of a current to a magnetic distribution established above leads of course to the conclusion that whatever process, or function, is available for the calculation of the forces acting on the magnetic shell, is also available for the calculation of the action on the current when in the field. This will be manifested as certain forces acting on the conductor which we have now to investigate. Effects of the electromagnetic action on the current itself will be discussed later.

Action of a Magnetic System on a Current.

The function from which we determine the force acting on a magnetic distribution in a magnetic field is the expression for the potential energy which the system possesses in virtue of its being in the field. We have found for this in the case of a shell of strength ϕ

Calculation of Force from Potential Energy of Shell in Field.

$$\mathsf{E} = \phi \iint \left(l\frac{dV}{dx} + m\frac{dV}{dy} + n\frac{dV}{dz} \right) dS. \quad \ldots \quad (6)$$

where V is the magnetic potential (due to the distribution producing the field and not at all to the shell itself), at the element dS the coordinates of which are x, y, z, and the integral is taken over the surface of the shell. But, if there be none of the magnetism producing the field at the shell itself, we have for the components of magnetic induction at (x, y, z) $a, b, c = -dV/dx$, $-dV/dy$, $-dV/dz$; and therefore, writing for ϕ the current strength γ, we get instead of (6)

$$\mathsf{E} = -\gamma \iint (la + mb + nc) dS. \quad \ldots \quad (6')$$

If the surface, as supposed here, do not pass through magnetized matter, α, β, γ coincide in value with a, b, c; but it is easy to see that a, b, c ought to be used in the integral in the general case. For let the surface bounded by the circuit be taken so as to pass through a portion of another medium. Then since

$$\iint (la + mb + nc) dS$$

has the same value for all surfaces having the same bounding edge, it is an expression which gives the same value of E for all positions of the surface.

The integral in this equation is the value of the magnetic induction through the shell, and here and in what follows we denote it by N. It is to be taken positive or negative according as it passes through the shell from the negative to the positive or from the positive to the negative side, that is, according as its direction agrees with or is opposite to that in which a right-handed screw would move through the circuit if the handle were turned round in the direction of the current. Hence

$$\mathsf{E} = -\gamma N \quad \ldots \ldots \quad (7)$$

Case of Permeability different from Unity. If the circuit is imbedded in a medium of magnetic permeability differing from unity, the magnetization of the medium must be taken into account in finding the potential energy of the system. We have simply as above to calculate the value of N for the circuit. It will not be necessary however to deal practically here with any such case. Those in which movable coils containing iron cores have to be dealt with do not cause any difficulty, since the magnetism of the core forms in each case part of the distribution producing the field.

Now the magnetic forces acting on the shell are such as to diminish its potential energy; and hence, if $d\psi$ be any small change of position or configuration of the shell, and Ψ the corresponding force producing it, we have for the work done by this force $\Psi d\psi$. The sum of this and the change in the value of the potential energy is zero, that is

$$\Psi d\psi + d\mathsf{E} = 0 \quad \ldots \ldots \quad (8)$$

or, γ remaining constant,

$$\Psi = \gamma \frac{dN}{d\psi}. \quad \ldots \ldots \quad (8')$$

Force on Element of Circuit. The direction of the electro-magnetic force is therefore to increase N; that is, the circuit if free to move as a rigid whole will change its position so as to increase N, and, what is here of great importance, if flexible, will alter its form so as to include a greater value of N. It is clear, then, that no force acts on an element of the circuit in the direction parallel to the magnetic force, for a displacement in that direction would not alter the value of E, and the resultant electro-magnetic force on each element is therefore at right angles to the magnetic force.

But the element itself, in the general case, is inclined to the direction of the magnetic induction. Let the angle between the latter direction (taken as that in which a north-seeking pole tends to move through the circuit) and that of the current in an element ds of the circuit be θ; and let the element be moved through any displacement $d\psi$ at right angles to the line of magnetic induction at its centre. The change in N is $\gamma \mathsf{B} \sin\theta\, ds\, d\psi$. Thus we have for the force on the element

$$\Psi = \gamma \mathsf{B} \sin\theta\, ds \quad \ldots \ldots \quad (9)$$

COMPONENTS OF ELECTROMAGNETIC FORCE

Rule for Direction of Force on Element.

The direction in which the element tends to move may be remembered by the following rule. Let, as supposed above, a human figure stand on the magnetic shell which replaces the circuit, so that, when the face of the figure is turned in the direction in which the current is flowing, the positive direction of the magnetic induction is from the feet of the figure towards the head. Then the element, if free to move, will do so towards the figure's right hand. Or, if the figure swim in the circuit so that the current enters at the feet and leaves at the head, and look in the positive direction of magnetic induction, the element will tend to move towards the *left* hand.

The direction of the force on an element of the circuit is shown in Fig. 27. The corresponding reaction is discussed below (Section II.).

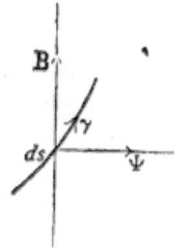

FIG. 27.

Denoting by l, m, n, the direction cosines of ds, we have

$$\sin \theta = \{(mc - nb)^2 + (na - lc)^2 + (ma - lb)^2\}^{\frac{1}{2}}/\mathbf{B}.$$

Hence (9) becomes

$$\Psi = \gamma\{(mc - nb)^2 + (na - lc)^2 + (ma - lb)^2\}^{\frac{1}{2}} ds \quad . \quad (9')$$

If σ denote the area of cross-section of the conductor at the element ds, taken at right angles to the direction of γ, then u, v, w, the components of current, are defined by the equations

$$l\gamma/\sigma, \; m\gamma/\sigma, \; n\gamma/\sigma = u, v, w.$$

Substituting in (9') and resolving Ψ/σ along the axes, denoting the components by X, Y, Z, we find instead of (9')

Equations of Electromagnetic Force.

$$\left. \begin{array}{l} X = vc - wb \\ Y = wa - uc \\ Z = ub - va \end{array} \right\} \quad . \quad . \quad . \quad . \quad . \quad (9'')$$

X, Y, Z, are the component electromagnetic forces per unit of volume acting on the conductor: we shall find them useful in considering action on non-linear conductors.

Distinction between Potential Energies of Current and Shell in Field.

With regard to the potential energy of the shell and field, care must be taken, while using this expression for the calculation of the force on the circuit (a procedure the legitimacy of which follows from the theorem of equivalence as regards forces), not to allow it to cause any misconception as to the energy of the current in the field. It is not the case that there is any sensible mutual potential energy of the current and the magnetic distribution, such that, when the circuit moves in the field in obedience to magnetic force, exhaustion of this potential energy takes place in the same way as when the shell moves in the field. The shell and field remaining each unchanged, the magnets are set in relative motion, and kinetic energy is acquired, or work is done against external resistance at the expense of potential energy, which so far as our knowledge goes at present may be regarded as a function of the configuration of the system. On the other hand the fact, as illustrated by the experiments of Joule referred to below (Chapter V.), and all experience of the motion of conductors in magnetic fields, is that the kinetic energy acquired, or external work done, in the case of motion of the circuit, is obtained at the expense of the battery or electrical generator maintaining the current, and no available energy is gained or lost in virtue of geometrical displacement *per se*. This subject will be further discussed under induction of currents in Chapter IV.

SECTION II.

ACTION OF CURRENTS ON CURRENTS.

Mutual Action of Two Circuits:

IT is a result of experiment that the equivalence of a current and a magnetic shell which enables the action of a current on a magnet, or of a magnet on a current, to be calculated, is also available for the determination of the action of currents on one another.* Experiments which prove this were made first by

---

* It is to be clearly understood that electrostatic action due to difference of potential between adjacent conductors is not here taken into account. We shall have examples later of combined electrostatic and electrodynamic action.

AMPÈRE'S EXPERIMENTS

Ampère, Weber, and others; but the best experimental proof of the truth of this proposition is to be found in the uniformly consistent results obtained by means of measuring instruments made and graduated to give absolute determinations by applying it. Weber's electro-dynamometer was the first instrument of this kind constructed, and with it the inventor accurately verified the laws of electromagnetic action which had previously been announced by Ampère, as a deduction from his celebrated series of four experiments.

Deducible from Equivalent Magnetic Shells.

Ampère however, besides giving the theorem of the equivalence of currents and magnetic shells, took another view of the subject, in which he regarded every element of a conductor carrying a current as acted on by every element of the other conductor, and the law of action which he gave was a law for the mutual action between two elements. This law agrees with experiment in so far as it gives when applied over the whole circuit of each conductor exactly the electromagnetic action observed; but it is only one of several laws of action between elements which do the same thing. The actions in all cases which have been investigated have been actions between parts of different closed circuits, or between different parts of one closed circuit, and no difference in result has been found between these two cases. We are in ignorance of how two unclosed conductors, or two parts of an unclosed circuit, carrying currents (if such an arrangement can really be obtained) act upon one another, but, though this be true, it is allowable in the case of closed circuits to establish and use any formula for the mutual action of each pair of elements, which is mathematically true in the sense of giving the actual forces observed between the circuits. A simple expression of this kind is that found by Ampère. We shall here give first some account of Ampère's experiments, and show how by means of a certain assumption the law given by him can be deduced.

These experiments were made by means of apparatus invented by Ampère himself, copies of which are now to be found in almost all collections of apparatus. The chief piece is one for enabling a part of a closed circuit (in itself generally nearly a closed circuit), to turn freely round a vertical axis. The arrangement with the movable conductor in position is shown in the diagram (Fig. 28).

Ampère's Experiments:

Two metallic cups containing mercury are arranged close together in the same vertical line at the extremities of two projecting arms, and in these rest the turned down extremities of the movable conductor. This has different forms according to the effect to be tested or measured. The two arms carrying

Apparatus.

the cups are in conducting contact with the mercury, and one of them is generally attached to a vertical metallic tube fixed to a heavy sole plate, the other is a continuation of a wire or rod which, insulated from the tube, passes up within it from the sole plate. The current is thus led to one cup and from the other without the conveying wires themselves producing any sensible action.

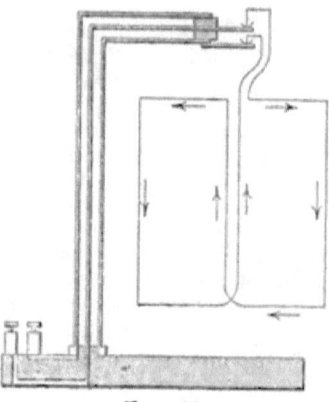

Fig. 28.

The portion of the circuit suspended in the cups in the first two experiments was (as shown in Fig. 28) a double rectangular frame of wire, the wires of which are insulated from one another at the points of crossing. This frame gives two nearly closed circuits of equal area; and round these the current flows in opposite directions, so that the suspended conductor does not experience any action in the earth's magnetic field.

First Experiment.
In Ampère's first experiment a wire (Fig. 29) carrying a current

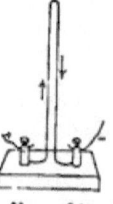

Fig. 29.

AMPÈRE'S EXPERIMENTS

was doubled on itself, and the two portions were kept from touching by insulating material between them. This double wire being brought near and parallel to one side of the suspended frame, the latter did not experience any sensible deflecting force, showing that the effect of the current in one direction in one portion of the doubled conductor neutralized almost exactly the effect of the opposite current in the other part. Exact experiments show that this neutralization is complete, if one conductor be a tube containing the other.

In the second experiment one of the two portions of the doubled wire was not straight, but (Fig. 30) contained a series of small and

Second Experiment.

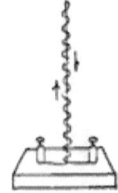

FIG. 30.

rather sharp bends, no part of any one of which was far from the straight conductor. The suspended conductor was still found unaffected. The conclusion from this experiment is that the effect of an element of a straight conductor may be replaced by that of a small crooked conductor having the same beginning and end as the element has, if the same current flow in both cases. In other words the effect of any element may be considered as the resultant in the ordinary sense of any number of component elements at the same place.

In Ampère's third experiment a conductor which formed an arc of a horizontal circle was made movable round a vertical axis through the centre of the circle. This was done by supporting the arc of wire on the convex surface of mercury projecting above a horizontal plane from troughs cut in it, and attaching it to a light radial arm of insulating material moving about the vertical axis. The current passed through the arc from one trough to the other. It was found that no magnet, or circuit carrying a current, produced any effect in moving the conductor in the direction of its length, that is the resultant force upon it was normal to the element.

Third Experiment.

Fourth Experiment.

In the fourth experiment currents were made to pass through three similar and nearly closed conductors A, B, C (Fig. 31), the middle one of which B was attached to the stand and was movable round a vertical axis. The currents in A and C were of equal strength and in the same direction; the direction and strength of the current in B were indifferent. The three circuits were similar in form, and the two, A, C, which were on opposite sides of the movable conductor B, were of very different dimensions, but so chosen that each dimension of the circuit B was n times the corresponding dimension of A, and $1/n$ of the corresponding dimension of C. The position of the conductor B relatively to C was similar to that of A relatively to B, and therefore the distance of any element of C from any element of

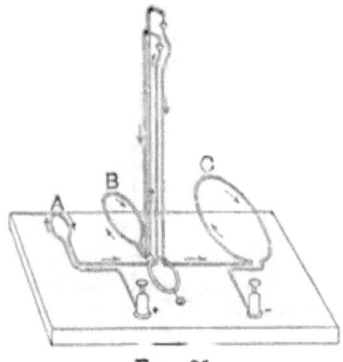

Fig. 31.

B was n times the distance of the corresponding elements in B and A.

Action between Two Elements inversely as Square of Distance.

The movable circuit B was thus subjected to two opposite force-systems from A and C, and was found to remain in equilibrium under that action. From this it follows that if we assume the action on the whole of the movable conductor to be made up of the actions on each of its elements of all the elements of the other two conductors, the action between any pair of elements varies inversely as the square of the distance between them. To prove this let r_1 be the distance between an element b_1 in B and an element a_2 in A, and r_2 the distance between two similarly situated elements, c_1 and b_2, in C and B; and let $f(r_1)$, $f(r_2)$ be the forces between the elements of the respective pairs per unit of length and per unit of current in each case. Now if ds be the length of each of the elements of B chosen, those of the elements a_2 and c_1 of A and C are respectively ds/n, nds.

From the equilibrium of B it is clear that the forces for corresponding pairs of elements are equal, and therefore we have, if γ be the current in A and C and γ_1 that in B,

$$\frac{ds^2}{n}\gamma\gamma_1 f(r_1) = nds^2 \gamma\gamma_1 f(r_2)$$

or

$$\frac{f(r_2)}{f(r_1)} = \frac{1}{n^2} \quad \ldots \ldots \ldots (10)$$

that is the law of force is the inverse square of the distance.

Now, since by the second experiment each element can be replaced by its components, we may first resolve each into two components parallel to and at right angles to the line joining the centres of the elements. Also by the first experiment the forces are as the lengths of the elements and as the strengths of the currents. Let ds, ds' be the lengths of the elements AB, $A'B'$, θ, θ', the angles which they make with the line joining their

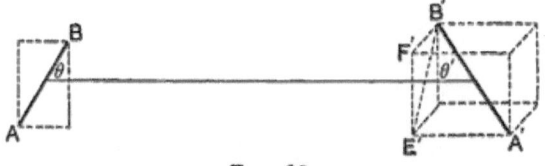

FIG. 32.

centres, as shown in Fig. 32, then the components are $ds \cos \theta$, $ds' \cos \theta'$ along the line, and $ds \sin \theta$, $ds' \sin \theta'$ perpendicular to it. The last ($B'E'$ in the Figure) is not in the same plane with $ds \sin \theta$, and gives, if η be the angle $B'E'F'$ between the two planes, the components $ds' \sin \theta' \cos \eta$, $ds' \sin \theta' \sin \eta$, parallel to and at right angles to $ds \sin \theta$. We now consider the actions (couples excluded) between the different pairs of these elements.

Theoretical Results of Ampère's Experiments.

In the first place we have the two elements $ds \cos \theta$, $ds' \cos \theta'$ in the same straight line. The only determinate direction of any action between these two elements is the straight line in which they lie. We suppose, therefore, that when θ, θ' are both acute the force between the elements is an attraction. It has the value $A\gamma\gamma' \cos \theta \cos \theta' ds ds'/r^2$, where A is a constant, and r is the distance between the centres of the elements.

Action of Elements in the same Straight Line.

Action of Parallel Elements Perpendicular to joining Line.

We take next the two elements $ds\sin\theta$, $ds'\sin\theta'\cos\eta$, which are parallel to one another and at right angles to the line joining their centres. This force has for value $B\gamma\gamma'\sin\theta\sin\theta'\cos\eta\,ds\,ds'/r^2$ where B is a constant, and must act in the plane of the elements; for there is no reason why a component at right angles to this plane should act towards one side or the other of it. Further it must act in the line joining the elements, for to change the sign of one of the currents reverses the action, and to change the sign of both must leave the action unchanged. We shall suppose that it is also an attraction.

Action of Elements Perpendicular to one another,

Lastly we have the four pairs of elements at right angles to one another. Of these the pair $ds\sin\theta$, $ds'\sin\theta'\sin\eta$ are at right angles to one another and likewise to the line joining their centres. Now if we make the assumption that the force between two components at right angles to one another is in the line joining their centres (an assumption necessary for Ampère's theory in the cases of the other two pairs of elements), we can easily prove that the force between the pair of elements now being considered is zero. Suppose that we have such a pair of elements a, β at right angles to one another, and let the force act as shown in Fig. 33 from left to right. Now if the whole system

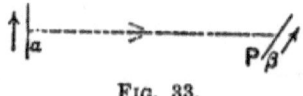

Fig. 33.

be turned through 180° round the direction of a as an axis (or be looked at from the other side of the paper), the direction of β will be reversed, and the force will now act from right to left. The system then turned from its new position through 90° about the line joining a, β, gives the original arrangement of the elements, with the force between them reversed. Hence no such force can exist.

The assumption made above is not really necessary for this case, since, if there exist a component at right angles to the line of centres, it must act in the plane of one of the elements and the line of centres, or in a plane bisecting the angle between the planes of the elements and the line of centres, and there is nothing to determine in which plane it must act.

There remain the three pairs of elements $ds\cos\theta$, $ds'\sin\theta'\cos\eta$, $ds\cos\theta$, $ds'\sin\theta'\sin\eta$, and $ds\sin\theta$, $ds'\cos\theta'$, the constituents of each of which are in one plane. Making the assumption stated above for these, we see that between the elements of each pair there can be no force, since if a force did exist it would not be reversed by the reversal of the current in $ds \sin \theta$, $ds' \sin \theta' \sin \eta$, or $ds' \sin \theta' \cos \eta$, for this would merely be equivalent to turning the whole system through 180° round the line joining the elements. We find therefore collecting these results a total force of attraction between the two elements ds, ds' of amount

Attraction between Two Elements.

$$dF = \gamma\gamma' \, ds \, ds' \frac{1}{r^2}(A \cos\theta \cos\theta' + B \sin\theta \sin\theta' \cos\eta) \quad (11)$$

and it remains to determine the coefficients A and B.

Now applying the result of Ampère's third experiment we resolve the force on ds into two components, one along ds, the other at right angles to it, and equate the integral of the former, taken round the circuit of ds', to zero. Hence

Determination of Constants.

$$\gamma\gamma' \, ds \left\{ A \int \frac{1}{r^2}\cos^2\theta \cos\theta' \, ds' + B \int \frac{1}{r^2}\sin\theta\cos\theta\sin\theta'\cos\eta \, ds' \right\} = 0 \quad (11')$$

This expression can be transformed as follows. We have by geometry, if the coordinates of the centres of the elements be x, y, z, x', y', z',

$$-\cos\theta = \frac{dr}{ds}; \quad -\cos\theta' = \frac{dr}{ds'} \quad \ldots \quad (12)$$

and $r^2 = (x - x')^2 + (y - y')^2 + (z - z')^2$.

The last gives

$$r\frac{dr}{ds} = -r\cos\theta = (x - x')\frac{dx}{ds} + (y - y')\frac{dy}{ds} + (z - z')\frac{dz}{ds};$$

and differentiating this with respect to s' we get

$$\frac{dr}{ds}\frac{dr}{ds'} + r\frac{d^2r}{ds\,ds'} = -\frac{dx}{ds}\frac{dx'}{ds'} - \frac{dy}{ds}\frac{dy'}{ds'} - \frac{dz}{ds}\frac{dz'}{ds'} = -\cos\epsilon \quad (13)$$

where ϵ is the angle between the elements ds, ds'.

But by (12)

$$r\frac{d^2r}{ds\,ds'} = r\sin\theta\frac{d\theta}{ds'} = r\sin\theta'\frac{d\theta'}{ds} = -\sin\theta\sin\theta'\cos\eta \quad (14)$$

since by geometry

$$-r\frac{d\theta}{ds'} = r\frac{d\theta'}{ds}\cos\eta = \sin\theta'\cos\eta.$$

Therefore by (12) and (13)

$$\cos \epsilon = - \cos \theta \cos \theta' + \sin \theta \sin \theta' \cos \eta \quad . \quad . \quad (15)$$

Substituting from (12) and (14) in (11′) and rearranging, we find

$$\gamma\gamma' ds \left[\left(A - \tfrac{1}{2}B \right) \int \frac{1}{r^2} \left(\frac{dr}{ds} \right)^2 dr - B \int \frac{d}{ds'} \left\{ \frac{1}{2r} \left(\frac{dr}{ds} \right)^2 \right\} ds' \right] = 0 \; . \; (16)$$

The discussion of the mutual actions of the pairs of elements may be avoided and equation (16) proved in the following manner. We have seen that the mutual action of ds and ds' is equal to $\gamma\gamma' \, ds \, ds' f(\theta, \theta', \epsilon)/r^2$, where $f(\theta, \theta', \epsilon)$ is a function of the relative positions to be determined. But by Ampère's second experiment it is the sum of the forces between the elements ds', and the three components dx, dy, dz, into which ds may be resolved. Hence $f(\theta, \theta', \epsilon) \, ds \, ds' = P \, dx \, ds' + Q \, dy \, ds' + R \, dz \, ds'$, or

$$f(\theta, \theta', \epsilon) = P \frac{dx}{ds} + Q \frac{dy}{ds} + R \frac{dz}{ds},$$

where P, Q, R depend only on the position of ds'. Thus $f(\theta, \theta', \epsilon)$ is a linear homogeneous function of the direction cosines of ds; and similarly it must be a linear homogeneous function of the direction cosines of ds'. To fulfil these conditions and involve ϵ it must be made up of two parts, $A \, dr/ds \cdot dr/ds'$, $Br \, d^2r/ds \, ds'$, where A and B are constants. Hence Ampère's third experiment gives

$$\gamma\gamma' ds \int \frac{1}{r^2} f(\theta, \theta', \epsilon) \frac{dr}{ds} ds' = \gamma\gamma' ds \int \frac{1}{r^2} \left\{ -A \left(\frac{dr}{ds} \right)^2 \frac{dr}{ds'} + Br \frac{d^2r}{ds \, ds} \frac{dr}{ds} \right\} ds' = 0,$$

which is equivalent to (16).

The second integral in (16) vanishes when taken round the circuit of s', and we are left with the equation

$$\gamma\gamma' ds \left(A - \tfrac{1}{2}B \right) \int \frac{1}{r^2} \left(\frac{dr}{ds} \right)^2 dr = 0 \quad . \quad . \quad . \quad (16')$$

ACTION BETWEEN TWO CURRENT ELEMENTS

Now the integral in this equation does not in general vanish,* and therefore $A = \frac{1}{2}B$. Hence substituting in (11), and using (15) we get

$$dF = B\gamma\gamma' ds ds' \frac{1}{r^2}\left(\cos\epsilon + \frac{3}{2}\cos\theta\cos\theta'\right) \quad . \quad . \quad (17)$$

Ampère's expression for the action between the elements.†

Ampère's Final Result.

Ampère assumed B to be equal to 1, which amounted to defining the unit of current as that current which flowing in the same direction in each of two parallel elements at unit distance apart gives unit force of attraction between them. We shall show that for agreement with the definition of unit current adopted above the value 2 must be given to B. Thus Ampère's unit of current is $1/\sqrt{2}$ of the electromagnetic unit of current now in ordinary use.

Returning to equation (74) above for the mutual potential energy of two magnetic shells we are led, by the theorem of equivalence of currents and magnetic shells, to write for the mutual potential energy of two closed circuits

$$\mathsf{E} = -\gamma\gamma'\int\frac{\cos\epsilon}{r}ds ds' \ddagger \quad . \quad . \quad . \quad . \quad (18)$$

Ampère's Expression deduced from the Magnetic Shell Theory.

We shall inquire what expression for the action between two elements can be deduced from the quantity on the right, and compare it with that given in (17). Substituting for $\cos\epsilon$ its value given by (13) and noticing that the integral of the complete differential $d^2r/ds ds' \cdot ds$ is zero, we get

$$\mathsf{E} = \gamma\gamma'\int\int\frac{1}{r}\frac{dr}{ds}\frac{dr}{ds'}ds\,ds' \quad . \quad . \quad . \quad . \quad (19)$$

---

* This may be seen by considering the particular case of a circuit, formed by two perpendicular straight lines, and a circular arc joining their extremities, acting on an element ds at the centre of the circular arc and in line with one of the straight lines. If the radius of the circle be c, and the distance of ds from the straight line perpendicular to it be a, the portion of the integral contributed by the latter straight line is $1/3a - a^2/3c^3$, by the other straight line $1/c - 1/a$, and by the circle zero. Hence the total integral is not zero.

† In Ampère's own formula the minus sign is prefixed to the second term. This arises from his using the supplementary angle to θ'.

‡ F. Neumann gave $-\gamma\gamma' ds ds' \cos\epsilon/r$ as the mutual potential of the two elements. The corresponding expression in (19) is due to Weber. Either gives the same result for closed circuits as does Ampère's formula. Thus the forces between the circuits may be found from (17), (18), or (19). The energy of the system is further discussed in Chap. IV. below.

Ampère's Expression deduced from the Magnetic Shell Theory.

Now let the circuit of ds be slightly deformed in any way while that of ds is kept unchanged: r, dr/ds, dr/ds', ds will be affected by the deformation. The change in $\mathbf{E}$ is $\delta\mathbf{E}$, and by taking the variation of the right-hand side of (19), remembering that

$$\delta\frac{dr}{ds} = \frac{d\delta r}{ds} - \frac{dr}{ds}\frac{d\delta s}{ds}, \quad \delta\frac{dr}{ds'} = \frac{d\delta r}{ds'},$$

we find

$$\delta\mathbf{E} = -\gamma\gamma'\left\{\iint \frac{1}{r}\frac{dr}{ds}\left(\frac{1}{r}\frac{dr}{ds'}\delta r - \frac{d\delta r}{ds'} + \frac{dr}{ds'}\frac{d\delta s}{ds}\right)dsds'\right.$$
$$\left. - \iint \frac{1}{r}\frac{dr}{ds'}\frac{d\delta r}{ds}dsds' - \iint \frac{1}{r}\frac{dr}{ds}\frac{dr}{ds'}d\delta s \cdot ds'\right\}$$

The last integral and the last term of the first integral cancel one another, and we have

$$\delta\mathbf{E} = -\gamma\gamma'\iint\left(\frac{1}{r^2}\frac{dr}{ds}\frac{dr}{ds'}\delta r - \frac{1}{r}\frac{dr}{ds}\frac{d\delta r}{ds'} - \frac{1}{r}\frac{dr}{ds'}\frac{d\delta r}{ds}\right)dsds' \quad (20)$$

Integrating the last two terms by parts and rejecting the integrals, round the circuits, of perfect differentials, we get

$$2\gamma\gamma'\iint\frac{1}{r^2}\left(\frac{dr}{ds}\frac{dr}{ds'} - r\frac{d^2r}{dsds'}\right)\delta r\,ds\,ds'$$

for the corresponding part of δE. Hence finally by (20)

$$\delta\mathbf{E} = \gamma\gamma'\iint\frac{1}{r^2}\left(\frac{dr}{ds}\frac{dr}{ds'} - 2r\frac{d^2r}{dsds'}\right)\delta r\,ds\,ds' \quad . \quad . \quad (21)$$

which by (15) becomes

$$\delta\mathbf{E} = 2\gamma\gamma'\iint\frac{1}{r^2}(\cos\epsilon + \tfrac{3}{2}\cos\theta\cos\theta')\delta r\,ds\,ds' \quad . \quad (22)$$

The interpretation of this result is an attraction of amount $2\gamma\gamma'(\cos\epsilon + \tfrac{3}{2}\cos\theta\cos\theta')/r^2$ between the elements ds, ds' in the line joining them. This agrees with Ampère's result and shows that the value of B in (17) is 2.

Having thus shown the equivalence of the two modes of regarding the mutual action of currents, we now give a very short account of the apparatus and experiments by which Weber investigated the subject.

WEBER'S EXPERIMENTS

Weber made his measurements of electromagnetic action by means of his electrodynamometer. This consisted of two circular coils, suspended by bifilar wires (which also conveyed the current) so as to be free to turn round a vertical axis, the other coil fixed and arranged so that by levelling the planes of its windings could be made vertical. The apparatus was in two forms: (1) with the movable coil suspended within the fixed coil, with the centres as nearly as may be coincident; (2) with the fixed and movable coils distinct so that they could be placed at any required distance from one another, and in any relative positions. Deflections of the movable coil were measured by the mirror and telescope method described above (Vol. I., p. 214).

By the first experiment made by Weber it was proved that the electromagnetic action between the two currents varied as the square of the current strength. Apparatus (1) was used, and the fixed coil was set up with its axis perpendicular to, while that of the suspended coil was in, the magnetic meridian. Currents of different strengths were sent through the coils, and to prevent too great a deflection, the current through the suspended coil was reduced to $1/246\cdot26$ of the whole current by a shunt of thick wire inserted between the terminals to which the bifilar wires were attached. A magnetometer with magnetized steel mirror in a damping covering of copper was set up north of the fixed coil, at a distance of $58\cdot3$ centimetres, and the tangents of the deflections of this mirror (read by a telescope as in the other case) gave a comparative measure of the different currents used. The results shown in the following table were obtained; and from these it will be seen that the mutual action between the systems was proportional to the square of the current, that is, to the product of the strengths of the two (equal) magnetic shells.

No. of cells used.	Comparative values of force between coils $= A$.	Force on magnetometer needle in arbitrary units $= B$.	Force on needle found by formula $5\cdot15534\sqrt{A}$.	Diff. $B - 5\cdot15534\sqrt{A}$.
3	440·038	108·426	108·144	− 0·282
2	198·255	72·398	72·389	+ 0·191
1	50·915	36·332	36·786	+ 0·454

In another series of experiments Weber used the apparatus (2). The axis of the suspended coil was placed horizontal and parallel

Weber's Experiments

to the magnetic meridian, while the fixed coil was placed with its axis at right angles to the magnetic meridian, and its centre (1) in the magnetic north and south horizontal line, (2) in the magnetic east and west line through that of the suspended coil. Experiments were made in each case with distances between the centres, of respectively 0, 30, 40, 50, 60 centimetres. The current from eight Bunsen's cells was sent through both coils, and also through a coil set up about 8 metres from the fixed coil so as to act on the magnetometer referred to above, and through a reversing key, so arranged that the current through the suspended coil could be sent first in one and then in the opposite direction without altering its direction in the rest of the circuit. The object of thus reversing the current was to determine and allow for the turning moment of the earth's magnetic field, when the axis of the suspended coil was deflected from the magnetic meridian. The corrected results of the experiments are shown in the table below, in which the second column for each series of positions gives the corresponding numerical values calculated by Ampère's formula (17) above.

Distance of centres of coils apart.	Position of centres of coils.			
	In magnetic east and west line.		In magnetic north and south line.	
	Couple observed.	Couple calculated.	Couple observed.	Couple calculated.
cms. 0	22960	22680	22960	22680
30	189·93	189·03	−77·11	−77·17
40	77·45	77·79	−34·77	−34·74
50	39·27	39·37	−18·24	−18·31
60	22·46	22·64	—	—

Here the results for the greater distances agree very fairly well with calculation from Ampère's formula, and we have shown that Ampère's formula and the magnetic shell theory give identical results.

FORCES BETWEEN STRAIGHT CONDUCTORS

It is to be remarked that in these experiments the two coils are not independent circuits; but that they may be so regarded is plain from the fact that the remaining portion of the circuit, if the wires are close or twisted together, is of no effect since it can be altered at pleasure without affecting the action between the coils, provided the current be maintained constant. *(Formulas Applicable to Action between Two Elements of same Current.)*

But the **deflections** θ, θ' **in the two** cases agree closely for the greater distances with the **formulas**

$$\tan \theta = \frac{2MM'}{d^3}\left(1 + \frac{a}{d^2}\right), \quad \tan \theta' = \frac{MM'}{d^3}\left(1 + \frac{b}{d^2}\right)$$

which express the action between two magnets of moments M, M', in the "end-on" and "side-on" positions and at distances d apart, great in comparison with the dimensions of the magnets.*

Elaborate experiments have also been made by Cazin, Boltzmann, and others in verification of the theory. For these the student should consult Wiedemann, *Elektricität*, Vol. III.

It is an experimental fact that the action between two long parallel conductors carrying currents is an attraction when the currents are in the same direction, and a repulsion when the currents are in opposite directions, and that if the conductors are not parallel there is attraction between them if the directions of the currents in the portions forming equal acute angles with one another are both towards or both from the shortest line joining the conductors, and repulsion if the direction of one is towards that line, and of the other from it. We have not space here to go into calculations regarding such cases, but their general nature may easily be seen by considering the magnetic fields produced by the currents, and the consequent motions of the conductors according to the rules given above. In both cases the lines of force are closed curves surrounding each conductor, and it is obvious that if we consider each circuit completed by a return wire at a great distance, the magnetic induction through each will be increased or diminished by the approach of the conductors if the currents are in the same direction, and diminished if they are in the opposite direction. The same will clearly be the case if the two conductors considered be parts of the same circuit. The action of a current in a straight conductor, on an element of a parallel conductor, is shown in Fig. 34, which with the statements made above explains itself. *(Forces between Straight Conductors.)*

---

* See pp. 72—75 above

Application of Ampère's Formula,

to find Action of a Thin Solenoid.

We give here the application made by Ampère * of his formula to the calculation of the force on an element of a conductor, and the turning moment on a given finite conductor produced by a simple solenoidal electromagnet, that is, a succession of infinitely small circuits arranged equidistantly at infinitely short distances apart with their centres on, and their planes at right angles to a given curve, and carrying currents such that the product of the area of the circuit and the current strength is the same in each case. The solution of this problem is of the greatest importance in Ampère's theory of magnetism, in which he supposes all effects of magnets to be produced by currents flowing in molecular circuits within the body. We shall see that the arrangement specified above is equivalent to a uniformly magnetized

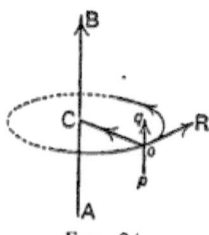

FIG. 34.

magnet, having a strength of magnetic pole equal to the sum of the products of current and area for the circuits round unit length of the given curve forming their common axis.

Let ds be an element of the closed circuit, and consider its action on an element ds' of another conductor, the current being unity in each case. If the coordinates of ds be x, y, z, and the origin be taken at the centre of ds', the direction cosines of ds and r are dx/ds, dy/ds, dz/ds, and x/r, y/r, z/r respectively. The expression for the action between the elements may be written

$$2d \cdot ds' \frac{1}{r^2}\left(-r\frac{d^2r}{ds\,ds'} + \tfrac{1}{2}\frac{dr}{ds}\frac{dr}{ds'}\right),$$

---

* "Théorie des Phénomènes électro-dynamiques," *Mémoires de l'Institut*, VI., 1823. The proof of Ampère's formula, and the applications here given, have been very elegantly treated by quaternion methods by Professor Tait: see his *Quaternions*, 2nd edition, p. 249.

ACTION OF ELEMENT ON CLOSED CIRCUIT

which by (14) becomes

$$2ds\,ds'\frac{1}{r^2}\left(-r\sin\theta'\frac{d\theta'}{ds} - \tfrac{1}{2}\cos\theta\frac{dr}{ds'}\right)$$

$$= 2ds\,ds'\,r^{-\tfrac{1}{2}}\frac{d}{ds}(r^{-\tfrac{1}{2}}\cos\theta').$$

Hence the component of this action along the axis of x or dX is given by *Components of Action of a Closed Circuit.*

$$dX = 2ds\,ds'\,r^{-\tfrac{3}{2}}x\frac{d}{ds}(r^{-\tfrac{1}{2}}\cos\theta') \quad\ldots\quad (23)$$

But if λ, μ, ν be the direction cosines of ds'

$$\cos\theta' = \lambda\frac{x}{r} + \mu\frac{y}{r} + \nu\frac{z}{r}.$$

Hence (23) becomes

$$dX = 2ds\,ds'\frac{x}{r^2}\left[\frac{d}{ds}\left\{\frac{1}{r^2}(\lambda x + \mu y + \nu z)\right\}\right]$$

$$= ds\,ds'\left\{\lambda\frac{d}{ds}\left(\frac{x^2}{r^3}\right) + \mu\frac{x}{y}\frac{d}{ds}\left(\frac{y^2}{r^3}\right) + \nu\frac{x}{z}\frac{d}{ds}\left(\frac{z^2}{r^3}\right)\right\} \quad\ldots\quad (24)$$

But

$$\frac{x}{y}\frac{d}{ds}\left(\frac{y^2}{r^3}\right) = \frac{d}{ds}\left(\frac{xy}{r^3}\right) + \frac{1}{r^3}\left(x\frac{dy}{ds} - y\frac{dx}{ds}\right),$$

and

$$\frac{x}{z}\frac{d}{ds}\left(\frac{z^2}{r^3}\right) = \frac{d}{ds}\left(\frac{xz}{r^3}\right) - \frac{1}{r^3}\left(z\frac{dx}{ds} - x\frac{dz}{ds}\right).$$

Substituting in (24) we find

$$dX = ds\,ds'\left[\frac{d}{ds}\left\{\frac{x}{r^3}(\lambda x + \mu y + \nu z)\right\} + \frac{\mu}{r^3}\left(x\frac{dy}{ds} - y\frac{dx}{ds}\right)\right.$$

$$\left. - \frac{\nu}{r^3}\left(z\frac{dx}{ds} - x\frac{dz}{ds}\right)\right] \quad\ldots\quad (25)$$

The first term disappears when integrated round the circuit of ds. Hence

$$X = ds'\left\{\int \frac{\mu}{r^3}\left(x\frac{dy}{ds} - y\frac{dx}{ds}\right)ds - \int \frac{\nu}{r^3}\left(z\frac{dx}{ds} - x\frac{dz}{ds}\right)ds\right\}$$

Similarly we obtain

$$Y = ds'\left\{\int \frac{\nu}{r^3}\left(y\frac{dz}{ds} - z\frac{dy}{ds}\right)ds - \int \frac{\lambda}{r^3}\left(x\frac{dy}{ds} - y\frac{dx}{ds}\right)ds\right\} \quad . \quad (26)$$

$$Z = ds'\left\{\int \frac{\lambda}{r^3}\left(z\frac{dx}{ds} - x\frac{dz}{ds}\right)ds - \int \frac{\mu}{r^3}\left(y\frac{dz}{ds} - z\frac{dy}{ds}\right)ds\right\}$$

Ampère's Directrix of Electro-Dynamic Action.

Denoting the integrals (divested of the multipliers λ, μ, ν) in these expressions by A, B, C, we have

$$\left.\begin{array}{l} X = ds'(\mu C - \nu B) \\ Y = ds'(\nu A - \lambda C) \\ Z = ds'(\lambda B - \mu A) \end{array}\right\} \quad . \quad . \quad . \quad . \quad (27)$$

These equations give $\lambda X + \mu Y + \nu Z = 0$, as they ought, since the component force along ds' is zero. Their form also shows that the resultant force on ds' is at right angles to the line the direction cosines of which are proportional to A, B, C, that is, its direction is at right angles to the plane through ds' and that line. The resultant of A, B, C, Ampère called the *directrix*. By comparison with (9) above we see that it is the magnetic induction at ds' produced by the circuit.

Equations of precisely the same form as (27) hold of course for any assemblage of circuits. In that case however A, B, C, are sums of integrals of the form given in (26).

Component Action in any Plane.

The component force in any plane may be found as follows. Let ϕ be the angle between the given plane and the plane containing ds' and the directrix. Then clearly the angle which the resultant force, R, makes with the given plane is $\pi/2 - \phi$, and the component is $R \sin \phi$. Squaring equations (27) and adding we find $R = ds'D \sin \omega$, where ω is the angle between ds' and the directrix, and $D = \sqrt{A^2 + B^2 + C^2}$. If ψ be the angle between the directrix and the given plane, we get, by projecting unit distance along the directrix on a line at right angles to ds', and then at right angles to the given plane, for the final projection the length $\sin \omega \sin \phi$. But the same line projected directly gives $\sin \psi$. Hence $\sin \omega \sin \phi = \sin \psi$. The component force in the given plane is therefore $ds'D \sin \omega \sin \phi = ds'D \sin \psi$. If a, b, c be the direction cosines of the normal to the given plane, $\sin \psi = aA/D + bB/D + cC/D$, and the component is $ds'(aA + bB + cC)$, or $ds'U$ where

$$U = aA + bB + cC. \quad . \quad . \quad . \quad . \quad (28)$$

ACTION OF ELEMENT ON SMALL PLANE CIRCUIT

From this we obtain the remarkable result that the action in the given plane is independent of the direction of ds' if only the element lie in that plane.

To apply the results found above to the problem of the solenoid, let the circuit be small and plane. The values of the components A, B, C can be calculated approximately for this case as follows. Let $MPQN$, Fig. 35, represent the circuit, and let it

Calculation of Result for a Small Circuit.

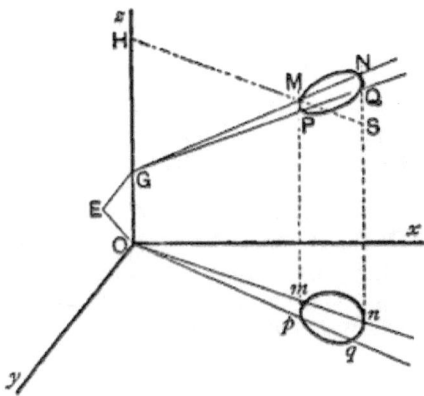

Fig. 35.

be cut by planes passing through the axis of z. Let two of these planes meet the circuit in MN, PQ, and Omn, Opq be their traces on the plane of x, y, meeting the projection of the circuit in mn, pq, we have for C the equation

$$C = \int \frac{1}{r^3}\left(x\frac{dy}{ds} - y\frac{dx}{ds}\right)ds \quad \ldots \quad (29)$$

taken round the circuit. Clearly this may be written

$$C = \int \frac{u^2 da}{r^3} \quad \ldots \quad \ldots \quad (30)$$

if a be the angle which Omn makes with Ox, and u the distance from O of the element of the projection corresponding to ds.

Now since the circuit is small we may suppose the elements mp, nq, intercepted by the planes, to be at a small distance δu

Calculation of Result for a Small Circuit

apart, corresponding to a small distance δr between the actual elements MP, NQ. We then find the value of C by calculating the sum of the contributions to it corresponding to the pairs of elements mp, nq of the projection on the plane of xy. Now taking the area swept out by the radius vector as positive when the end of the radius moves from m to p, and therefore negative when it moves from q to n, and taking a between the extreme tangents drawn from O to the projection, we get

$$C = \int\left\{\frac{u^2}{r^3} - \frac{(u+\delta u)^2}{(r+\delta r)^3}\right\}da = \int \frac{1}{r^3}\left(\frac{3u^2}{r}\delta r - 2u\delta u\right)da \ . \quad (31)$$

Now $r^2 = u^2 + z^2$, and therefore $\delta r = (u\delta u + z\delta z)/r$. Letting fall a perpendicular OE from O, on the plane of the circuit, and calling its length h, and its direction cosines l, m, n, we have $OG = h/n$. But if HMS be drawn parallel to the plane of x, y, we have by the similar triangles MSN, MHG, $\delta z/(z - OG) = \delta u/u$. Hence $\delta r = \{u^2 + z(z - h/n)\}\delta u/ur = (r^2 - zh/n)\delta u/ur$. Substituting in (31), and taking mean values of r and z, which since the circuit is small, may be those for the mean point of the area, we have, if S be the area of the circuit

$$C = \frac{1}{r^3}\left(1 - \frac{3zh}{nr^2}\right)nS$$

and similarly

$$A = \frac{1}{r^3}\left(1 - \frac{3xh}{lr^2}\right)lS \qquad \cdots \cdots (32)$$

$$B = \frac{1}{r^3}\left(1 - \frac{3yh}{mr^2}\right)mS$$

Application of Result to Solenoid

Now consider a solenoid, as defined above (p. 134), made up of such circuits uniformly arranged along a common axis, and such that the current and the area in each case have a constant infinitely small product; then taking the circuits as infinitely close, denoting by g the sum of the products per unit of length of the axis, and by ds an element of the axis, we have gds for the sum of the products for the element ds. Hence, to find A, B, C for this part of the assemblage of circuits, we have to substitute gds for S in (32). Doing this, and denoting by A', B', C', the values of A, B, C, for the whole assemblage of circuits, we have

$$A' = g\int \frac{1}{r^3}\left(l - \frac{3hx}{r^2}\right)ds$$

$$B' = g\int \frac{1}{r^3}\left(m - \frac{3hy}{r^2}\right)ds \quad \right\} \quad \ldots \ldots \quad (33)$$

$$C' = g\int \frac{1}{r^3}\left(n - \frac{3hz}{r^2}\right)ds$$

Application of Result to Solenoid.

Now we have here

$l, m, n = dx/ds, dy/ds, dz/ds$, and $h = xdx/ds + ydy/ds + zdz/ds$.

Hence substituting and integrating from one end of the axis of the solenoid to the other, we find

$$A' = g\left(\frac{x_2}{r_2^3} - \frac{x_1}{r_1^3}\right), \quad B' = g\left(\frac{y_2}{r_2^3} - \frac{y_1}{r_1^3}\right), \quad C' = g\left(\frac{z_2}{r_2^3} - \frac{z_1}{r_1^3}\right) \quad (34)$$

where the suffixes distinguish the values of the quantities for the two ends of the solenoid.

These values of A', B', C' are proportional to the direction cosines of the directrix for this case, and substituted in (27) give the components of the force on ds'. It is evident that the force on ds' depends only on g and the positions of the ends of the solenoid.

If the axis be a closed curve or extend to infinity in both directions, the values of A', B', C' are zero, and hence by (27) the solenoid exerts no force on ds'.

Actions of Closed, Doubly Infinite, and Singly Infinite Solenoids.

If the axis extend to infinity in the direction of integration the first terms in (34) are zero, and we have

$$A' = -gx_1/r_1^3, \quad B' = -gy_1/r_1^3, \quad C' = -gz_1/r_1^3 \quad . \quad (35)$$

Substituting these values in (27) we see at once that the action is at right angles to the plane through ds' and the extremity of the solenoid. Let now the conductor be straight and infinitely extended in both directions. Then changing the origin to the extremity of the solenoid, taking as the plane of xy the plane through the conductor and the end of the solenoid, and the direction of the conductor as that of x, we find $A' = gx/r^3$, $B' = gy/r^3$, $C' = 0$, $\lambda = 1$, $\mu = \nu = 0$. Hence the resultant force is $Z = gds'a/(a^2 + x^2)^{\frac{3}{2}}$, where a is the constant value of y. Writing dx for ds', and integrating over the whole conductor from $-\infty$ to $+\infty$, we find the value $2g/a$ for the total force. This result shows that the action between the conductor and the

Singly Infinite Solenoid equivalent to Magnetic Pole.

140   ELECTROMAGNETIC THEORY

extremity of the solenoid is the same as that (p. 112) between the conductor and a magnetic pole of strength g.

Finite Solenoid equivalent to Two Singly Infinite Solenoids.
Returning to (35) it is clear that if we had another solenoid with value of g numerically the same but opposite in sign, and extending to infinity from the point x_2, y_2, z_2, we should have for it

$$A' = gx_2/r_2^3, \quad B' = gy_2/r_2^3, \quad C' = gz_2/r_2^3 \quad . \quad . \quad (36)$$

and these two solenoids acting together would be equivalent to the finite solenoid already discussed.

Action of a Solenoid on a Finite Conductor.
We can now consider the action of a finite solenoid on a conductor of finite length s' carrying a current of unit strength. The component forces on the element ds' are by (27) and (34)

$$\left. \begin{array}{l} X = gds'\left\{\mu\left(\dfrac{z_2}{r_2^3} - \dfrac{z_1}{r_1^3}\right) - \nu\left(\dfrac{y_2}{r_2^3} - \dfrac{y_1}{r_1^3}\right)\right\} \\ Y = \qquad \&c. \qquad\qquad \&c. \\ Z = \qquad \&c. \qquad\qquad \&c. \end{array} \right\} \quad . \quad . \quad (37)$$

If we no longer take the origin at ds', but transfer it to the extremity x_1, y_1, z_1 of the solenoid, and take the line joining its ends as axis of x, the calculation of the action on the conductor will be simplified. The coordinates of ds' are now $-x_1, -y_1, -z_1$, which we shall write x, y, z. If l be the numerical value of the distance between the ends of the solenoid, $x_2 = -x + l$. Substituting these in (37) with the values dx, dy, dz for $\lambda ds', \mu ds', \nu ds'$, we get equations adapted to the calculation of the force components for the whole conductor.

Turning Moment on Finite Conductor,
We shall apply these to find the moment tending to turn the conductor round the line joining the extremities of the solenoid. The moment, dM, of the forces on ds' is $Zy - Yz$. Calculating this by equations (37), modified as just described, we find after reduction

$$dM = g\left\{\dfrac{d}{ds'}\left(\dfrac{x-l}{r_2}\right) - \dfrac{d}{ds'}\left(\dfrac{x}{r_1}\right)\right\}ds' = g\dfrac{d}{ds'}(\cos\theta_2 - \cos\theta_1)ds' \quad (38)$$

depends only on Position of ends.
where θ_2, θ_1 are the angles which the radii drawn from the extremities of the solenoid make with the axis of x. Integrating from one end of the conductor to the other, and distinguishing by accents the angles for the end where the integration terminates from the angles for the end at which it begins, we get finally

$$M = g\,(\cos\theta'_2 - \cos\theta'_1 - \cos\theta_2 + \cos\theta_1) \quad . \quad (39)$$

ACTION OF SOLENOID

This result, derived from Ampère's formula, agrees with that which we should obtain from equation (8) above, by considering the solenoid as an infinitely thin uniformly magnetized bar magnet. The turning moment of such a magnet on the conductor may be obtained most simply as follows. The magnetic field of such a magnet may be regarded as produced by equal and opposite quantities of magnetism at its **extremities**. Take the line joining these as **axis**, and draw lines from the ends to an **element** ds of **the conductor** making with the positive direction **of the axis the** angles θ_1, θ_2, and let the element make angles ϕ_1, ϕ_2 **with these** lines. First suppose the conductor wholly in a plane through the axis. Let the strength of each pole be m, then considering the action first of the positive pole (distant r_1 from the element), the force on the element is $mds \sin \phi_1/r_1^2$, and its **direction** is at right angles to the plane in which the conductor lies. **The** moment of this force round the axis **is** therefore $mds \sin \theta_1 \sin \phi_1/r_1$. Similarly the other **pole** gives a moment $- mds \sin \theta_2 \sin \phi_2/r_2$. The total **moment is** therefore $mds(\sin \theta_1 \sin \phi_1/r_1 - \sin \theta_2 \sin \phi_2/r_2$. But $r d\theta_1/ds = \sin \phi_1$, $r_2 d\theta_2/ds = \sin \phi_2$. Hence the moment may be written $m (\sin \theta_1 d\theta_1 - \sin \theta_2 d\theta_2)$, and the total moment is therefore

$$m (\cos \theta'_2 - \cos \theta'_1 - \cos \theta_2 + \cos \theta_1) \quad . \quad . \quad . \quad (40)$$

which if m be taken equal to g agrees with the value given in (39).

If the conductor be not in a plane through the axis, we may resolve any element ds into two components—one in such a plane, the other at right angles to it. The latter **component** will be acted on by a force passing through the axis, and therefore having no moment round it. Each element having been thus dealt with, all the components in planes through the axis may by rotation round the axis be transferred without alteration of turning moment to one such plane. They will therefore give a **continuous curve in that** plane, the **values** of $\theta_1, \theta_2, \theta'_1, \theta'_2$ for which **are** the same as for the actual conductor.

By (40) and the equation of the lines of force of such a magnet, (18) Chap. I. above, we obtain the following interesting result. Let **the lines of force** be revolved round **the** magnet so as to generate **coaxial surfaces.** Then the turning moment **on a** conductor carrying a given current in the field of the magnet, is the same whatever be the length **and** position of the conductor provided it terminate in the same two of these surfaces. The moment is **equal to the** difference of the parameters of these surfaces, and is **therefore zero** when both ends are on the same surface.

Action of Solenoid Compared with that of a Uniform Magnet.

Theorem with respect to Moment on Conductor.

Moment on a Conductor in Field of Single Pole.

By the process just employed we can show that the moment on a conductor in the field of a single magnetic pole, tending to turn it round any axis through the pole, depends only on the position of its ends. For each element can be resolved into two components, one in a plane through the element and the axis, the other at right angles to that plane. The force on the latter passes through the axis, and gives no moment. The former gives (supposing current and pole both unity) a moment $-\sin\theta d\theta$, where θ is the angle which the line drawn to it from the pole makes with the axis, and $d\theta$ is the change in θ between the ends of the element, for this is the same both for the element and its component in the plane through the axis. Hence integrating along the conductor from θ_1 to θ_2 we get $\cos\theta_2 - \cos\theta_1$ for the total moment. This is zero if $\theta_1 = \theta_2$, that is if the circuit be closed.

Moment on a Conductor on Field of Magnetic Poles in Straight Line.

We obtain the same result for each of any number of magnetic poles, and hence if we have any number of magnetic poles in one line, the moment which they exert on an unclosed conductor in their field is $\Sigma dm \cos(\theta_2 - \theta_1)$, the sum being taken for every element, dm, of magnetism in the magnet. Hence for a closed circuit this sum is zero round the line in which the elements lie.

It follows by equality of action and reaction that the couple which a straight linear distribution of magnetism experiences round its own line as axis, or that on a single pole round any axis, in consequence of the action of the current in a closed conductor in the corresponding field is zero.

Application of Results, to find Equation of Lines of Force of any Straight Magnet.

We can employ the result obtained above to find the equation of the lines of force of a uniformly magnetized magnet, or indeed of any straight linear distribution of magnetism. For let there be a single magnetic pole at a given point. Let the conductor be supposed made of flexible material, and to be held fixed at one end and laid along a line of force. Then let the other end be carried round the axis on which lies the magnetism, and be stretched and guided so as always to rest on the surface swept out by the line of force. No work is done against or by the action of the field, since the conductor is nowhere made to cut across lines of force. Hence for a single pole we have the equation, $\cos\theta_1 - \cos\theta_2 = 0$, that is the line of force is a straight line through the pole. In the same way we find for any assemblage of poles in a straight line the equation of the lines of force

$$\Sigma (\cos\theta_1 - \cos\theta_2) = \text{const.} \quad \ldots \quad (41)$$

It has been shown (p. 118 above) that an element of a circuit carrying a current in a magnetic field, in which the induction at

the element is **B**, is acted on by a force at right angles to the element and the direction of the magnetic induction, of amount **B**γ sin θ*ds* where *ds* is the length of the element, and θ the angle between its direction and that of the magnetic induction. We may of course suppose the induction at the element to be due to a single magnetic pole of proper strength, and properly situated in a field of uniform permeability. The reaction of the element on the magnetic system in the general case, and in this on the pole will therefore be **B**γ sin θ*ds*.
Reaction of Elements of Circuit on Magnetic System.

If we suppose the action on the element to be a single force of magnitude **B**γ sin θ*ds* applied at the element itself, the reaction on the pole will be an equal and opposite force *at the element*, and this is equivalent to an equal force at the pole and a couple. The moment of the integral couple due to the whole circuit is zero as has just been seen. We may therefore use the elementary forces on unit pole to calculate the magnetic action of the circuit upon it, that is the magnetic field-intensity which the circuit there produces, with certainty that no action between the pole and the circuit will be neglected, although these, or any other terms which integrated round the circuit give a zero result, are left out of account.
Reaction of each Element of Circuit

Thus we take as the intensity of the field produced at the pole by the element the expression **B**γ sin θ*ds*. But **B** is μ**H** where **H** is the intensity of the field produced at the element by the pole, and if a spherical surface of radius r equal to the distance between the pole and element be described round the pole as centre, we shall have $4\pi r^2 \mathbf{B} = 4\pi$, or

$$\mathbf{B} = \frac{1}{r^2}, \quad \mathbf{H} = \frac{1}{\mu r^2} \quad \ldots \ldots \quad (42)$$

and **B**γ sin θ*ds* = γ sin θ*ds*/r^2. The direction of this force is at right angles to the plane through the element and the line joining its centre to the pole. To specify the direction of the field-intensity due to an element let an observer be supposed immersed in the current in the element, so that it flows from his feet to his head, and have his face turned towards the pole, then the latter, if positive or north-seeking, will tend to move towards his right hand.
Magnetic Field-Intensity produced by Element of Circuit.

It is to be observed that this result illustrates the fact stated at p. 116 above, that the action of a circuit on a magnetic pole and therefore on any other distribution of magnetism is independent of the nature of the medium occupying the field.

It also gives another definition of unit current (which may be compared with that given at p. 105), as that current which

Second Definition of Unit Current. flowing in a thin wire forming a circle of unit radius acts on a magnetic pole placed at the centre with unit force per unit length of the circumference. Since the forces due to all the elements are in the same direction, the force at the centre of a circle of radius r carrying a current of strength γ is $2\pi\gamma/r$. Unit current is therefore that current which flowing in a circle of unit radius produces a magnetic field intensity at the centre of 2π units.

CHAPTER IV

GENERAL THEORY OF CURRENT INDUCTION AND ELECTROMAGNETIC ACTION

Section I

LAGRANGE'S DYNAMICAL METHOD

SINCE between a conductor carrying a current and a magnet, or between two conductors carrying currents, mutual forces exist, we are constrained to admit that the system in each case is a dynamical system, that is to say is subject to ordinary dynamical laws. One general principle which we find fulfilled by all natural systems is the law of Conservation of Energy. This is however not generally sufficient by itself to completely explain the properties of a material system, and we are compelled to have recourse to the results of experience to supplement it. A case in point is the theory of thermodynamics. Here the foundation consists of the two so-called laws of thermodynamics, *viz.* the principle of the Conservation of Energy applied to a system taking in and giving out heat, and doing work against external pressure; and the famous second law founded on an axiom derived from experience. Whether the attempt to dispense with the second law of thermodynamics, by an appeal to the dynamical principle of Least Action, be successful or not, it will remain true that the law of Conservation of Energy is by itself insufficient to account for thermal phenomena. The

Laws fulfilled by Natural System. Conservation of Energy, and Second Law of Thermodynamics.

same holds in other parts of physical science. The principle of Conservation of Energy does not in general suffice to enable the relations of which it is a consequence in any particular case to be deduced from it. In the electrical application we have to appeal to experience to guide us in forming an expression for the kinetic energy (or that part of the kinetic energy which comes within our cognizance by the outward effects of its variation) in each of the cases we have to consider; and hence, by assuming that we have to deal with the mutual actions of the particles of a material system, deduce the actions between two mutually influencing circuits, or between different parts of a system of mutually influencing circuits and magnets. In the case of a circuit and a magnet this proceeding is valuable rather for the sake of comparison with that of two mutually influencing circuits; since in the former case we have to reach the expression for the energy from the observed forces, instead of as in the latter finding the kinetic energy from general considerations, and deducing from that the forces between different parts of the system, and the electromotive forces in the circuits.

A system of Currents or of Currents and Magnets to be regarded as a Dynamical System.

Motion of a System of Mutually Influencing Particles.

In considering the motion of a material system we have to deal with a body or number of bodies, the parts of which vary in configuration within certain limits, determined by the conditions which the connections of the system fulfil. Thus, for example, a simple disconnected particle is free to move in all directions, while another, constrained to remain in a straight groove in a fixed support, can only be displaced along the groove. These limitations of the motions of the different parts are expressed by certain equations, commonly called the kinematical equations, which lay down the geometrical conditions which the coordinates of the different parts fulfil. From these equations the same number of coordinates can be determined in terms of the remaining coordinates, so that the former are given in terms of the latter, which thus become the independent variables. The system is said to have as many degrees of freedom as there are of independent coordinates. For example, the kinematical conditions imposed on a rigid body prevent the different parts of the body from receiving any change of relative configuration. The body is however left free to move as a whole in any one of three mutually rectangular directions in space, or to revolve round any one of these lines as axis, or to undergo any displacement compounded of these motions of translation and of rotation. It has therefore six degrees of freedom. Again, if a given point in the body, say, its centre of mass, be constrained to remain in a certain plane, this single kinematical condition limits the motion to rotation round some axis through the centre of mass, and to translation of the body as a whole along the given plane. Thus the equation of condition removes one of the degrees of freedom. *Degrees of Freedom.*

Now Lagrange has given a method of expressing the motion of the body by just as many equations as there are degrees of freedom. The possible displacements of the body, whether translations or rotations, are indicated by variations of certain coordinates ψ, ϕ, &c., along each of which acts a "force" of corresponding nature given, as explained below, by proper resolution in that direction of the actual force system. The word "coordinate" has here an enlarged or generalised meaning, and the word "force" has a corresponding range of interpretation. Its meaning is thus defined:—Let $d\psi$ denote a displacement of the type ψ, and let Ψ denote the corresponding force. Then $\Psi d\psi$ is the work (in the ordinary dynamical sense) done in the displacement supposed taking place alone. Thus if $d\psi$ is an elementary rotation Ψ is a couple, if $d\psi$ is a quantity of fluid past a cross-section in a pipe, Ψ is the flowing pressure at the cross-section, and so on. *Lagrange's Method. Generalised Coordinates.*

This system of generalised coordinates and forces has been extended to phenomena outside the range of ordinary kinetics, for example to electricity and magnetism. Thus we are led to regard quantity of electricity, conducted past a given cross-section of a conductor, as a coordinate, and the electromotive force producing the current as the corresponding force. Thus current is regarded as the rate of variation of the electric coordinate with the time. For this idea we are already prepared from what we know of the phenomena and energetics of electric currents. We shall return to its discussion after giving a short account of Lagrange's dynamical method. *Generalised Signification of "Force." Application of Method of Generalised Coordinates to Electricity.*

L 2

Lagrange's Dynamical Method. We shall denote by ψ, ϕ, θ, &c., $\dot{\psi}$, $\dot{\phi}$, $\dot{\theta}$, &c., the generalised coordinates and velocities of a material system at any time, and suppose these connected with the Cartesian coordinates $x_1, y_1, z_1, x_2, y_2, z_2$, &c., and the corresponding velocities $\dot{x}_1, \dot{y}_1, \dot{z}_1, \dot{x}_2, \dot{y}_2, \dot{z}_2$, of different points of the system, by the equations

$$\left.\begin{aligned}\dot{x}_1 &= \frac{\partial x_1}{\partial \psi}\dot{\psi} + \frac{\partial x_1}{\partial \phi}\dot{\phi} + \text{\&c.}\\ \dot{y}_1 &= \frac{\partial y_1}{\partial \psi}\dot{\psi} + \frac{\partial y_1}{\partial \phi}\dot{\phi} + \text{\&c.}\\ \dot{z}_1 &= \frac{\partial z_1}{\partial \psi}\dot{\psi} + \frac{\partial z_1}{\partial \phi}\dot{\phi} + \text{\&c.}\end{aligned}\right\} \quad \ldots \ldots (1)$$

with similar equations for the other points. Here and in what follows the symbol ∂ denotes partial differentiation with respect to any specified variable supposed explicitly appearing in given expressions for the quantities differentiated, while d is used to denote total differentials, and total differentiation with respect to any variable. Thus equations (1) presuppose that $x_1, y_1, z_1, x_2, y_2, z_2$, &c. can be expressed in terms of the generalised coordinates ψ, ϕ, θ, &c. *without involving* $\dot{\psi}, \dot{\phi}$, *&c.* Now if m_1, m_2, &c. be the masses of particles having the coordinates $x_1, y_1, z_1, x_2, y_2, z_2$, &c. we have for the kinetic energy

$$T = \tfrac{1}{2}\Sigma m(\dot{x}^2 + \dot{y}^2 + \dot{z}^2). \quad \ldots \ldots (2)$$

Expression for Kinetic Energy in terms of Generalised Coordinates and Velocities. Substituting from (1) in this equation we get

$$T = \tfrac{1}{2}\{(\psi,\psi)\dot{\psi}^2 + (\phi,\phi)\dot{\phi}^2 + \ldots + 2(\psi,\phi)\dot{\psi}\dot{\phi} + \text{\&c.} \quad (3)$$

where $(\psi,\psi), (\phi,\phi)$, &c. denote coefficients which are functions of the coordinates ψ, ϕ, &c. Thus the kinetic energy is a homogeneous quadratic function of the generalised velocities.

Let $X_1, Y_1, Z_1, X_2, Y_2, Z_2$, &c. denote forces, according to the ordinary signification, which act on the particles m_1, m_2, &c. ; then in any possible displacements $\delta x_1, \delta y_1, \delta z_1, \delta x_2, \delta y_2, \delta z_2$, &c. of these particles, the work done by the forces is

$$\delta W = \Sigma(X\delta x + Y\delta y + Z\delta z) \quad \ldots \ldots (4)$$

Now by (1)

$$\left.\begin{aligned}\delta x_1 &= \frac{\partial x_1}{\partial \psi}\delta\psi + \frac{\partial x_1}{\partial \phi}\delta\phi + \text{\&c.}\\ \text{\&c.} \quad & \quad \text{\&c.} \quad \quad \text{\&c.}\end{aligned}\right\} \quad \ldots (5)$$

GENERALISED COORDINATES

Substituting from (5) in (4) and collecting terms we find

$$\delta W = \Sigma\{(X\frac{\partial x}{\partial \psi} + Y\frac{\partial y}{\partial \psi} + Z\frac{\partial z}{\partial \psi})\delta\psi + (X\frac{\partial x}{\partial \phi} + Y\frac{\partial y}{\partial \phi} + Z\frac{\partial z}{\partial \phi})\delta\phi$$
$$+ \&c.\}$$

Generalised Components of Forces.

$$= \Psi\delta\psi + \Phi\delta\phi + \&c. \quad \ldots \ldots \quad (6)$$

where

$$\left. \begin{array}{l} \Psi = \Sigma(X\dfrac{\partial x}{\partial \psi} + Y\dfrac{\partial y}{\partial \psi} + Z\dfrac{\partial z}{\partial \psi}) \\[6pt] \Phi = \Sigma(X\dfrac{\partial x}{\partial \phi} + Y\dfrac{\partial y}{\partial \phi} + Z\dfrac{\partial z}{\partial \phi}) \\[6pt] \&c. \quad\quad \&c. \quad\quad \&c. \end{array} \right\} \quad \ldots \ldots \quad (7)$$

The time integral of any force Ψ over any interval τ we shall denote by $\overline{\Psi}$, so that we have

Time Integral of a Force.

$$\overline{\Psi} = \int_0^\tau \Psi dt.$$

Thus $\overline{\Psi}$ is equal to the total momentum generated by the force in the time τ. The use of the time-integral is, in general, confined to cases in which a finite change of momentum is produced in what may in the particular circumstances of the case be regarded as an infinitely short time. It is usual to say that then the force is an *impulsive* one; and the time-integral is called the *impulse* of the force in the interval τ. The work W done by an impulse is of course given by

Impulsive Forces and Impulse Work done by an Impulse.

$$W = \int_0^\tau \Psi \frac{d\psi}{dt} dt \quad \ldots \ldots \quad (8)$$

Now let $(\dot\psi)$ denote a value of $\dot\psi$ intermediate between the velocity at the beginning of τ and the velocity at the end, such that the work done is $(\dot\psi) \cdot \overline{\Psi}$. If τ be very small, $(\dot\psi)$ is the arithmetic mean between these two velocities. For let t be the time from the beginning to any instant of the interval τ, the velocity $\dot\psi$ at that instant is given by

$$\dot\psi = \dot\psi_0 + C\int_0^t \Psi dt$$

where $\dot\psi_0$ is the initial velocity, and C is a constant depending upon the conditions to which the coordinate is subject. It is to

150   CURRENT INDUCTION AND ELECTROMAGNETIC ACTION

be remarked that the effect of the impulsive force in altering the velocity is given by the second term on the right, only if the interval t is exceedingly short. If the interval is finite the system will after time t have moved from its initial configuration, and C in this case will be a variable quantity coming within the integration. Denoting the time-integral in the last equation by Ψ_t, we have for the rate of working at time t, the value $\Psi \times (\dot{\psi}_0 + C\Psi_t)$. Hence the whole work done in the interval τ is

$$\int_0^\tau \Psi(\dot{\psi}_0 + C\Psi_t)\,dt = \Psi\dot{\psi}_0 + C\int_0^\tau \Psi_t \cdot d\Psi_t$$

$$= \Psi\dot{\psi}_0 + \frac{C}{2}\Psi^2 = \tfrac{1}{2}\Psi(\dot{\psi} + \dot{\psi}_0). \quad (9)$$

which proves the statement as to the value of $(\dot{\psi})$.

The work done by the impulse may thus be written in the form $\Psi(\dot{\psi} + \tfrac{1}{2}\delta\dot{\psi})$. Similarly if the other generalised component forces act, the whole work done will be

$$\Psi(\dot{\psi} + \tfrac{1}{2}\delta\dot{\psi}) + \Phi(\dot{\phi} + \tfrac{1}{2}\delta\dot{\phi}) + \&c.$$

Work done in Originating a given Motion from Rest. Hence the work done in setting the system into motion from rest is

$$\tfrac{1}{2}(\Psi\dot{\psi} + \Phi\dot{\phi} + \&c.)$$

and this is equal to the kinetic energy which the system has, viz.

$$T = \tfrac{1}{2}\{(\dot{\psi},\dot{\psi})\dot{\psi}^2 + (\dot{\phi},\dot{\phi})\dot{\phi}^2 + \ldots + 2(\dot{\phi},\dot{\psi})\dot{\phi}\dot{\psi} + \&c.\} \quad (10)$$

If Ψ, Φ, &c., as here, be such as to produce from rest the given motion of the system, it is usual to denote the momenta which they produce by ξ, η, ζ, &c. Thus we have the equations

$$\left.\begin{aligned}\xi &= (\dot{\psi},\dot{\psi})\dot{\psi} + (\dot{\phi},\dot{\psi})\dot{\phi} + (\dot{\theta},\dot{\psi})\dot{\theta} + \&c.\\ \eta &= (\dot{\psi},\dot{\phi})\dot{\psi} + (\dot{\phi},\dot{\phi})\dot{\phi} + \quad \&c.\\ &\&c. \qquad\quad \&c. \qquad\quad \&c.\end{aligned}\right\} \quad (11)$$

Generalised Components of Momentum. where of course $(\dot{\psi},\dot{\phi}) = (\dot{\phi},\dot{\psi})$; &c. The quantities on the right are called the components of momentum; and are obtained from (10) by differentiation with respect to $\dot{\psi}, \dot{\phi}$, &c. Thus we have

$$\xi = \frac{\partial T}{\partial \dot{\psi}}, \quad \eta = \frac{\partial T}{\partial \dot{\phi}}, \quad \zeta = \frac{\partial T}{\partial \dot{\theta}}, \&c. \quad \ldots \quad (12)$$

By means of these equations another mode of expressing T which is very useful is obtained. Clearly $\partial T/\partial \dot\psi$, &c. are linear functions of $\dot\psi$, $\dot\phi$, &c. and the coefficients (ψ, ψ), (ψ, ϕ), &c. Hence by equations (2) $\dot\psi$, $\dot\phi$, &c. can be expressed as linear functions of ξ, η, ζ, &c. and the coefficients. Thus by substitution in (10) the kinetic energy can be expressed as a homogeneous quadratic function of the components of momentum ξ, η, ζ, &c. Denoting the kinetic energy, when thus expressed, by T_m, we have

$$T_m = \tfrac{1}{2}\{[\psi, \psi]\xi^2 + [\phi, \phi]\eta^2 + \ldots + 2[\psi, \phi]\xi\eta + \ldots\} \quad (13)$$

in which the coefficients are distinguished from those in the other expression by square brackets. T_m has been called the *reciprocal function* of T.

The Kinetic Energy expressed in terms of the Momentum Components.

Multiplying equations (12) by $\dot\psi$, $\dot\phi$, $\dot\theta$, &c. respectively, adding and remembering that T is a homogeneous function of the velocities of the second degree, we have

$$2T = \xi\dot\psi + \eta\dot\phi + \zeta\dot\theta + \text{&c.} \quad \ldots \quad (14)$$

Hence denoting for clearness the kinetic energy expressed in terms of the velocities by T_v we may write

$$T_m = -T_v + \xi\dot\psi + \eta\dot\phi + \zeta\dot\theta + \text{&c.} \quad \ldots \quad (15)$$

Differentiating this expression for T_m, remembering that ξ, η, &c. are functions of $\dot\psi$, $\dot\phi$, &c. given by (10), we find

$$\frac{\partial T_m}{\partial \xi}\frac{\partial \xi}{\partial \dot\psi} + \frac{\partial T_m}{\partial \eta}\frac{\partial \eta}{\partial \dot\psi} + \text{&c.} = -\frac{\partial T_v}{\partial \dot\psi} + \xi + \dot\psi\frac{\partial \xi}{\partial \dot\psi} + \text{&c.}$$

$$= \dot\psi\frac{\partial \xi}{\partial \dot\psi} + \dot\phi\frac{\partial \eta}{\partial \dot\psi} + \text{&c.}$$

since $\xi = \partial T_v/\partial \dot\psi$. The equation just obtained gives

$$\frac{\partial T_m}{\partial \xi} = \dot\psi\,;\quad \frac{\partial T_m}{\partial \eta} = \dot\phi, \text{&c.} \quad \ldots \quad (16)$$

Velocities expressed in terms of Momenta or Impulses.

These equations give the velocities produced by impulses equal to ξ, η, ζ, &c. when the kinetic energy is known in terms of the latter quantities.

Again, (15) gives by differentiation with respect to any coordinate, ψ say,

$$\frac{\partial T_m}{\partial \psi} + \frac{\partial T_m}{\partial \xi}\frac{\partial \xi}{\partial \psi} + \text{&c.} = -\frac{\partial T_v}{\partial \psi} + \dot\psi\frac{\partial \xi}{\partial \psi} + \dot\phi\frac{\partial \eta}{\partial \psi} + \text{&c.}$$

which by (16) becomes

Relation between the Reciprocal Functions.

$$\frac{\partial T_m}{\partial \psi} = -\frac{\partial T_v}{\partial \psi} \quad . \quad . \quad . \quad . \quad . \quad (17)$$

This relation, which is very important, holds of course for every coordinate.

It may be noted that the differentiation of (15), by which (17) is obtained, is performed upon T_m with respect to ψ regarded as occurring explicitly in T_m, and in ξ, which is a function of the velocities and coefficients of T_v given by (10). This amounts to partial differentiation of T_v with respect to ψ, and therefore ψ, ϕ, &c. on the right of (15) are in this operation not regarded as functions of ψ.

Proof of Lagrange's Equations of Motion.

We may now establish Lagrange's equations of motion. By the Second Law of Motion for a system of particles of masses m_1, m_2, &c. we have the equations

$$\left. \begin{array}{ll} m_1\ddot{x}_1 = X_1, & m_2\ddot{x}_2 = X_2, \text{ &c.} \\ m_1\ddot{y}_1 = Y_1, & m_2\ddot{y}_2 = Y_2, \text{ &c.} \\ m_1\ddot{z}_1 = Z_1, & m_2\ddot{z}_2 = Z_2, \text{ &c.} \end{array} \right\} \quad . \quad . \quad . \quad (18)$$

where $\ddot{x}$, $\ddot{y}$, $\ddot{z}$, denote the component accelerations. If the particles are free X, Y, Z, &c. are the components of the forces applied to the individual particles; if they are not free, then these are the forces which actually produce the accelerations, that is the forces due to the externally applied forces and the constraints introduced by the connections of the system. From (18) and (7) we obtain

$$\left. \Sigma m \left(\ddot{x} \frac{\partial x}{\partial \psi} + \ddot{y} \frac{\partial y}{\partial \psi} + \ddot{z} \frac{\partial z}{\partial \psi} \right) = \Psi \right\} \quad . \quad . \quad . \quad (19)$$
&c. &c. &c.

Now

$$\ddot{x} \frac{\partial x}{\partial \psi} = \frac{d}{dt}\left(\dot{x} \frac{\partial x}{\partial \psi} \right) - \dot{x} \frac{d}{dt} \frac{\partial x}{\partial \psi}$$

and by (1) we have

$$\frac{\partial x}{\partial \psi} = \frac{\partial \dot{x}}{\partial \dot{\psi}}, \quad \frac{d}{dt}\frac{\partial x}{\partial \psi} = \frac{\partial \dot{x}}{\partial \psi}, \text{ (see below).}$$

Hence

$$m\ddot{x} \frac{\partial x}{\partial \psi} = m \frac{d}{dt}\left(\dot{x} \frac{\partial \dot{x}}{\partial \dot{\psi}} \right) - \dot{x} \frac{\partial \dot{x}}{\partial \psi} \quad . \quad . \quad . \quad (20)$$

LAGRANGE'S EQUATIONS

Dealing in the same way with the other quantities on the left in (19), and adding we find

$$\Sigma m \left(\ddot{x} \frac{\partial \dot{x}}{\partial \psi} + \&c. \right) = \Sigma m \left\{ \frac{d}{dt}\left(\dot{x}\frac{\partial \dot{x}}{\partial \psi} + \&c.\right) - \left(\dot{x}\frac{\partial \dot{x}}{\partial \psi} + \&c.\right) \right\}$$

$$= \frac{d}{dt}\frac{\partial T}{\partial \dot{\psi}} - \frac{\partial T}{\partial \psi}, \text{ by (2)}$$

&c. &c.

We have therefore finally

$$\left. \begin{array}{c} \dfrac{d}{dt}\dfrac{\partial T}{\partial \dot{\psi}} - \dfrac{\partial T}{\partial \psi} = \Psi \\[4pt] \dfrac{d}{dt}\dfrac{\partial T}{\partial \dot{\phi}} - \dfrac{\partial T}{\partial \phi} = \Phi \\[4pt] \&c. \quad \&c. \end{array} \right\} \quad \ldots \ldots \quad (21) \quad \text{Lagrange's Equations.}$$

which are Lagrange's celebrated equations of motion.*

If the forces Ψ, Φ, &c., have a potential V then $\Psi = -\partial V/\partial \psi$, $\Phi = -\partial V/\partial \phi$, &c., and since $\partial V/\partial \dot{\psi} = 0$, &c., we may, putting $L = T - V$, write (21) in the form

$$\left. \begin{array}{c} \dfrac{d}{dt}\dfrac{\partial L}{\partial \dot{\psi}} - \dfrac{\partial L}{\partial \psi} = 0 \\[4pt] \dfrac{d}{dt}\dfrac{\partial L}{\partial \dot{\phi}} - \dfrac{\partial L}{\partial \phi} = 0 \\[4pt] \&c. \quad \&c. \end{array} \right\} \quad \ldots \ldots \quad (22)$$

The meaning of the differential coefficients is to be here most carefully observed, otherwise the reader to whom this subject is new may very readily fall into error. If the process given above be attentively examined it will be seen that the substitution of $\partial \dot{x}/\partial \dot{\psi}$ for $\partial x/\partial \psi$, &c., in obtaining (20) presupposes $\dot{x}$, &c. to be expressed as in (1). Thus we replace differentiation of any variable x, with respect to any one of the new independent variables, ψ say, by *partial* differentiation of $\dot{x}$ with respect to $\dot{\psi}$ supposing $\dot{x}$ expressed as in (1). This gives $\partial T/\partial \dot{\psi}$ as the

---

* This simple mode of deducing the equations of Lagrange from the equations of motion of a free particle was given first I believe in Thomson and Tait's *Natural Philosophy* (second edition), vol. i. part i. p. 302.

equivalent of $m\dot{x}\partial\dot{x}/\partial\psi + $ &c., where T is expressed, as in (3), as a homogeneous quadratic function of the velocities obtained by taking the squares of equations (1).

Again the reader should mark the substitution

$$\frac{\partial \dot{x}}{\partial \psi} \equiv \frac{d}{dt}\frac{\partial x}{\partial \psi}.$$

This replaces the expression on the right by the partial differential coefficient with respect to ψ of $\dot{x}$ expressed as in (1). This can be justified as follows. Writing w for $\partial x/\partial \psi$, we have

$$\frac{d}{dt}\frac{\partial x}{\partial \psi} = \frac{dw}{dt} = \frac{\partial w}{\partial \psi}\dot{\psi} + \frac{\partial w}{\partial \phi}\dot{\phi} + \&c.$$

But by (1)

$$\frac{\partial \dot{x}}{\partial \psi} = \frac{\partial w}{\partial \psi}\dot{\psi} + \frac{\partial w}{\partial \phi}\dot{\phi} + \&c.$$

the value just found for dw/dt. Also by multiplying $\partial\dot{x}/\partial\psi$, as just found, by the value of $\dot{x}$ as given in (1) we get the result of differentiating with respect to ψ the expression given in (3) for T.

Meaning of Terms in Lagrange's Equations. It is important to notice that $\partial T/\partial\psi$, &c. in (21) are forces depending on the appearance of the coordinates in the coefficients of the terms of the value of T as given in (3) and therefore depend on the relations to which the coordinates are subjected. If the particles are perfectly free and are referred to Cartesian coordinates these forces are zero, since then the coefficients do not involve the coordinates. The second terms on the left in (21) express generally the external forces required on account of the geometrical relations of the system.

It is also to be noted that these substitutions are not affected by the explicit appearance of the time t in the equations connecting the Cartesian with the generalised coordinates, and that the equations therefore hold for this case also.

As stated above Lagrange's equations are as many in number as there are coordinates left undetermined by the kinematical conditions, and the determination of the state of the system is reduced to the solution of these equations. It is very remarkable that these coordinates, as is evident from the process of derivation given above, may be of any physical nature whatever, and connect the corresponding physical state with the motions of particles composing the system. If we are given a sufficient number of kinematical equations Lagrange's equations enable us

LAGRANGE'S EQUATIONS APPLIED TO ELECTRICITY

to determine any required physical state of the system, although we have no means of obtaining any knowledge of the actual motions of the parts of the system, from general considerations regarding which these equations have been deduced.

Lagrange's equations can be put into a remarkable form, due to Sir W. Rowan Hamilton, which is frequently more convenient. Recurring to the components of momentum ξ, η, &c. given by (11) we see that

$$\frac{d}{dt}\frac{dT_v}{d\dot\psi} = \frac{d\xi}{dt} \\ \&c. \quad \&c.$$ . . . . (23)

Hamilton's Equations of Motion.

Hence using (21) and (17) we have

$$\frac{d\xi}{dt} + \frac{\partial T_m}{\partial \psi} = \Psi \\ \frac{d\eta}{dt} + \frac{\partial T_m}{\partial \phi} = \Phi \\ \&c. \quad \&c.$$ . . . . . (24)

These are Lagrange's equations as modified by Hamilton. They express that the force applied in each case is equal to the time-rate of increase of the corresponding component of momentum, plus the rate of increase of the kinetic energy per unit of increase of the corresponding displacement. The term $\partial T_m/\partial \psi$ is in another form the force required to balance the force of type ψ arising from the kinematical conditions to which the system is subject, and the time-rate of gain of momentum is equal to the difference between this term and the applied force.

Meaning of the Terms in Hamilton's Equations.

To apply Lagrange's equations to electrical and magnetic phenomena we have to consider what terms enter into the expression for the total kinetic energy T of a system of current-carrying conductors. If we denote by p, q, r, &c. the generalised coordinates of the system of conductors irrespective of currents, and by ψ, ϕ, &c. the electrical coordinates, we might expect T to be made up of three parts T_1, T_2, T_3, given by the equations

Application of Lagrange's Equations to Electricity and Magnetism.

$$T_1 = \tfrac{1}{2}(p, p)\dot p^2 + \ldots + (p, q)\dot p\dot q + \ldots \\ T_2 = \tfrac{1}{2}(\psi, \psi)\dot\psi^2 + \ldots + (\psi, \phi)\dot\psi\dot\phi + \ldots \\ T_3 = (\psi, p)\dot\psi\dot p + \ldots$$

so that T_1 denotes the ordinary molar kinetic energy of the conductors, T_2 the electrokinetic energy of the system, and T_3 the energy depending on products of velocities corresponding to the ordinary coordinates, and velocities corresponding to the electrical coordinates. There is no reason to believe that the part T_3 exists,* and as we are not concerned with the part T_1 in electrical investigations, we have only to consider T_2.

Electrical Coordinates. The question at once arises: what are to be considered the electrical coordinates of the system? Now all experience shows that the state of the system (that is the electric condition of the conductors and the electric and magnetic states of the field) remains unchanged, when the currents are kept constant, and the configuration of the conductors is maintained unaltered. Hence we are led at once to take the currents in the several conductors as ψ, ϕ, &c., and to conclude that the coefficients $(\psi, \psi), \ldots (\psi, \phi) \ldots$ in T_2 depend only on the coordinates p, q, &c., which fix the configuration.

Potential Energy of Magnetic Field now regarded as Kinetic Energy. In what has gone before we have regarded the energy of a magnetic field as potential energy; but we shall now find that energy appearing, with changed sign, as the electrokinetic energy, or that part of it determined for us by the coordinates which we have adopted as fixing the state of the system. There is not here any contradiction, only a change of point of view. The fact is, that everything points to the conclusion that what we are in the habit of regarding as potential energy is really kinetic energy. Every increase of scientific knowledge of matter furnishes additional proof that all its properties have their explanation in motion, and the conviction is being more and more forced upon every physical student that the ordinary division of energy into potential and kinetic results from our incomplete knowledge of the material system considered. If we had perfect knowledge of the coordinates of all the parts of the system at a given instant, and further knew fully the conditions to which these coordinates are subject, we should *ipso facto* be able to define the configuration at any time of any portion of the system, and to state how at that time the whole kinetic energy is divided between that portion and the rest of the system. Thus the ordinary transformation of potential into kinetic energy, and *vice versa*, is only a process of redistribution of kinetic energy between the different parts of the system.

Ignoration of Coordinates. Apparent Potential Energy. It can in fact be proved that if a dynamical system be specified partly by a certain group of position coordinates of type ϕ, and partly by velocities of other coordinates of type ψ, so that the kinetic energy is the sum of two parts $T\phi$, $T\psi$, the first of which is expressed as a quadratic function of the velocities of type ϕ, and the other as a quadratic function of the momenta corre-

---

* On this point see Maxwell, *El. and Mag.* vol. ii. § 573.

sponding to the velocities of type ψ, the alteration of the position coordinates will take place just as if the system had a quantity of kinetic energy T_ϕ, and a quantity of potential energy T_ψ. We suppose that there is no potential energy, so that if $T_{\dot\psi}$ be the energy T_ψ expressed in terms of the velocities of type $\dot\psi$, we have by Lagrange's equations, for the motion corresponding to any ϕ-coordinate

$$\frac{d}{dt}\frac{\partial (T_\phi + T_{\dot\psi})}{\partial \dot\phi} - \frac{\partial (T_\phi + T_{\dot\psi})}{\partial \phi} = 0 \quad . \quad . \quad . \quad (25)$$

Now by (17) above $\partial T_\psi/\partial \phi = -\partial T_{\dot\psi}/\partial \phi$, and since $\partial T_\psi/\partial \dot\phi = \partial T_{\dot\psi}/\partial \dot\phi = 0$, we may write (25) in the form

$$\frac{d}{dt}\frac{\partial (T_\phi - T_\psi)}{\partial \dot\phi} - \frac{\partial (T_\phi - T_\psi)}{\partial \phi} = 0 \quad . \quad . \quad . \quad (26)$$

Ignoration of Coordinates. Form of Lagrange's Equations.

or

$$\frac{d}{dt}\frac{\partial T_\phi}{\partial \dot\phi} - \frac{\partial T_\phi}{\partial \phi} = -\frac{\partial T_\psi}{\partial \phi} \quad . \quad . \quad . \quad . \quad (27)$$

which agrees exactly with the equation given by (22) above for the motion corresponding to a ϕ coordinate when the potential energy of the system is T_ψ.

It may be remarked here that if a quantity of potential energy V existed, then writing

$$L' = T_\phi - T_\psi - V = L - 2T_\psi$$

we should have instead of (26)

$$\frac{d}{dt}\frac{\partial L'}{\partial \dot\phi} - \frac{\partial L'}{\partial \phi} = 0 \quad . \quad . \quad . \quad . \quad . \quad (28)$$

Ignoration of Coordinates. Routh's Modified Form of the Lagrangian Function.

Hence in the case in which the kinetic energy is expressed in terms of velocities corresponding to certain coordinates and the momenta corresponding to certain others, we can use the modified Lagrangian function L' instead of L, and in precisely the same way. This very important theorem is due to Routh.*

The above investigation shows how without knowing one set of coordinates we can still express the energy, and find the equations of motion for those coordinates we do know, and well illustrates the view suggested that ordinary potential energy is

---

* *Stability of Motion*, p. 61.

merely the kinetic energy corresponding to the ignored or unknown coordinates.

Electrokinetic Energy. We shall find however that the electrical and configurational coordinates of the system give an expression for the electrokinetic energy in terms of the current strengths, which completely accounts for all ordinary phenomena of the electromagnetic field. We assume according to the theory given above that the electrokinetic energy T is given by the equation

$$T = \tfrac{1}{2}(L_1\dot{y}_1^2 + 2M_{12}\dot{y}_1\dot{y}_2 + \ldots + L_2\dot{y}_2^2 + 2M_{23}\dot{y}_2\dot{y}_3 + \ldots) \quad (29)$$

in which we have adopted the notation $\dot{y}_1, \dot{y}_2$, &c. for the currents in the circuits specified by the suffixes, and L_1, M_{12}, &c. for the coefficients, which as stated above, involve only the configurational coordinates. We shall see presently that L_1, L_2, &c. are what have been called the coefficients of self-induction for the different single circuits, and M_{12}, M_{13}, M_{23}, &c., the coefficients of mutual induction for the pairs of circuits indicated by the suffixes.

If we identify, as we shall presently, the electrokinetic energy with the intrinsic energy of the magnetic field as defined in Chap. III., but with its sign changed, and denote the magnetic inductions through the several circuits by N_1, N_2, &c., we shall have the equation

Generalised Components of Electromagnetic Momentum:

$$T = \tfrac{1}{2}(N_1\dot{y}_1 + N_2\dot{y}_2 + \ldots \&c. + N_k\dot{y}_k + \ldots) \quad (30)$$

Hence by (70) we get

$$\left.\begin{aligned} N_1 &= \frac{\partial T}{\partial \dot{y}_1} = L_1\dot{y}_1 + M_{12}\dot{y}_2 + \ldots + M_{1k}\dot{y}_k + \ldots \\ N_2 &= \frac{\partial T}{\partial \dot{y}_2} = M_{21}\dot{y}_1 + L_2\dot{y}_2 + \ldots + M_{2k}\dot{y}_k + \ldots \\ N_k &= \frac{\partial T}{\partial \dot{y}_k} = M_{k1}\dot{y}_1 + M_{k2}\dot{y}_2 + \ldots + L_k\dot{y}_k + \ldots \end{aligned}\right\} \quad (31)$$

where $M_{pq} = M_{qp}$. N_1, N_2, &c. are the generalised components of electromagnetic momentum for the system.

Or, Magnetic Inductions through the Circuits. If all the currents be zero except $\dot{y}_k$ for example, then N_k becomes $L_k\dot{y}_k$, so that L_k is the induction through the circuit numbered k, per unit of its own current. Through the other circuits the inductions are respectively $M_{k1}\dot{y}_k, M_{2k}\dot{y}_k$, &c. so that any one, say $M_{ik}\dot{y}_k$, is the induction through the circuit i, in consequence of the existence of the current $\dot{y}_k$ in the circuit k. Thus

DYNAMICAL THEORY OF TWO CIRCUITS

M_{ik} is the magnetic induction through the circuit i per unit of current in the circuit k, that is, it is what is called the coefficient of mutual induction between the two circuits. An expression by which M_{ik} can be calculated is given by (18), p. 129; this is also available for the calculation of L_i or L_k according as both ds and ds' are supposed to belong to the circuit i or the circuit k. Equation (18) shows that $M_{pq} = M_{qp}$, which is also evident from the simple consideration, that when there are only two circuits p, and q, the mutual electrokinetic energy is proportional to each of the current strengths, and therefore to their product, and to a coefficient depending on the geometrical arrangement of the circuits.

Using Lagrange's equations of motion and introducing Lord Rayleigh's Dissipation Function F,* we have, since the coefficients in T do not contain the coordinates y, for the impressed electromotive forces in the different circuits the typical equation

$$\frac{dN_k}{dt} + \frac{\partial F}{\partial \dot{y}_k} = E_k \qquad \ldots \ldots \quad (32)$$

But
$$F = \tfrac{1}{2}\Sigma R_k \dot{y}_k^2$$

where R_k is the resistance in the circuit indicated by the suffix k. Hence

$$\frac{dN_k}{dt} = E_k - R_k \dot{y}_k \qquad \ldots \ldots \quad (33)$$

As an example let the system be composed of two circuits only; and denote the currents in them by γ_1, γ_2, the respective coefficients of self-induction (or the *inductances* of the circuits as it is now usual to call them) by L_1, L_2, and the coefficient of mutual induction by M. For T we have the equation

Case of Two Circuits.

$$T = \tfrac{1}{2}(L_1 \gamma_1^2 + 2M\gamma_1\gamma_2 + L_2\gamma_2^2) \qquad \ldots \quad (29')$$

---

* See Lord Rayleigh's *Theory of Sound*, vol. i. p. 78. The Dissipation Function is such that when differentiated with respect to a velocity taken as independent variable it gives the applied force required to overcome the dissipative resistance to motion. The time rate of dissipation of energy due to any dissipative force is the product of that force into the corresponding velocity. Since we know that the rate of dissipation of energy in any circuit is $R_k \dot{y}_k^2$ the form of the function is in the present case obvious.

If E_1, E_2, be the electromotive forces, R_1, R_2 the resistances, equation (33) gives for this case

$$\left. \begin{aligned} E_1 - \frac{d}{dt}(L_1\gamma_1 + M\gamma_2) &= R_1\gamma_1 \\ E_2 - \frac{d}{dt}(L_2\gamma_2 + M\gamma_1) &= R_2\gamma_2 \end{aligned} \right\} \quad \ldots \quad (33')$$

Electromotive Forces of Induction. The electromotive forces of the type dN_k/dt are the parts of the impressed electromotive forces which are employed in working against the electromotive forces due to induction. Thus $-dN_k/dt$ is the type of the actual electromotive force due to induction.

The inductive electromotive force is thus equal to the time-rate of diminution of the electromagnetic momentum, and is a reaction against the applied electromotive force analogous to the inertia reaction of a particle against an applied force producing acceleration. It is such an inductive electromotive force which renders the rise or fall of the current in a circuit gradual when an external electromotive force is applied or removed.

Again the forces which must be applied to work against the similarly reacting internal electromagnetic forces are

$$-\frac{\partial T}{\partial x_1}, \quad -\frac{\partial T}{\partial x_2}, \quad \&c.$$

where x_1, x_2, &c. are coordinates determining the geometrical positions of the circuits. Thus the mutual electromagnetic forces having equal and opposite values to these, are equal to the corresponding space-rates of increase of the electrokinetic energy.

Dynamical Illustrations of Current Induction. Maxwell's Illustration. The following dynamical illustrations of inductive action between two circuits have been suggested by Clerk Maxwell and Lord Rayleigh. They are exceedingly valuable as helping the mind to picture the kind of dynamical action which must go on in the medium surrounding the circuits.

Maxwell constructed a dynamical model which is shown in Fig. 36. Two equal vertical wheels a and b having a common axis are bevelled and toothed so that an equal third wheel c is geared with them, and mounted as described below with its plane at right angles to that of a or b. a is fast to the horizontal axle, which also has rigidly attached to it a cross of four bars of which two only are shown in the figure. These are furnished with sliding weights so that the moment of inertia of the cross may be altered at will. One arm of the cross forms an axle round which the wheel c turns freely. The wheel b is mounted

DYNAMICAL ILLUSTRATIONS OF CURRENT INDUCTION

on a loose sleeve turning on the axle of a. A and B are two pulleys grooved and fitted with rope brakes so that any required resistance to their motion may be applied. A is fast to the axle of the wheel a, B to the sleeve also carrying b. In what follows we shall by the "wheel A" designate the whole system rigidly connected with the pulley A, and so on the other side by the "wheel B."

The theory of this machine will form a good illustration of Lagrange's dynamical method. The kinetic energy T is easily found. If A and B have angular velocities ω_1 and ω_2 in the same

Dynamical Theory of Maxwell's Machine.

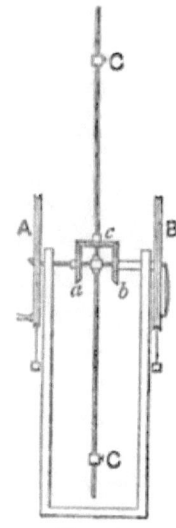

FIG. 36.

direction, and a be the radius of two equal circles round them which have contact with c, the tangential velocities of the two points of contact will be $\omega_1 a$, $\omega_2 a$. The arithmetical mean of these must clearly be the tangential velocity of the centre of the circle of contact of c. The angular velocity of the cross is thus $\frac{1}{2}a(\omega_1 + \omega_2)/a$, or $\frac{1}{2}(\omega_1 + \omega_2)$, the mean of the angular velocities of the wheels. The angular velocity of the wheel c round its axis is clearly $\frac{1}{2}(\omega_1 - \omega_2)$ since a is its radius. Putting then $m_1 k_1^2$, $m_2 k_2^2$, $m_3 k_3^2$, mk^2, for the respective moments of inertia round their **axes of the three wheels** A, B, c, and the cross, we have

$$T = \tfrac{1}{2}\{(m_1k_1^2 + \tfrac{1}{4}m_3k_3^2 + \tfrac{1}{4}mk^2)\omega_1^2 + (m_2k_2^2 + \tfrac{1}{4}m_3k_3^2 + \tfrac{1}{4}mk^2)\omega_2^2 + \tfrac{1}{2}(mk^2 - m_3k_3^2)\omega_1\omega_2\}.$$

Replacing the **coefficients of** ω_1^2, ω_2^2, $\omega_1\omega_2$ by L_1, L_2, $2M$ respectively, we get

$$T = \tfrac{1}{2}(L_1\omega_1^2 + 2M\omega_1\omega_2 + L_2\omega_2^2)$$

which is of precisely **the same form as that given above for the electrokinetic energy** of the currents **in two circuits.**

The wheels A and B correspond **to the two mutually influencing circuits**, and their angular **velocities to the currents in them,** while L_1, L_2, M represent **the coefficients of induction.** The rotating arms and attached masses, carried round by the wheel c, and forming **the connecting link between A and B,** correspond to **the medium in which the circuits are situated,** and through which their mutual action is propagated.

Dynamical Equations of Model analogous to Equations of E.M.F. for Two Circuits.

To make the analogy between the action of the mechanism perfect, we shall suppose the resisting forces applied to A and B to be proportional to the angular velocities. We shall therefore denote the resisting couple on A by $R_1\omega_1$, and that on B by $R_2\omega_2$, where R_1, R_2 are constants. If then Θ_1, Θ_2 denote the external couples applied to A and B, we have by Lagrange's equations

$$\left.\begin{aligned}\frac{d}{dt}(L_1\omega_1 + M\omega_2) &= \Theta_1 - R_1\omega_1 \\ \frac{d}{dt}(L_2\omega_2 + M\omega_1) &= \Theta_2 - R_2\omega_2\end{aligned}\right\} \quad \ldots \quad (33'')$$

which *mutatis mutandis* are **precisely the equations of inductive electromotive force (33')** given **above for two circuits.**

The interpretation of these equations is very simple. For example $Md\omega_2/dt$ is the applied force necessary to overcome the reaction on A produced by the acceleration $d\omega_2/dt$ in B; or in other words if this acceleration exist in B, a force $Md\omega_2/dt$ must be applied to A to **prevent** it from moving. Similarly for the other wheel.

Analogue of "Make" of Primary Circuit.

Interesting conclusions can easily **be** deduced from equations (33″), and the reader will do well **to** verify them by careful consideration of the mechanism. First let $\Theta_2 = 0$, and let the **system be started** from rest by imparting a positive angular acceleration by means of a couple Θ_1 to A. At the beginning $\omega_2 = 0$ and therefore

$$L_2 \frac{d\omega_2}{dt} + M \frac{d\omega_1}{dt} = 0.$$

Thus the angular acceleration of B must be opposite to that of A, and a negative angular velocity will begin to be generated in B. This will continue to increase so long as $M d\omega_1/dt$ is greater than $R_2 \omega_2$ in numerical value, since $-R_2\omega_2$ is now positive. At the beginning when $\omega_1 = 0$, $d\omega_1/dt$ is greatest, and $d\omega_2/dt$ has its greatest negative value. The changes afterwards may best be studied by integrating equations (33″), which may be done by alternate elimination of ω_1 and ω_2. Precisely similar equations [(65)] for the electric problem, are integrated on p. 180 below, and may be there studied. It will be seen that if Θ_1 be maintained constant, ω_2 rises to a maximum, and gradually dies away to zero as ω_1 approaches its steady value. The wheel A then rotates at constant speed while B remains at rest.

This corresponds precisely to the gradual rise of the current in the primary circuit to its final steady value while the backward induced current in the secondary rises to a maximum and thereafter dies away to zero.

If after the steady state has been reached A be retarded, it is clear from the differential equations that B will receive a forward acceleration $d\omega_2/dt$, and will acquire a velocity in the direction of motion of A, which, as A comes gradually to rest, will rise to a maximum, and then gradually die away to nothing. If A be suddenly arrested the forward acceleration of B will, in consequence of the angular momentum of the rotating mass, be correspondingly great, and a great resisting force applied to B may be overcome.

Dynamical Analogue of "Break" of Primary Circuit.

This corresponds precisely to what takes place when the primary circuit is broken. The electromotive force in the secondary circuit is greater the more sudden the break in the primary, and depends also on the nature of the circuits and their surroundings, in a manner which suggests some motion of the medium analogous to the angular momentum of the system set into motion by the bevel wheel c.

Again, let the coefficient R_2 be so great that even with a very small value of ω_2 the value of $R_2\omega_2$ is considerable, then $d\omega_2/dt$ is comparatively small, and consequently by the first equation $d\omega_1/dt$ has a much smaller positive value for the same applied couple than in the case in which B has a considerable backward acceleration. The reason is of course that now the wheel

Dynamical Analogue of High Resistance or Air Break in Secondary.

M 2

164 CURRENT INDUCTION AND ELECTROMAGNETIC ACTION

A cannot turn without carrying the massive cross with it. On the other hand, if the wheel A be stopped suddenly the stress thrown on B will be very great, and it will be forcibly moved forward.

This corresponds to a very high resistance or air-break in the secondary, so that a spark is prevented when the circuit is completed, while the sudden breaking of the circuit sends a current through a high resistance, and a spark across an air-space as in the secondary coil of an inductorium.

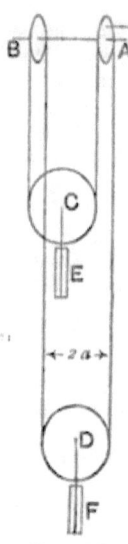

FIG. 37.

Dynamical Analogue of Leyden Jar in Secondary. The charging and discharging of a Leyden jar in the secondary circuit may be represented by the mechanism. An arm attached to the wheel B bears at each turn upon a spring attached to the framework. This spring resists the motion of the wheel until the arm is released after turning through a certain angle. Thus if the motion of A is sufficient the wheel B will be turned against the spring until the latter slips and recoils, while B runs on, to come round and repeat the same process so long as the acceleration of A endures.

The bending of the spring is the analogue of the charging of the Leyden jar, the slip of a sudden discharge by a spark. The amount of bending which the spring is capable of sustaining before slipping answers to the capacity of the jar. If it is so great as to entirely prevent the slip, the deflection will attain a maximum, and then be gradually undone as the couple on B diminishes and the spring brings the wheel back. This is analogous to the silent discharge of the jar by backward conduction through the secondary coil connecting its coatings.

Lord Rayleigh* has employed Huyghens' gearing (represented in Fig. 37) instead of the differential gearing described above. The two equal pulleys A, B, are loose on a round steel spindle. Over them passes an endless cord which carries at its two bights two equal movable pulleys C, D as shown. To these are attached equal weights E, F. This gives a system practically without potential energy, since whatever motion takes place, one of the weights must rise just as much as the other falls, and the weight of the cord may be neglected. *Lord Rayleigh's Illustration by Huyghens' Gearing.*

The expression for the kinetic energy is of the same form as before. Let m be the mass of each movable pulley and attached weight, $m'k^2$ the moment of inertia round its axis of each of the four pulleys, supposed here for simplicity all equal in every respect. Then if ω_1, ω_2 be the angular velocities of A and B, and a their common radius, the linear velocity of each weight is $\tfrac{1}{2}a(\omega_1 + \omega_2)$, and the angular velocity of each of the pulleys C and D is $\tfrac{1}{2}(\omega_1 - \omega_2)$. Hence for the kinetic energy of the whole system we have *Dynamical Theory of Illustration.*

$$T = \tfrac{1}{2}m'k^2\{\omega_1^2 + \omega_2^2 + \tfrac{1}{2}(\omega_1 - \omega_2)^2\} + \tfrac{1}{4}ma^2(\omega_1 + \omega_2)^2$$

or

$$T = \tfrac{1}{2}\{(\tfrac{1}{2}ma^2 + \tfrac{3}{2}m'k^2)(\omega_1^2 + \omega_2^2) + (ma^2 - m'k^2)\omega_1\omega_2\}.$$

Thus we have

$$L_1 = L_2 = \tfrac{1}{2}ma^2 + \tfrac{3}{2}m'k^2, \quad M = ma^2 - m'k^2,$$

and the same equations as in (33″) above.

The action of the apparatus is very similar to that of the differential gearing, and we have the same analogies. Angular velocities of A and B represent currents in the primary and secondary circuits respectively. Frictional resistances may be applied to A and B to resist their motion as before, and thus to play the part of the resistances of the circuits. Whatever friction there is at the bearings serves this purpose so far as it goes.

---

* *Phil. Mag.* July 1890, p. 30.

Dynamical Analogues of "Make" and "Break."

If the system be at rest a sudden motion of A in either direction produces in consequence of the inertia of the masses E, F, simply a turning of their pulleys and a moving round of the cord, and therefore a backward motion of the pulley B. As A goes on moving however, one weight rises and the other falls, until on the attainment of sufficient velocity of the system as much cord is gained by the rise of one weight as is required for the fall of the other, and there is no motion of the cord on B, which has come to rest. Thus we have the rise and gradual dying away of the induced current in the secondary as that in the primary attains its steady state.

Again if A is suddenly stopped the masses C and D will go on moving, and the motion of the cord will be shifted to the pulley B, which will now turn in the same direction as A did formerly. This illustrates the induced current in the secondary on break of the primary.

A condenser in the secondary may be represented as before by a spring pulled out by the pulley B as it turns, and let slip when a certain amount of stretching (representing the capacity of the condenser) has been attained.

Relation of Induction Coefficients to Energy of System.

We shall now investigate the relation between the coefficients of self and mutual induction and the intrinsic energy (supposed in Chap. III. above to be potential) of the system, as found by the analogy between a circuit carrying a current and a magnetic shell. The value to be given to the energy on the assumption that it is wholly kinetic will then immediately appear.

Electromotive Forces of Induction.

It is a well-known experimental result that any variation of the current in a conductor, or any displacement or deformation of the circuit, produces an electromotive force in the circuit itself and in any neighbouring circuit. We shall suppose that these electromotive forces are, in the case of variation of the current, proportional to the rate of variation, that is, considering any circuit in which the current at any instant has strength γ_1, we suppose the electromotive force in the circuit itself to be $A d\gamma_1/dt$, and in a neighbouring circuit $C d\gamma_1/dt$. A and C are coefficients which do not depend on the currents, but only on the configuration of the system. If γ_2 be the current in the neighbouring circuit at the same instant, then the electromotive forces due to the variation of the current in the second circuit are $B d\gamma_2/dt$ in the second circuit itself, and $C' d\gamma_2/dt$ in the first.

Self-Induction.

The proof of the validity of all these assumptions is to be found in the cumulative evidence afforded by all the experiments which have been made on the effects of varying currents, and the agreement of the consequences of the theory with observed facts.

THEORY OF TWO CIRCUITS

Again, it is found that if the first circuit be deformed so that A varies, an electromotive force of amount $\gamma_1 dA/dt$ will, during the change, exist in the first, and, in like manner, if the second circuit be deformed, an electromotive force of amount $\gamma_2 dB/dt$ will exist in the second.

Let now the first circuit be held fixed and the second be displaced relatively to it, while the currents are not allowed to vary. The circuit being changed in position without alteration of the current, the coefficients C, C' are altered in a short interval of time, dt, by the respective amounts dC, dC'. We shall suppose that this gives rise to electromotive forces of amount $\gamma_2 dC'/dt$, $\gamma_1 dC/dt$, in the first and second circuits respectively. *Mutual Induction.*

The existence of the two last electromotive forces might be inferred with a certain amount of probability from the first, by supposing the circuits and currents to be first completely given in some specified arrangement, then the currents to fade away in the given system while currents replacing them grow up in another system of circuits representing the former in its displaced or deformed state.

Thus the inductive electromotive forces are:—in the first circuit *Calculation of Inductive E.M.Fs. for two Circuits.*

$$A\frac{d\gamma_1}{dt} + \gamma_1 \frac{dA}{dt} + C\frac{d\gamma_2}{dt} + \gamma_2 \frac{dC'}{dt}$$

or

$$\frac{d}{dt}(A\gamma_1 + C'\gamma_2);$$

in the second circuit

$$\frac{d}{dt}(B\gamma_2 + C\gamma_1).$$

These electromotive forces fall to be added to the impressed electromotive forces E_1, E_2 in the circuits. Thus we get, applying Ohm's law, which we here assume as holding whatever the origin of the electromotive forces may be,

$$\left. \begin{array}{l} E_1 + \dfrac{d}{dt}(A\gamma_1 + C'\gamma_2) = R_1\gamma_1 \\[6pt] E_2 + \dfrac{d}{dt}(B\gamma_2 + C'\gamma_1) = R_2\gamma_2 \end{array} \right\} \quad \ldots \quad (34)$$

Now in the general case deformations and movements of the circuits take place under the action of external ponderomotive forces applied to the system, so that work is done by these forces *Work done by External Forces.*

168   CURRENT INDUCTION AND ELECTROMAGNETIC ACTION

Analysis of Energy of System. on the system, or by the system against these forces. The amount of this work is to be found as follows. Let E denote the quantity, called (p. 129 above) the **potential** energy **of the** system, which gives the internal ponderomotive forces on the circuits; then regarding the two currents as two mutually acting magnetic shells, we see, as in the case of the two mutually acting magnetic systems discussed at pp. 35, 36, that E is made up of three parts, one depending on the first circuit alone, another on the second circuit alone, and the **third** on the mutual action of the circuits. By the expression for the mutual potential **energy given** on p. 129 above we can easily see (and the proper **expressions** will be developed presently), **that** the parts of the energy **due** to the existence of the separate **circuits** apart from their mutual action, may be written $\tfrac{1}{2}\gamma_1^2 L'_1$, $\tfrac{1}{2}\gamma_2^2 L'_2$, where L'_1, L'_2 are coefficients depending only upon the **configura**tion of the **system**. The mutual potential **energy** is $\gamma_1\gamma_2 M'$ where $-M'$ denotes **the** integral in (19), p. 129. The value of E is thus given by

$$\mathsf{E} = \tfrac{1}{2}\gamma_1^2 L'_1 + \tfrac{1}{2}\gamma_2^2 L'_2 + \gamma_1\gamma_2 M' \quad . \quad . \quad . \quad (35)$$

Calculation of Work of External Forces and Energy furnished by Generators in Circuits. The work done by external forces in time dt is equal to the increase of potential energy, depending on deformation or displacement, which takes place in that time, and is

$$\tfrac{1}{2}\gamma_1^2 dL'_1 + \tfrac{1}{2}\gamma_2^2 dL'_2 + \gamma_1\gamma_2 dM'.$$

(Of course if this expression is negative it really expresses work done by the system against external forces.) Thus there is furnished by the generators and external forces energy amounting to

$$(E_1\gamma_1 + E_2\gamma_2)dt + \tfrac{1}{2}\gamma_1^2 dL'_1 + \tfrac{1}{2}\gamma_2^2 dL'_2 + \gamma_1\gamma_2 dM'.$$

Application of Principle of Conservation of Energy. This is used in generating heat of which the dynamical equivalent is $(R_1\gamma_1^2 + R_2\gamma_2^2)dt$, and in increasing the total intrinsic (potential) energy of the system by an amount $d\mathsf{E}$. Thus we get

$$(E_1\gamma_1 + E_2\gamma_2)dt + \tfrac{1}{2}(\gamma_1^2 dL'_1 + \gamma_2^2 dL'_2 + 2\gamma_1\gamma_2 dM')$$
$$= (R_1\gamma_1^2 + R_2\gamma_2^2)dt + d\mathsf{E}.$$

But substituting in this the value of $(R_1\gamma_1^2 + R_2\gamma_2^2)dt$ given by (34) we find

$$\tfrac{1}{2}(\gamma_1^2 dL'_1 + \gamma_2^2 dL'_2 + 2\gamma_1\gamma_2 dM')$$
$$= \gamma_1 \frac{d}{dt}(A\gamma_1 + C'\gamma_2)dt + \gamma_2 \frac{d}{dt}(B\gamma_2 + C\gamma_1)dt + d\mathsf{E} \quad (36)$$

ENERGY OF SYSTEM OF TWO CIRCUITS

If the system be put through a complete cycle of changes of configuration and current, so that the physical state of the system at the end is the same as at the beginning, E must have the same value as before, that is, $d\mathsf{E}$ is a perfect differential. We are at liberty to suppose that the cycle does not include any change of configuration, and in this case we have by the last equation

Proof of Existence of only one Induction Coefficient between Two Circuits.

$$- d\mathsf{E} = (\gamma_1 A + \gamma_2 C) d\gamma_1 + (\gamma_2 B + \gamma_1 C') d\gamma_2 . \quad . \quad (37)$$

The quantity on the left is a perfect differential, and therefore by the ordinary criterion

$$C = C' \quad . \quad . \quad . \quad . \quad . \quad . \quad . \quad (38)$$

that is, the coefficient of induction of the first circuit on the second is equal to that of the second circuit on the first. Hence (37) integrated on the supposition that the configuration is not altered, gives

$$- \mathsf{E} = \tfrac{1}{2} A \gamma_1^2 + \tfrac{1}{2} B \gamma_2^2 + C \gamma_1 \gamma_2 \quad . \quad . \quad . \quad (39)$$

there being no constant or arbitrary function of integration, since E vanishes when γ_1 and γ_2 are zero.

If now the configuration only varies

$$- d\mathsf{E} = \tfrac{1}{2}(\gamma_1^2 dA + \gamma_2^2 dB + 2\gamma_1 \gamma_2 dC).$$

Equating this value of $- d\mathsf{E}$ to that given by (36) for the case in which γ_1, γ_2 are constant, we find

Relation of Inductive E.M.Fs. to Energy of System.

$$\gamma_1^2 dA + \gamma_2^2 dB + 2\gamma_1 \gamma_2 dC = \gamma_1^2 dL_1^2 + \gamma_2^2 dL_2^2 + 2\gamma_1 \gamma_2 dM'.$$

Hence

$$dA = dL'_1, \; dB = dL'_2, \; dC = dM'.$$

Thus we conclude

$$A = L'_1, \; B = L'_2, \; C = M' \quad . \quad . \quad . \quad . \quad (40)$$

since if we diminish the area embraced by each of the circuits A, L_1 and B, L'_2 vanish together, and as we separate the circuits towards an infinite distance apart C and M decrease together towards zero.

We see now by equations (33) and (34) that if we suppose the energy electrokinetic we have only to put

Relation of Electrokinetic Energy to

$$T = - \mathsf{E} + \text{constant} . \quad . \quad . \quad . \quad . \quad (41)$$

170   CURRENT INDUCTION AND ELECTROMAGNETIC ACTION

Magnetic Energy of System: $T = -E$.
and the typical Lagrangian equations (21) or (24) with the potential energy taken as zero will give the electromotive and electromagnetic forces of the system. The value of the constant is not required and we may take it as zero if we please. Thus the value of the electrokinetic energy of the system may be regarded as simply $-E$.

Equation (41) is in form apparently a statement of the conservation of the energy of the system; but it is to be carefully noticed that it only furnishes a value of T which used in the Lagrangian equations with the value of the potential energy taken as zero, gives dynamical equations expressing the results of experiment.

Relation of Induction Coefficients to Magnetic Energy of System.
It is clear from equations (33′) and (34) and the results expressed in (40) that

$$L_1 = -L'_1\,;\ L_2 = -L'_2\,;\ M = -M' \quad . \quad . \quad (42)$$

Magnetic Theorems introduced on Basis of Equivalence of Circuit and Magnetic Shell.
By the equivalence of a circuit and a magnetic shell, the expression N_1 for the magnetic induction through any circuit A due to a current γ_2 in any other circuit B is given by

$$N_1 = \int \left(F \frac{dx}{ds_1} + G \frac{dy}{ds_1} + H \frac{dz}{ds_1} \right) ds_1 \quad . \quad . \quad (43)$$

where F, G, H are the components of vector potential produced by the current in B at an element ds_1 of A, and the integration is taken round the circuit of A.

If u, v, w, be the components of current in the first circuit as these are defined on p. 114 above, this equation may be written

$$N_1 = \frac{1}{\gamma_1} \int (Fu + Gv + Hw) d\varpi \quad . \quad . \quad . \quad (43')$$

where $d\varpi$ is an element of volume of the first conductor and the integration is extended throughout the whole space in which the current is not zero.

By (50) and (51) of Chap. I. above, (43) and therefore also (43′) can be written

$$N_1 = \gamma_2 \int\int \frac{\cos\theta}{r} ds_1 ds_2 \quad . \quad . \quad . \quad . \quad (44)$$

where θ is the angle between two elements ds_1, ds_2 of the respective circuits, r the distance of these elements apart, and the integration is taken round both circuits. Similarly we should find

COEFFICIENTS OF INDUCTION

for the magnetic induction N_2 through B due to the current in A the equation

$$N_2 = \gamma_1 \iint \frac{\cos\theta}{r} ds_1 ds_2 \quad \ldots \ldots (45)$$

These equations may be written

$$N_1 = M\gamma_2, \; N_2 = M\gamma_1 \quad \ldots \ldots (46)$$

if

$$M \equiv \iint \frac{\cos\theta}{r} ds_1 ds_2 \quad \ldots \ldots (47)$$

General Expressions for Coefficients of Induction.
1. Mutual Induction.

M is what has been called above the coefficient of mutual induction of the two circuits, and may be interpreted as the magnetic induction through *either* of the circuits due to unit current in the other.

The part of the electrokinetic energy which depends upon the existence of one current-carrying circuit in presence of the other is thus

$$M\gamma_1\gamma_2 = \gamma_1\gamma_2 \iint \frac{\cos\theta}{r} ds_1 ds_2. \quad \ldots \ldots (48)$$

But the current in any one circuit produces a certain magnetic induction through its own circuit which may be calculated thus. Suppose the circuit to be built up of coincident filaments each containing an element of current $d\gamma$, and consider the increase dN of magnetic induction through the previously existing circuit due to the addition of a single such filament, we have

$$dN = \int \left(dF \frac{dx}{ds} + dG \frac{dy}{ds} + dH \frac{dz}{ds} \right) ds$$

or, since $u, v, w = (\gamma dx/ds, \gamma dy/ds, \gamma dz/ds)/$ *area of cross-section*

$$\gamma dN = \int (u dF + v dG + w dH) d\varpi. \quad \ldots \ldots (49)$$

Since all the filaments are coincident the components of vector potential at any element of the added filament due to the current in the previously existing circuit are F, G, H. Hence the induction through the filament is N, and we have

$$N = \int \left(F \frac{dx}{ds} + G \frac{dy}{ds} + H \frac{dz}{ds} \right) ds$$

or
$$N d\gamma = \int (F du + G dv + H dw) d\varpi \quad . \quad . \quad . \quad . \quad (50)$$

But if T_1 be the electrokinetic energy so far as it depends upon this one circuit alone, and dT_1 be the increase of T_1 due to the addition of the filament carrying the current $d\gamma$, then since by (29′)

$$T_1 = \tfrac{1}{2} L \gamma^2 = \tfrac{1}{2} N \gamma$$

we find from (49) and (50)

$$\begin{aligned} dT_1 &= \tfrac{1}{2}(N d\gamma + \gamma dN) \\ &= \tfrac{1}{2}\int (F du + G dv + H dw + u dF + v dG + w dH) d\varpi \\ &= \tfrac{1}{2}\int d(Fu + Gv + Hw) d\varpi. \end{aligned}$$

Hence

$$T_1 = \tfrac{1}{2}\int (Fu + Gv + Hw) d\varpi$$

$$= \tfrac{1}{2}\gamma \int \left(F \frac{dx}{ds} + G \frac{dy}{ds} + H \frac{dz}{ds} \right) ds$$

Here F, G, H are the components of vector potential at any element ds of the circuit produced by the current in the circuit itself and therefore we can write

$$T_1 = \tfrac{1}{2}\gamma^2 \int\int \frac{\cos\theta}{r} ds\, ds' \quad . \quad . \quad . \quad . \quad . \quad (51)$$

2. Self-Induction. where ds, ds', belong to the same circuit. Thus for the coefficient of self-induction we get

$$L = \int\int \frac{\cos\theta}{r} ds\, ds' \quad . \quad . \quad . \quad . \quad . \quad (52)$$

In Chapter VI. we shall proceed to the calculation of the values of L and M for circuits of different forms and arrangement.

The coefficient of self-induction for any circuit may be defined as the coefficient of mutual induction between that circuit and a coincident circuit. It has therefore the value just given in (52) in which ds, ds' are now to be taken as elements of the two coincident circuits and the integrations are performed round these circuits.

VECTOR POTENTIAL OF SYSTEM OF CURRENTS

This idea of two coincident circuits gives an easy method of deducing from the expression for the mutual energy of two circuits the energy due to the existence of a single circuit alone. For let ds, ds' be any two coincident elements of the two circuits and ds_1, ds'_1 any two other coincident elements, and γ, γ', the currents in the circuits. Then if the circuits be deformed, and still kept coincident, the alteration of the mutual energy will be $\gamma\gamma'\delta m$, in consequence of the action between the two elements ds, ds'_1, and $\gamma\gamma'\delta m_1$, in consequence of the action between the two elements ds', ds_1, where $m \equiv \cos\epsilon\, ds ds'_1/r$, $m_1 \equiv \cos\epsilon\, ds'\, ds_1/r_1$. The whole alteration of energy is therefore $\gamma\gamma'\delta(m + m_1) = 2\gamma\gamma'\delta m$, since obviously $m = m_1$. This gives $2\gamma^2\delta m$ if the currents be γ in each case.

On the other hand the alteration of energy due to the deformation of *either* of these coincident circuits alone is $\gamma^2\delta m$, or half the former amount. Hence if the currents be equal **the work done in establishing** either circuit alone, **or its intrinsic energy, is** half that spent in establishing either in presence of the other. But the latter is

$$\gamma^2 \int\int \frac{\cos\epsilon}{r} ds\, ds' = \gamma^2 L.$$

Hence the energy in the former case is $\tfrac{1}{2}\gamma^2 L$.

From the value of N given in the equations

$$N = \gamma' \int\int \frac{1}{r}\left(\frac{dx}{ds}\frac{dx'}{ds'} + \frac{dy}{ds}\frac{dy'}{ds'} + \frac{dz}{ds}\frac{dz'}{ds'}\right)ds ds'$$

$$= \int\left(F\frac{dx}{ds} + G\frac{dy}{ds} + H\frac{dz}{ds}\right)ds \quad\ldots\ldots\quad (53)$$

Calculation of the Values of F, G, H in the General Case.

which applies to two linear circuits (not necessarily of uniform cross-section) very important expressions for F, G, H can be deduced. Denoting by u', v', w', and $d\sigma'$ components of current and the cross-section at (x', y', z') of the conductor carrying the current γ', we can write the first equation in the form

$$N = \int\int\left(\frac{u'}{r}\frac{dx}{ds} + \frac{v'}{r}\frac{dy}{ds} + \frac{w'}{r}\frac{dz}{ds}\right)ds d\varpi'$$

where $d\varpi'\, (= ds'\, d\sigma')$ is an element of volume of the conductor, and the integration with respect to that conductor may therefore be taken as one throughout its volume. If then the circuit giving rise to F, G, H at the point (x, y, z) of the other be non-linear, and we assume that it can be regarded as an assemblage of elementary circuits bounded by lines of flow, we have

$$N = \Sigma \int\int \left(\frac{u'}{r}\frac{dx}{ds} + \frac{v'}{r}\frac{dy}{ds} + \frac{w'}{r}\frac{dz}{ds}\right) ds\, d\varpi'$$

where Σ denotes summation of the integrals obtained for each elementary circuit. But clearly by integrating with respect to $d\varpi'$, taking the corresponding values of u', v', w' throughout the non-linear conductor, we get rid of the summation and find

$$N = \int\int \left(\frac{u'}{r}\frac{dx}{ds} + \frac{v'}{r}\frac{dy}{ds} + \frac{w'}{r}\frac{dz}{ds}\right) ds\, d\varpi'$$

By comparison of this with the second value of N given in (53) above, we obtain

$$F = \int \frac{u'}{r} d\varpi', \quad G = \int \frac{v'}{r} d\varpi', \quad H = \int \frac{w'}{r} d\varpi' \quad . \quad (54)$$

where the integration is extended throughout all space in which the current (u', v', w') differs from zero. To each of the expressions under the integral sign in (54) might of course be added any term which gives a zero result when integrated round the circuit; but it will not in any of the following developments of the subject of electromagnetism be found necessary to do so.

In the case of a system of currents, however complex, producing the components F, G, H of vector potential at the point (x, y, z), the expressions (54) will evidently still hold, for we have only to sum the values of each of these expressions as found for each circuit; and this is done equally by extending the integration throughout all space, taking at each point the actual values of the components u', v', w' there, due to the whole complex system. This system will include any circuit coinciding with that through which the magnetic induction is taken, and carrying any current, and this includes the latter circuit with the current which actually exists in it. Hence we drop the accents in (54) and write

$$F = \int \frac{u}{r} d\varpi, \quad G = \int \frac{v}{r} d\varpi, \quad H = \int \frac{w}{r} d\varpi \quad . \quad . \quad (55)$$

General Values of F, G, H for any where u, v, w denote the components of current at the element of volume $d\varpi$, and r the distance of that element from x, y, z, and the integrals are taken throughout the whole space in which currents exist.

The medium has been here supposed to have unit magnetic inductive capacity. If however the magnetic inductive capacity be μ (and be uniform throughout all space not at an infinite distance from the system) the equations become

$$F = \int \frac{\mu u}{r} d\varpi, \quad G = \int \frac{\mu v}{r} d\varpi, \quad H = \int \frac{\mu w}{r} d\varpi \quad . \quad . \quad (55')$$

If the value of μ be different in different parts of the field these equations must be modified by the addition of integrals over the separating surfaces, but into these cases we cannot here enter.* It may be remarked however that in the case of continuous variations of μ from point to point, if the magnetization of the medium be lamellar and the medium fill the whole field, equations (55') still hold, while they must be corrected by certain simple integrals over the separating surfaces when different parts of the field are filled by different media.

By equations (5) p. 116, we have by differentiation and addition

$$\frac{\partial u}{\partial x} + \frac{\partial v}{\partial y} + \frac{\partial w}{\partial z} = 0 \quad . \quad . \quad . \quad . \quad . \quad (56)$$

From this we can show that the values of F, G, H given in (55) fulfil the equation

$$\frac{\partial F}{\partial x} + \frac{\partial G}{\partial y} + \frac{\partial H}{\partial z} = 0 \quad . \quad . \quad . \quad . \quad . \quad (57)$$

To find the value of F at the point $(x + dx, y, z)$ we may suppose the whole system of currents displaced through a distance dx in the opposite direction. This will bring the point $(x + dx, y, z)$ to coincidence with (x, y, z), and the value of F at the latter point will then be that previously existing for the former. But the values of u and F at any point will have become $u + \partial u/\partial x \cdot dx$, and $F + \partial F/\partial x \cdot dx$ respectively. In the same way we may deal with G and H. Hence we have

$$\frac{dF}{dx} = \int \frac{\frac{\partial u}{\partial x}}{r} d\varpi, \quad \frac{dG}{dy} = \int \frac{\frac{\partial v}{\partial y}}{r} d\varpi, \quad \frac{dH}{dz} = \int \frac{\frac{\partial w}{\partial z}}{r} d\varpi,$$

---

* See a paper by Prof. Schuster in the *Phil. Mag.* for July, 1891.

$$\frac{\partial F}{\partial x} + \frac{\partial G}{\partial y} + \frac{\partial H}{\partial z} = \int \frac{1}{r}\left(\frac{\partial u}{\partial x} + \frac{\partial v}{\partial y} + \frac{\partial w}{\partial z}\right)dx = 0 \quad (58)$$

It is to be noticed that, if (56) do not hold in any space, (57) will not either. In general (56) does hold (and the current therefore is analogous to the flow of an incompressible fluid), but there are exceptional cases in which it does not. For example it does not hold throughout any portion of space enclosing a critical point, such as a *source*, or place where electricity is supposed to be generated, or a *sink*, where electricity is supposed to be destroyed. The corrections however in such cases are easily made.

We shall now illustrate the dynamical theory set forth above by applying it to a few important particular cases.

First take the case of two circuits specified above by equations (29′) and (33′).

Two Undeformable Circuits: Change of Electrokinetic Energy.
We shall suppose the circuits rigid so that L_1, L_2 are invariable. Then if the circuits be subjected only to those changes which take place from their mutual action, and dT be the change in T which takes place in a small interval of time dt

$$dT = L_1\gamma_1 d\gamma_1 + M(\gamma_2 d\gamma_1 + \gamma_1 d\gamma_2) + \gamma_1\gamma_2 dM + L_2\gamma_2 d\gamma_2 \quad (59)$$

Electromagnetic work done.
If a typical coordinate x, fixing the relative positions of the circuits (say in the case of two plane circuits maintained parallel to one another, the distance between them) be altered at the same time by an amount dx, the work done by electromagnetic forces has the value $\partial T/\partial x \cdot dx$. Calling this dW, and putting dM for the corresponding change in M, we obtain

$$dW = \gamma_1\gamma_2 dM \quad \ldots \quad \ldots \quad (60)$$

This work is spent in producing kinetic energy in the displaced conductors, or if these are not free, in moving them against external resistance, or in both ways.

Whole Energy spent in any change accounted for.
The work done by the impressed electromotive forces over and above that dissipated is

$$(E_1 - R_1\gamma_1)\gamma_1 dt + (E_2 - R_2\gamma_2)\gamma_2 dt$$

which by (33′) has the value

$$L_1\gamma_1 d\gamma_1 + M(\gamma_1 d\gamma_2 + \gamma_2 d\gamma_1) + 2\gamma_1\gamma_2 dM + L_2\gamma_2 d\gamma_2$$

or by (59) and (60) the value $dT + dW$, so that the energy is all accounted for.

MUTUAL ACTION OF CIRCUIT AND MAGNETIC SHELL

Thus the impressed electromotive forces working against the inductive electromotive forces do an amount of work which is accounted for by an increase dT of the electrokinetic energy, and an amount of work dW done in displacing the circuits. The former quantity dT may be separated into two parts, (1) the work done against the inductive electromotive forces which depend only on changes of the currents, and (2) an increase of electrokinetic energy amounting to $\gamma_1\gamma_2 dM$ arising through alteration of the relative positions of the circuits. If the conductors are displaced from rest to rest again, so that γ_1, γ_2, have resumed their steady values $d\gamma_1 = 0$, $d\gamma_2 = 0$, and the energy furnished by the batteries is $2\gamma_1\gamma_2 dM$ of which one-half is accounted for in dT, the other in dW which has its equivalent in work done against the external resistance by which the conductors were brought to rest. This result was published by Sir William Thomson in 1851. *Increase of Electrokinetic Energy. Electromagnetic work done in circuits.*

According to Ampère's theory of magnetism each molecule of a magnetized body is supposed to be the seat of a current of electricity which flows in an infinitesimal circuit. When the positive faces of these circuits are turned in a common direction, or, if one direction preponderates, the circuits produce by their combined influence a magnetic field, the intensity of which is finite in amount. If the arrangement of the molecules of the body remains unchanged, and the currents in them are unaltered, the resulting magnetic field is constant. *Ampère's Theory of Magnetism.*

Since there is supposed to be no progressive change in the internal state of a body permanently magnetized, and there is no experimental evidence of such a change, there is neither consumption nor production of energy in any of these circuits, so that both the electromotive force and the resistance in each are supposed to be zero. A theory of electric currents will be developed presently, which will throw some light on the state of the medium within the magnetized body in relation to these currents: at present we shall take Ampère's theory for granted. *Electromotive Force and Resistance in Molecular Circuits of Permanent Magnet both zero.*

Let us, then, consider as another application of the dynamical theory, the case of the mutual action of an ordinary circuit and a magnetic shell. As we have seen (p. 104) a magnetic shell is equivalent to a current of strength equal to that of the shell circulating round its edge. Hence putting γ_2 for this current, and $E_2 = 0$, $R_2 = 0$, we have instead of (33') above *Case of Circuit and Magnetic Shell*

$$\left. \begin{array}{l} E_1 - \dfrac{d}{dt}(L_1\gamma_1 + M\gamma_2) = R_1\gamma_1 \\[6pt] \dfrac{d}{dt}(L_2\gamma_2 + M\gamma_1) = 0 \end{array} \right\} \quad \ldots \quad (61)$$

VOL. II.

As before we get, supposing L_1, L_2 invariable,

$$\left. \begin{array}{l} dT = L_1\gamma_1 d\gamma_1 + M(\gamma_2 d\gamma_1 + \gamma_1 d\gamma_2) + \gamma_1\gamma_2 dM + L_2\gamma_2 d\gamma_2 \\ dW = \gamma_1\gamma_2 dM. \end{array} \right\} \quad (62)$$

The energy furnished by the battery is

$$L_1\gamma_1 d\gamma_1 + M\gamma_1 d\gamma_2 + \gamma_1\gamma_2 dM$$

and by the other circuit no energy is given. But, multiplied by γ_2, the second of (61) gives

$$L_2\gamma_2 d\gamma_2 + M\gamma_2 d\gamma_1 + \gamma_1\gamma_2 dM = 0.$$

Thus

$$dT = L_1\gamma_1 d\gamma_1 + M\gamma_1 d\gamma_2$$

and $dT + dW$ is again exactly equal to the whole energy furnished by the battery.

Direct Inductive Change in the Molecular Currents is probably very small.
If the changes are estimated for the system when brought to rest, $d\gamma_1 = 0$. The other current γ_2 does not, however, remain constant, and we have to inquire what is its effect upon the magnet. We know that the moment of a hard magnet is not seriously altered by a displacement produced by the mutual forces when γ_1 is moderate in amount; and the small alteration generally observed is opposite to that which would be produced by the inductive change in γ_2. We are led to conclude that γ_2, on which, according to Ampère's theory, the intensity of magnetization depends, must remain practically constant. Writing, then, the second of (61) in the form

A Consequence of this Supposition.
$$L_2\gamma_2 + M\gamma_1 = L_2\gamma_2'$$

where γ_2' is the initial current before the magnet was brought into the field of the circuit, we see that if the dynamical theory is applicable to this case, $M\gamma_1/L_2$ must be a quantity small in comparison with γ_2'. This gives $d\gamma_2$ a small quantity of the second order, and makes the value of dT, the change in the electrokinetic energy, depend only on the term $L_1\gamma_1 d\gamma_1$, which is zero if the system is not in relative motion. Thus the fact, that the mechanical value of a current in a conductor is not affected by bringing permanent magnets into its neighbourhood, is not contradicted by the dynamical theory, if the supposition here made as to the value of $M\gamma_1/L_2$ is actually true.

Physical Reasons for the Supposition.
In support of this supposition we have the consideration that it leads to the result, probable on other grounds, that the number of molecular circuits in an ordinary magnet is exceedingly great. Let us suppose that, instead of a simple magnetic shell, we have

a solenoid made up of equal distinct circuits, in each of which a current γ flows. If the number of circuits per unit of length be n, the coefficient of self-induction is (Chap. V.) $4\pi n^2 A$, where A is the area of the circuit, and the total induction through the solenoid (neglecting the effect of its ends) is $4\pi n^2 l \gamma A$, if l be its length. If the circuit carrying the current γ_1 consist of n' turns of wire closely surrounding the solenoid, the induction through it and the solenoid is (Chap. VI.) $4\pi n n' \gamma A$. Thus the maximum value of M is $4\pi n n' A$. Thus we get $M/L_2 = n'/nl$, that is the ratio, which the theory indicates must be vanishingly small, is equal to the ratio of the number of turns in the circuit to the number of circuits in the solenoid.

It is also necessary, as we shall see below, in order that the Ampèrean currents may give an inductive magnetization in iron agreeing with experiment, that the self-induction of each molecular circuit be great, that is A/L_2 must be small. If the current flow in a ring channel this condition, as Maxwell* has pointed out, may be fulfilled by supposing the radius R of the mean line of the channel great in comparison with r the radius of the channel, since L_2 depends on log R/r (see Chap. VI.).

Considering next the case of two mutually acting magnetic shells; we have the same expressions for dT and dW as in (62), but equation (61) becomes, since now $E_1 = 0$, $R_1 = 0$.

Case of Two Magnetic Shells.

$$L_1 \gamma_1 d\gamma_1 + M \gamma_1 d\gamma_2 + \gamma_1 \gamma_2 dM = 0$$
$$L_2 \gamma_2 d\gamma_2 + M \gamma_2 d\gamma_1 + \gamma_1 \gamma_2 dM = 0.$$

Thus substituting in (62) we find

$$dT = -\gamma_1 \gamma_2 dM . \quad \ldots \ldots \quad (63)$$

which is equal and opposite to the work done by the electromagnetic forces in moving the conductors. Thus the electromagnetic work done on the conductors is at the expense of the electrokinetic energy of the system.

We consider next the march of the current in two mutually influencing circuits invariable in form and position. The equations are by (33')

Currents in Two Mutually Influencing Circuits. Differential Equations.

$$\left. \begin{array}{l} L_1 \dfrac{d\gamma_1}{dt} + M \dfrac{d\gamma_2}{dt} + R_1 \gamma_1 - E_1 = 0 \\ L_2 \dfrac{d\gamma_2}{dt} + M \dfrac{d\gamma_1}{dt} + R_2 \gamma_2 - E_2 = 0 \end{array} \right\} \ldots (64)$$

---

* *El. and Mag.* vol. ii. § 844.

These may be written

$$\left.\begin{aligned}(L_1\frac{d}{dt} + R_1)\gamma_1 + M\frac{d\gamma_2}{dt} - E_1 = 0 \\ M\frac{d\gamma_1}{dt} + (L_2\frac{d}{dt} + R_2)\gamma_2 - E_2 = 0\end{aligned}\right\} \quad \ldots \quad (65)$$

Operating on the first equation by $L_2 d/dt + R_2$, and on the second by $M d/dt$, and subtracting we get

Solution of Differential Equations.
$$(L_1L_2 - M^2)\frac{d^2\gamma_1}{dt^2} + (L_2R_1 + L_1R_2)\frac{d\gamma_1}{dt} + R_2(R_1\gamma_1 - E_1) = 0 \quad (66)$$

The solution of this equation is

$$R_1\gamma_1 - E_1 = A_1\epsilon^{at} + B_1\epsilon^{\beta t} \ldots \ldots (67)$$

where a, β, are the roots of the quadratic

$$(L_1L_2 - M^2)x^2 + (L_2R_1 + L_1R_2)x + R_1R_2 = 0 \quad (68)$$

By eliminating in the same way γ_1 from (65) we should obtain an equation for γ_2 precisely similar to (66). Hence

$$R_2\gamma_2 - E_2 = A_2\epsilon^{at} + B_2\epsilon^{\beta t} \ldots \ldots (69)$$

The roots of the quadratic (68) are both real, since

$$(L_1R_2 - L_2R_1)^2 > -4M^2R_1R_2$$

or which is the same

$$(L_1R_2 + L_2R_1)^2 > 4(L_1L_2 - M^2)R_1R_2.$$

Neither of the currents can ever become infinite, so that a and β must both be negative as well as real. The necessary condition for this is

$$L_1R_2 + L_2R_1 > \{(L_1R_2 + L_2R_1)^2 - 4(L_1L_2 - M^2)R_1R_2\}^{\frac{1}{2}}$$

which is true if $L_1L_2 > M^2$. This we shall see later is always the case.

We suppose now that both circuits are closed at the same instant, which is taken as the zero of reckoning for t. Thus we must have when $t = 0$, $\gamma_1 = 0$, $\gamma_2 = 0$, and therefore (67) and (69) become

$$-E_1 = A_1 + B_1, \quad -E_2 = A_2 + B_2.$$

PRIMARY AND SECONDARY INDUCTION COILS

Hence (67) and (69) may be written

$$R_1\gamma_1 = E_1(1 - \epsilon^{\beta t}) + A_1(\epsilon^{\alpha t} - \epsilon^{\beta t})$$
$$R_2\gamma_2 = E_2(1 - \epsilon^{\beta t}) + A_2(\epsilon^{\alpha t} - \epsilon^{\beta t}) \quad \ldots \quad (70)$$

Equations of Currents.

By calculating $d\gamma_1/dt$, $d\gamma_2/dt$ from these last equations, substituting in (65) and putting $t = 0$ we could determine A_1 and A_2. It will however be simpler to determine these coefficients for each particular case to which we apply the equations obtained.

One very important case is that of a primary and secondary circuit, the former of which includes a battery of electromotive force E, the latter no electromotive force. We shall suppose that the first of equations (65), (70) refer to the primary current, and the second of these, with $E_2 = 0$, to the secondary circuit.

Primary and Secondary Circuits of an Induction Coil without Iron.

Putting $E_2 = 0$ in (70) differentiating and substituting in (65) we get, for $t = 0$

$$(\alpha - \beta)\left(\frac{L_1}{R_1}A_1 + \frac{M}{R_2}A_2\right) - E\left(1 + \beta\frac{L_1}{R_1}\right) = 0$$

$$(\alpha - \beta)\left(\frac{M}{R_1}A_1 + \frac{L_2}{R_2}A_2\right) - E\beta\frac{M}{R_1} = 0$$

which give

$$A_1 = \frac{E}{\alpha - \beta}\left\{\beta + \frac{L_2 R_1}{L_1 L_2 - M^2}\right\}$$
$$A_2 = -\frac{E}{\alpha - \beta}\frac{R_2 M}{L_1 L_2 - M^2} \quad \ldots \quad (71)$$

We can now find the whole quantity of electricity which flows through the secondary circuit at the completion of the primary. For by the second of (70)

Quantity of Electricity passing at "Make."

$$R_2\int_0^\infty \gamma_2 dt = A_2\int_0^\infty (\epsilon^{\alpha t} - \epsilon^{\beta t})dt = A_2\frac{\alpha - \beta}{\alpha\beta}.$$

But referring to the quadratic which determines α, β, we see that $\alpha\beta = R_1 R_2/(L_1 L_2 - M^2)$. Hence substituting the value of A_2 from (71) in the last equation we find

$$\int_0^\infty \gamma_2 dt = -\frac{E}{R_1}\frac{M}{R_2} \quad \ldots \ldots \quad (72)$$

It is to be noticed that the integration is taken from 0 to ∞, to cover the whole variable period of the current, although as a matter of fact the current in the primary has become constant, and the secondary current has ceased, in a very short time as compared with any ordinary interval.

The result in (72) can be at once obtained by direct integration of the second of (65). For if γ be the final constant value of γ_1 we have

$$M\gamma + L_2 \int_0^\infty \frac{d\gamma_2}{dt} dt + R_2 \int_0^\infty \gamma_2 dt = 0$$

and since $\gamma_2 = 0$ both when $t = 0$, and when $t = \infty$, and $\gamma = E/R_1$, this gives

$$\int_0^\infty \gamma_2 dt = -\gamma \frac{M}{R_2} = -\frac{E}{R_1} \frac{M}{R_2}$$

as before.

Quantity of Electricity passing at "Break." The whole quantity of electricity which passes in the secondary conductor when the primary circuit is broken can also be obtained by direct integration. In this case the primary current is initially γ and finally 0, the secondary both finally and initially 0, and instead of the last result we get

$$\int_0^\infty \gamma_2 dt = \gamma \frac{M}{R_2} = \frac{E}{R_1} \frac{M}{R_2} \quad . \quad . \quad . \quad . \quad (73)$$

Thus the quantity of electricity which passes at break is equal and opposite to that which passes when the circuit is made.

If the break of the primary circuit occupies a time τ we have

$$M \int_0^\tau \frac{d\gamma_1}{dt} dt + L_2 \int_0^\tau \frac{d\gamma_2}{dt} dt + R_2 \int_0^\tau \gamma_2 dt = 0$$

or $\quad -M\gamma + L_2\gamma_2 + R_2 \int_0^\tau \gamma_2 dt = 0 \quad . \quad . \quad . \quad (74)$

Instantaneous Break. If the break take place so quickly as to be practically instantaneous the third term will be zero in the last expression, and the value of the current in the secondary at the end of the break will be

$$\gamma_2 = \frac{M}{L_2} \gamma \quad . \quad . \quad . \quad . \quad . \quad . \quad (75)$$

The current thus rises quickly in the secondary to this value, and then gradually dies away, the energy being consumed in producing heat. By (29') above the energy initially is $\frac{1}{2} M^2/L_2 \cdot \gamma^2$,

ALTERNATE CURRENT TRANSFORMER WITHOUT IRON

and at any time when the current is γ_2 is $\frac{1}{2} L_2 \gamma_2^2$. The rate at which the energy is being dissipated in heat is $R \gamma_2^2$. Hence we have

$$-\frac{d}{dt}\left(\tfrac{1}{2} L_2 \gamma_2^2\right) = R \gamma_2^2,$$

or $\quad L_2 \dfrac{d\gamma_2}{dt} + R_2 \gamma_2 = 0.$

Integrating this and remembering that the initial value of γ_2 is $\gamma M/L_2$ we have

$$\gamma_2 = \frac{M}{L_2}\frac{E}{R_1}\epsilon^{-\frac{R_2}{L_2}t} \quad \ldots \ldots \quad (76)$$

which gives the mode in which the current dies out.

A very important practical case is that in which the electromotive force E in the primary circuit is a periodic function of the time. We shall suppose that the function is a single harmonic term $E_0 \sin nt$, and that there is, as before, no electromotive force in the other circuit. By (33') the equations are

Theory of Alternate Current Transformer without Iron.

$$\left. \begin{array}{l} L_1 \dfrac{d\gamma_1}{dt} + M \dfrac{d\gamma_2}{dt} + R_1 \gamma_1 = E_0 \sin nt \\[4pt] L_2 \dfrac{d\gamma_2}{dt} + M \dfrac{d\gamma_1}{dt} + R_2 \gamma_2 = 0 \end{array} \right\} \ldots (77)$$

Differential Equations for Currents in Primary and Secondary.

To eliminate γ_2 we operate as before on the first of these by $R_2 + L_2 d/dt$, and on the second by $M d/dt$, and subtract. The resulting equation is

$$\frac{d^2\gamma_1}{dt^2} + a\frac{d\gamma_1}{dt} + b^2 \gamma_1 = c_1 \sin(nt - e) \ldots \quad (78)$$

where

$$\left. \begin{array}{l} \sin e = -nL_2/(R_2^2 + n^2 L_2^2)^{\frac{1}{2}} \\ \cos e = R_2/(R_2^2 + n^2 L_2^2)^{\frac{1}{2}} \end{array} \right\} \ldots (79)$$

and for brevity a, b^2, c_1 are put for $(L_1 R_2 + L_2 R_1)/(L_1 L_2 - M^2)$, $R_1 R_2/(L_1 L_2 - M^2)$, $E_0 (R_2^2 + n^2 L_2^2)^{\frac{1}{2}}/(L_1 L_2 - M^2)$, respectively.

A particular solution of (78) is

Solution of Differential Equations.

$$\gamma_1 = f_1 \sin(nt - \theta_1) \ldots \ldots \quad (80)$$

184   CURRENT INDUCTION AND ELECTROMAGNETIC ACTION

where f_1 and θ_1 are given by the equations (found by substituting from (80) in (78) and equating coefficients of sin nt, cos nt, on the two sides of the resulting equation)

$$\left.\begin{aligned} f_1 &= c_1/\{(b^2-n^2)^2 + a^2n^2\}^{\frac{1}{2}} \\ \tan\theta_1 &= \frac{(b^2-n^2)\tan e + an}{b^2 - n^2 - an\tan e} \end{aligned}\right\} \quad \ldots \quad (81)$$

Hence if a, β be as before the roots of the quadratic

$$x^2 + ax + b^2 = 0,$$

the complete solution of (72) is

Current in Primary Circuit.

$$\gamma_1 = A_1 \epsilon^{at} + B_1 \epsilon^{\beta t} + \frac{c_1}{\{(b^2-n^2)^2 + a^2n^2\}^{\frac{1}{2}}} \sin(nt - \theta_1) \quad (82)$$

Similarly by operating on the second of (77) with $R_1 + L_1 d/dt$, and on the first with $M d/dt$, and subtracting the first result from the second, we get

$$\frac{d^2\gamma_2}{dt^2} + a\frac{d\gamma_2}{dt} + b^2\gamma_2 = c_2 \sin\left(nt - \frac{\pi}{2}\right). \quad \ldots \quad (83)$$

where a, b^2, have the same values as before, and $c_2 = E_0 M n/(L_1 L_2 - M^2)$. Since this equation is derivable from (78) by putting γ_2 for γ_1, c_2 for c_1, and $\pi/2$ for e, the solution is obtained at once from (80), (81), and (82) by making the corresponding changes. Thus we get

Current in Secondary Circuit.

$$\left.\begin{aligned} \gamma_2 &= A_2 \epsilon^{at} + B_2 \epsilon^{\beta t} + \frac{c_2}{\{(b^2-n^2)^2 + a^2n^2\}^{\frac{1}{2}}} \sin(nt - \theta_2) \\ \text{where} & \\ \tan\theta_2 &= -\frac{b^2 - n^2}{an} \end{aligned}\right\} \quad . \quad (84)$$

Since a and β are negative the exponential terms in (82) and (84) are only effective for a short time after the circuits are closed, or after some change takes place in the impressed electromotive force. Except in these circumstances, therefore, they may be neglected.

Phase-Difference between Primary

By inspection of equations (78), (82), (83), (84), or by direct calculation from (81) and (84), we see that the difference of phase, $\theta_2 - \theta_1$, between the primary and secondary currents, is $\pi/2 - e$.

If we put $M = 0$, in the first of (77), that is, if we suppose the primary and secondary currents without mutual influence, that equation becomes

$$L_1 \frac{d\gamma_1}{dt} + R_1\gamma_1 = E_0 \sin nt \quad . \quad . \quad . \quad . \quad (85)$$

of which the solution is obtained by the method pursued above, p. 183, for the solution of (78), and is

where
$$\left.\begin{array}{c} \gamma_1 = A e^{-\frac{R_1}{L_1}t} + \dfrac{E_0}{(R_1^2 + n^2L_1^2)^{\frac{1}{2}}} \sin(nt - \theta) \\ \\ \tan \theta = \dfrac{nL_1}{R_1} \end{array}\right\} \quad . \quad (86)$$

This result is of course also deducible from (82) by putting $M = 0$ in that equation. It is of great importance in the theory of alternating machines. It is hardly necessary to remark that if $M = 0$, the current is always zero in the secondary circuit.

In consequence of self-induction the phase of the current in (86) differs by θ from that of the electromotive force, and the resistance is virtually (though not really) increased from R_1 to $(R_1^2 + n^2L_1^2)^{\frac{1}{2}}$. The latter quantity is now very commonly called the impedance of the circuit.

The result of (86) enables us to state the effect of the presence of the secondary circuit on the current in the primary in a very simple manner, due to Clerk Maxwell.* Inserting in (82) the values of a, b^2, and c_1, and attending only to the harmonic term, we can write the equation in the form

$$\gamma_1 = \frac{E_0 \sin(nt - \theta_1)}{\left\{\left(R_1 + n^2 \dfrac{M^2 R_2}{R_2^2 + n^2 L_2^2}\right)^2 + n^2\left(L_1 - n^2 \dfrac{M^2 L_2}{R_2^2 + n^2 L_2^2}\right)^2\right\}^{\frac{1}{2}}} \quad (87)$$

Thus the resistance has been virtually increased by the amount $n^2M^2R_2/(R_2^2 + n^2L_2^2)$, and the self-induction virtually diminished by the amount $n^2M^2L_2/(R_2^2 + n^2L_2^2)$.

Similarly, we see by (84) and (86) that the current in the secondary circuit is the same as if the circuit were independent and contained a harmonic electromotive force of maximum

---

* "A Dynamical Theory of the Electromagnetic Field," Maxwell's *Collected Papers*, p. 547, or *Phil. Trans. R.S.* vol. clv. (1865).

amount $E_0 Mn/(R_1^2 + n^2 L_1^2)^{\frac{1}{2}}$, and had a resistance $R_2 + n^2 M^2 R_1/(R_1^2 + n^2 L_1^2)$, and a self-induction $L_2 - n^2 M^2 L_1/(R_1^2 + n^2 L_1^2)$.

Another important case is that of a number of conductors each having self-induction, but without mutual induction, joined so as to form a multiple arc between two points A, B, of a circuit in which is a simple harmonic electromotive force.*

Theory of Multiple Arc in Alternating Circuit.

We may suppose that a simple harmonic difference of potential is applied to the points A, B. If the inductances and resistances of the different circuits be $L_1, L_2, \ldots L_n, \ldots, R_1, R_2 \ldots R_n, \ldots$, the currents in the different conductors at an instant at which the impressed difference of potential in $V_0 \sin nt$ are given by the equations

$$L_1 \frac{d\gamma_1}{dt} + R_1 \gamma_1 = V_0 \sin nt$$

$$L_2 \frac{d\gamma_2}{dt} + R_2 \gamma_2 = V_0 \sin nt$$

$$\cdots \cdots \cdots \cdots \quad (88)$$

$$L_n \frac{d\gamma_n}{dt} + R_n \gamma_n = V_0 \sin nt$$

$$\cdots \cdots \cdots \cdots$$

By (86) the solution of the typical equation

$$L \frac{d\gamma}{dt} + R\gamma = V_0 \sin nt$$

is

$$\gamma = A e^{-\frac{R}{L}t} + \frac{V_0}{(R^2 + n^2 L^2)^{\frac{1}{2}}} \sin(nt - e) \quad . \quad . \quad (89)$$

where

$$\tan e = \frac{nL}{R}.$$

Using this solution in each of (88) and adding the values of γ, we get (neglecting the exponential terms) for the total current at time t the equation

$$\sum \gamma = V_0 \sum \frac{1}{(R^2 + n^2 L^2)^{\frac{1}{2}}} \sin(nt - e) \quad . \quad . \quad (90)$$

---

* Lord Rayleigh, *Phil. Mag.* 1886.

If now we put

$$R' = \frac{\sum \frac{R}{R^2 + n^2 L^2}}{\left(\sum \frac{R}{R^2 + n^2 L^2}\right)^2 + \left(\sum \frac{nL}{R^2 + n^2 L^2}\right)^2},$$

$$nL' = \frac{\sum \frac{nL}{R^2 + n^2 L^2}}{\left(\sum \frac{R}{R^2 + n^2 L^2}\right)^2 + \left(\sum \frac{nL}{R^2 + n^2 L^2}\right)^2},$$

(90) may be written in the form

$$\left. \begin{array}{c} \sum \gamma = \dfrac{V_0}{(R'^2 + n^2 L'^2)^{\frac{1}{2}}} \sin(nt - \theta) \\[2mm] \tan \theta = \dfrac{nL'}{R'} \end{array} \right\} \quad \ldots \quad (91)$$

where

By the principle of continuity this must be the current in the main conductor connected to A and B.

Comparing this result with (86) or (89) we see that the total current is the same as if the point A, B, were joined by a single conductor of resistance R' and inductance L'. We may call these the **equivalent resistance and inductance of the multiple arc.** The angle θ measures in angle the lag in phase of the total current in the main conductor behind the impressed difference of potential.

Another important case is that of a condenser charged to any given potential, and then discharged by having its plates connected by a wire of given inductance L and resistance R. Let the capacity of the condenser be C, the difference of potential between its plates at time t, V, and the current at the same instant γ. We shall suppose that the current has the same value at each instant at all parts of the conducting wire, although this will not be the case in a coil so arranged that the adjacent parts are at considerably different potentials, and have therefore sensible electrostatic capacities. The energy of the distribution on the plates is $\frac{1}{2}CV^2$, and that of the current in the connecting wire $\frac{1}{2}L\gamma^2$. Hence the total electrical energy is $\frac{1}{2}CV^2 + \frac{1}{2}L\gamma^2$. The rate of diminution of this must be equal to the rate at which

Discharge of Condenser through Inductive Resistance.

energy is dissipated in the form of heat in the circuit. Hence we have

$$\tfrac{1}{2}\frac{d}{dt}(CV^2 + L\gamma^2) + R\gamma^2 = 0,$$

or since $\gamma = - CdV/dt$

$$CL\frac{d^2V}{dt^2} + RC\frac{dV}{dt} + V = 0. \quad \ldots \quad (92)$$

Solution for Non-Oscillatory Discharge.

The solution of this equation is

$$V = \epsilon^{-\tfrac{R}{2L}t}\left\{A\epsilon^{\tfrac{1}{2L}\left(R^2 - \tfrac{4L}{C}\right)^{\tfrac{1}{2}}t} + B\epsilon^{-\tfrac{1}{2L}\left(R^2 - \tfrac{4L}{C}\right)^{\tfrac{1}{2}}t}\right\}. \quad (93)$$

which if the exponents within the brackets are real represents a discharge continually progressing towards equalization of the potentials of the plates.

Solution for Oscillatory Discharge.

If the exponents are imaginary, which will be the case if $R^2C < 4L$, the solution becomes

$$V = \epsilon^{-\tfrac{R}{2L}t} A \cos\left(\frac{1}{2L}\sqrt{\frac{4L}{C} - R^2} \cdot t - e\right) \quad \ldots \quad (94)$$

where A and e are constants to be determined from the initial circumstances. For example if $V = V_0$, and $dV/dt = 0$, when $t = 0$, we have

$$\left. \begin{aligned} A &= V_0\left\{\frac{4L}{4L - CR^2}\right\}^{\tfrac{1}{2}} \\ \tan e &= \frac{R\sqrt{C}}{(4L - CR^2)^{\tfrac{1}{2}}} \end{aligned} \right\} \quad \ldots \quad (95)$$

Thus we have the very remarkable result that the discharge is oscillatory with a period of magnitude $T = 4\pi L/\sqrt{4L/C - R^2}$ if the inductance L is sufficiently great, and with a diminishing amplitude the logarithmic decrement of which is $RT/4L$. The current is zero when the difference of potential is a maximum or minimum, since then $dV/dt = 0$.

That the discharge might be oscillatory in certain cases, that is, that the condenser might become charged alternately positive and negative many times in succession was suggested by Helmholtz in his famous essay *Die Erhaltung der Kraft* from certain anomalous phenomena of magnetization produced by Leyden jar discharges. The above theory is practically that given in 1857 by Sir William Thomson, who at the same time suggested, that it might be tested by examining in a rotating mirror the image of a spark produced by the discharge of a Leyden jar. This was done by Feddersen and the theory completely verified in 1859.* A short account of his and other experiments on transient currents, proving the validity of the theory stated above, will be given later.

It is to be noticed that in the investigation given above the energy of the initially charged condenser is supposed to be gradually dissipated in heat produced in the conductor joining the plates. It has been found by Hertz that when an oscillatory discharge thus takes place, there is a radiation of energy into the surrounding medium, which theory indicates must at a considerable distance from the condenser travel outwards with the velocity of light. No account of such radiation is taken here. The whole subject will be dealt with in a later chapter.

---

* *Pogg. Ann.* Vol. 108, (1859) and Vols. 112 and 113 (1861).

CHAPTER V

GENERAL ELECTROMAGNETIC THEORY

SECTION I

INDUCTIVE ELECTROMOTIVE FORCES IN A CONTINUOUS MEDIUM. ELECTROMAGNETIC THEORY OF LIGHT

Inductive Electromotive Forces. WE now proceed to the determination of inductive electromotive forces, first, in the circuit of a conductor, second, in a continuous medium. If N_k denote total magnetic induction through the circuit we have seen (p. 159 above) that the electromotive force in the circuit is $-dN_k/dt$. We have by (49) of Chap. I. above

$$N_k = \int (la + mb + nc)dS = \int \left(F\frac{dx}{ds} + G\frac{dy}{ds} + H\frac{dz}{ds} \right) ds \ . \quad (1)$$

Now the value of N_k may be altered in two ways:—(1) by motion of the circuit; (2) by variation of the currents in the circuits producing the magnetic induction. The currents being expressed in terms of the time, the time-rate of the latter variation has the value

E.M.F. due to Variation of Current.
$$\frac{\partial N_k}{\partial t} = \int \left(\frac{\partial F}{\partial t}\frac{dx}{ds} + \frac{\partial G}{\partial t}\frac{dy}{ds} + \frac{\partial H}{\partial t}\frac{dz}{ds} \right) ds \ . \ . \ . \quad (2)$$

since while it is taking place we suppose dx/ds, &c. to be constant.

E.M.F. due to Motion of the Circuit. To calculate the time-rate of variation of N_k due to motion of the circuit, let the resultant velocity of any element ds of the circuit be q, with components $\dot{x}, \dot{y}, \dot{z}$. If θ be the angle between the positive direction of **B** and the direction of motion of ds, the direction cosines of a line at right angles to both q and **B** are

EQUATIONS OF ELECTROMOTIVE FORCE

$$\frac{b\dot{z}-c\dot{y}}{q\mathbf{B}\sin\theta},\ \frac{c\dot{x}-a\dot{z}}{q\mathbf{B}\sin\theta},\ \frac{a\dot{y}-b\dot{x}}{q\mathbf{B}\sin\theta},$$

and the positive direction of this line is that in which a right-handed screw would advance, if the handle were turned round O from the direction $\mathbf{B}$ to the direction q. Hence the revolved

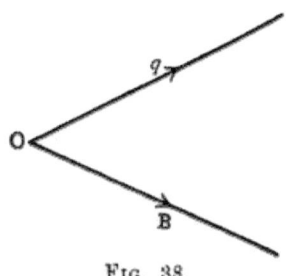

FIG. 38.

part ds' of ds, at right angles to the plane of $\mathbf{B}$ and q, is given by

$$ds' = \frac{ds}{q\mathbf{B}\sin\theta}\left\{\frac{dx}{ds}(b\dot{z}-c\dot{y})+\frac{dy}{ds}(c\dot{x}-a\dot{z})+\frac{dz}{ds}(a\dot{y}-b\dot{x})\right\}$$

The velocity with which this component cuts normally across the lines of magnetic induction is $q\sin\theta$; and hence for the time-rate of the variation of N_k due to the motion of ds, we have

$$\mathbf{B}q\sin\theta ds' = ds\left\{\frac{dx}{ds}(b\dot{z}-c\dot{y})+\frac{dy}{ds}(c\dot{x}-a\dot{z})+\frac{dz}{ds}(a\dot{y}-b\dot{x})\right\}. \quad (3)$$

Thus for the total time-rate of variation of N_k we get finally the equation

$$-\frac{dN_k}{dt} = \int\left(P\frac{dx}{ds}+Q\frac{dy}{ds}+R\frac{dz}{ds}\right)ds \qquad (4)$$

where

$$\left.\begin{array}{l}P = c\dot{y}-b\dot{z}-\dfrac{\partial F}{\partial t}-\dfrac{\partial\psi}{\partial x}\\[4pt]Q = a\dot{z}-c\dot{x}-\dfrac{\partial G}{\partial t}-\dfrac{\partial\psi}{\partial y}\\[4pt]R = b\dot{x}-a\dot{y}-\dfrac{\partial H}{\partial t}-\dfrac{\partial\psi}{\partial z}\end{array}\right\} \quad (5)$$

Equations of Electromotive Force.

in which for the sake of generality the components $-\partial\psi/\partial x$, $-\partial\psi/\partial y$, $-\partial\psi/\partial z$, of a single valued function $-\psi$ of the coordinates have been introduced. They contribute nothing to the integral in (4), which is taken round the circuit. ψ may be taken as the potential of any electrostatic distribution which may exist in the field.

$P, Q, R,$ are called the components of the electromotive force at the element ds. We shall denote their resultant by **E**, and ask the reader to carefully distinguish between it and the line integral of electromotive force round a circuit, which is commonly called the electromotive force in the circuit. We shall denote the latter by E. The direction of the electromotive force **E**, at ds, is that of the current which it tends to produce. Thus when a portion of a circuit is displaced in consequence of the mutual action of the current and the magnetic field the direction of the electromotive force is such that the current is diminished.

Simple Cases of Induction. 1. Rails and Slider in Uniform Field.

To take an example, let the circuit consist of a pair of parallel rails AB, CD, bridged across at one end by a fixed conductor, and elsewhere by a movable cross-bar or slider EF, and let a current flow round in the direction of the arrows. Let also, to fix the ideas, the plane of the rails be horizontal, and the direc-

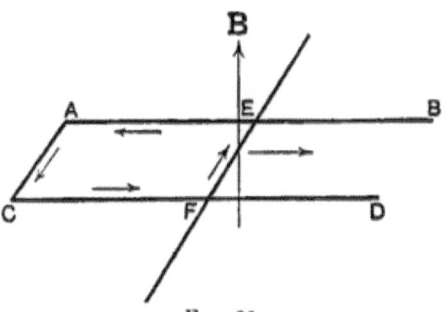

Fig. 39.

tion of the magnetic induction be upward. Then the slider if free to move will do so towards BD, and the resulting electromotive force will tend to produce a current in the opposite direction to that shown by the arrows.

If the slider be at right angles to the rails, and have a length l from rail to rail, and a velocity v, and **B** be the (uniform) value of the induction, the time-rate of change of the total induction through the circuit is **B**lv, and this is the line integral, E, of the

INDUCTIVE ELECTROMOTIVE FORCE

electromotive force round **the circuit**. **If** ds **be** an element of length of the slider, $\mathbf{B}ds.v$ is the integral electromotive force in ds, so that $\mathbf{B}v$ is the value of **E** at the element.

It is to be observed that the expression here given for the total electromotive force E is complete, and includes **the** effect **not** only of change of the part of the total magnetic induction through the circuit due to currents in other circuits, but also of the part of the induction due to the current in the circuit moved. This latter part will be zero if there be no current in the displaced circuit independent of induction. If then a portion of the circuit be moved by external force in any direction the electromotive force set up will produce a current in the opposite direction to that which flowing in the circuit would have produced by mutual action the motion of the element.

In the preceding statement the induction through a closed circuit has been regarded as undergoing variation. It is not however necessary for the production of electromotive **force at a** point, that there should be a closed conducting circuit, nor that any actual current should **flow**. Wherever a portion of matter, whether conducting **or** insulating, **is** in motion **in** a magnetic field, in such a manner that it cuts **across lines** of magnetic force, there inductive electromotive force is produced. If the moving substance is **a portion** of a linear conductor, a difference of potential between its extremities is produced, which it is always consistent with experience to **take as** the line-integral of **E** along the conductor. This difference of potential of course tends to produce a backward current in the conductor, and thus to annul itself. An electric field is thus produced which **is** maintained by the inductive action.

A simple but very important case of this kind is furnished by a **metal disk** spun round an axis so that its plane cuts across the lines of magnetic induction. If the disk be circular and spin round **its** centre with its plane at right angles to the magnetic induction $\mathbf{B}$, supposed uniform, a difference of potential will be produced between its centre and its edge, which can be measured by means of a sufficiently sensitive electrometer. If w be the angular velocity, and r the radius of the disc the value of **E** at a point on any radius at a distance x from the centre is $wx\mathbf{B}$. Hence the total difference of potential between the **centre** and **the** outer **extremity** of the radius is

Inductive Electromotive Force in Case of Zero Current.

Simple Cases of Induction. 2. Metal Disk Spinning in Magnetic Field.

$$w\mathbf{B}\int_0^r x\,dx = \tfrac{1}{2}w\mathbf{B}r^2.$$

Or, if v be the velocity of the **rim** of the disk, the difference of potential between the rim and the centre is $\tfrac{1}{2}vr\mathbf{B}$.

VOL. II.

With the disk in this state no work is spent in driving it beyond that expended in overcoming friction. If however the value of **B** is not the same at every point of the disk, internal currents will be set up in the metal, the mutual action between which and the magnetic field will oppose the motion. Thus energy will be consumed in driving the disk across the lines of induction in the field, and this will have its equivalent in heat produced in the metal. For example, very marked heating effects can be produced in a copper disk by placing it so that a portion of it only, between the centre and one part of the rim, is in an intense magnetic field, and rotating it rapidly in that position.

Application of Rotating Disk to Determination of E.M.F.

If an external circuit be arranged by connecting one end of a wire to the edge of the disk the other to its centre, a current will flow for which E will have the value $\frac{1}{2}vrB$ just found, and the resistance will be the sum of the resistances of the wire and the disk. The latter would in an actual case be difficult to reckon, as the current would spread laterally in the disk about the radial portion connecting the contacts, and would depend on the positions and nature of the contacts themselves.

But if an electromotive force equal and opposite to that due to the rotation be placed in the wire, the current will be reduced to zero, and the latter electromotive force will be obtainable in absolute units by calculation from the rotation and other circumstances of the disk. This process has been applied to the determination of resistances in absolute measure (see Chap. XII.).

Maxwell's Electromagnetic Theory of Light.

On the equations (5) above of electromotive force, Clerk Maxwell has based a very remarkable theory of the propagation of disturbance in the electromagnetic field. It will be convenient to give a short statement of this theory here, reserving an account of its recent experimental verification by Hertz to a later chapter.

We suppose the medium to be at rest and to be isotropic, and assume that the equations of electromotive force (5) hold at every point of it. Then at any point

$$\left. \begin{aligned} P &= -\frac{\partial F}{\partial t} - \frac{\partial \psi}{\partial x} \\ Q &= -\frac{\partial G}{\partial t} - \frac{\partial \psi}{\partial y} \\ R &= -\frac{\partial H}{\partial t} - \frac{\partial \psi}{\partial z} \end{aligned} \right\} \quad \ldots \ldots (6)$$

Maxwell assumes that the true current at any point of the medium, supposed partially conducting, is made up of two parts,

DISPLACEMENT CURRENTS IN DIELECTRIC

one the true conduction current (comparatively small in a good insulator), the other (comparatively small in a good conductor) the time-rate of variation there of the electric displacement. If **E** be the electromotive force at the point the electric displacement there is taken as in the direction of **E**, and as having the value $KE/4\pi$ where K is the specific inductive capacity of the medium. Thus the components of the electric displacement are

Total Current = Displacement Current + Conduction Current.

$$f = \frac{K}{4\pi} P, \quad g = \frac{K}{4\pi} Q, \quad h = \frac{K}{4\pi} R \quad \ldots \quad (7)$$

Hence the components of the displacement current are given by

Components of Displacement Currents.

$$\dot{f} = \frac{K}{4\pi} \dot{P}, \quad \dot{g} = \frac{K}{4\pi} \dot{Q}, \quad \dot{h} = \frac{K}{4\pi} \dot{R} \quad \ldots \quad (8)$$

where the Newtonian fluxional notation is adopted for brevity to denote time-rate of variation of a quantity *at a given point*. The components of the conduction current at the point in question are kP, kQ, kR where k is the specific conductivity of the medium. The components of the total current are therefore

$$\left. \begin{array}{l} u = kP + \dfrac{K}{4\pi} \dot{P} \\[4pt] v = kQ + \dfrac{K}{4\pi} \dot{Q} \\[4pt] w = kR + \dfrac{K}{4\pi} \dot{R} \end{array} \right\} \quad \ldots \quad (9)$$

By (6), and *assuming* that $\partial \psi/\partial x$, &c. are independent of the time, and therefore play no part in periodic phenomena, we may write

$$\left. \begin{array}{l} u = - k\dot{F} - \dfrac{K}{4\pi} \ddot{F} \\[4pt] v = - k\dot{G} - \dfrac{K}{4\pi} \ddot{G} \\[4pt] w = - k\dot{H} - \dfrac{K}{4\pi} \ddot{H} \end{array} \right\} \quad \ldots \quad (10)$$

But assuming that displacement currents produce magnetic effects according to the same laws as those which hold for ordinary conduction currents, we have by (5) of **Chapter II**, and (50) of Chapter I., since $a = \mu a$, &c.

$$4\pi\mu u = \frac{\partial c}{\partial y} - \frac{\partial b}{\partial z} = \frac{\partial}{\partial y}\left(\frac{\partial G}{\partial x} - \frac{\partial F}{\partial y}\right) - \frac{\partial}{\partial z}\left(\frac{\partial F}{\partial z} - \frac{\partial H}{\partial x}\right)$$
$$= \frac{\partial J}{\partial x} - \nabla^2 F \quad \ldots \ldots (11)$$

where J denotes $\partial F/\partial x + \partial G/\partial y + \partial H/\partial z$, and ∇^2 as usual the operator $\partial^2/\partial x^2 + \partial^2/\partial y^2 + \partial^2/\partial z^2$. Similar equations are found in the same way for v and w. Thus we have the three equations of currents

$$\left. \begin{aligned} u &= \frac{1}{4\pi\mu}\left(\frac{\partial J}{\partial x} - \nabla^2 F\right) \\ v &= \frac{1}{4\pi\mu}\left(\frac{\partial J}{\partial y} - \nabla^2 G\right) \\ w &= \frac{1}{4\pi\mu}\left(\frac{\partial J}{\partial z} - \nabla^2 H\right) \end{aligned} \right\} \quad \ldots \ldots (12)$$

Differential Equations of Propagation of an Electromagnetic Disturbance.

Equating these values of u, v, w, to those given in (10) we find

$$\left. \begin{aligned} 4\pi\mu k \frac{\partial F}{\partial t} + K\mu \frac{\partial^2 F}{\partial t^2} &= \nabla^2 F - \frac{\partial J}{\partial x} \\ 4\pi\mu k \frac{\partial G}{\partial t} + K\mu \frac{\partial^2 G}{\partial t^2} &= \nabla^2 G - \frac{\partial J}{\partial y} \\ 4\pi\mu k \frac{\partial H}{\partial t} + K\mu \frac{\partial^2 H}{\partial t^2} &= \nabla^2 H - \frac{\partial J}{\partial z} \end{aligned} \right\} \quad \ldots (13)$$

These are Maxwell's differential equations of propagation of an electromagnetic disturbance of any type.*

Discussion as to Value of J.

It is easy to see that the quantity J which appears in them must be zero in the case of a periodic disturbance. For equations (13) differentiated with respect to x, y, z, and added give

$$4\pi\mu k \frac{dJ}{dt} + K\mu \frac{d^2 J}{dt^2} = 0 \quad \ldots \ldots (14)$$

of which the solution is

$$J = C + C' \exp(-4\pi kt/K) \quad \ldots \ldots (15)$$

---

* *Phil. Trans. R. S.* vol. clv. (1865), or *Reprint of Scientific Papers*, vol. i. p. 578; also *El. and Mag.* vol. ii. p. 395 (Sec. Ed.).

DISPLACEMENT CURRENTS IN DIELECTRIC

one the true conduction current (comparatively small in a good insulator), the other (comparatively small in a good conductor) the time-rate of variation there of the electric displacement. If **E** be the electromotive force at the point the electric displacement there is taken as in the direction of **E**, and as having the value $KE/4\pi$ where K is the specific inductive capacity of the medium. Thus the components of the electric displacement are

Total Current = Displacement Current + Conduction Current.

$$f = \frac{K}{4\pi} P, \quad g = \frac{K}{4\pi} Q, \quad h = \frac{K}{4\pi} R \quad . \quad . \quad . \quad (7)$$

Hence the components of the displacement current are given by

Components of Displacement Currents.

$$\dot{f} = \frac{K}{4\pi} \dot{P}, \quad \dot{g} = \frac{K}{4\pi} \dot{Q}, \quad \dot{h} = \frac{K}{4\pi} \dot{R} \quad . \quad . \quad . \quad (8)$$

where the Newtonian fluxional notation is adopted for brevity to denote time-rate of variation of a quantity *at a given point*. The components of the conduction current at the point in question are kP, kQ, kR where k is the specific conductivity of the medium. The components of the total current are therefore

$$\left. \begin{array}{l} u = kP + \dfrac{K}{4\pi} \dot{P} \\[4pt] v = kQ + \dfrac{K}{4\pi} \dot{Q} \\[4pt] w = kR + \dfrac{K}{4\pi} \dot{R} \end{array} \right\} \quad . \quad . \quad . \quad . \quad (9)$$

By (6), and *assuming* that $\partial\psi/\partial x$, &c. are independent of the time, and therefore play no part in periodic phenomena, we may write

$$\left. \begin{array}{l} u = -k\dot{F} - \dfrac{K}{4\pi} \ddot{F} \\[4pt] v = -k\dot{G} - \dfrac{K}{4\pi} \ddot{G} \\[4pt] w = -k\dot{H} - \dfrac{K}{4\pi} \ddot{H} \end{array} \right\} \quad . \quad . \quad . \quad . \quad (10)$$

But assuming that displacement currents produce magnetic effects according to the same laws as those which hold for ordinary conduction currents, we have by (5) of **Chapter II**, and (50) of Chapter I., since $a = \mu a$, &c.

$$4\pi\mu u = \frac{\partial c}{\partial y} - \frac{\partial b}{\partial z} = \frac{\partial}{\partial y}\left(\frac{\partial G}{\partial x} - \frac{\partial F}{\partial y}\right) - \frac{\partial}{\partial z}\left(\frac{\partial F}{\partial z} - \frac{\partial H}{\partial x}\right)$$
$$= \frac{\partial J}{\partial x} - \nabla^2 F \quad \ldots \ldots \quad (11)$$

where J denotes $\partial F/\partial x + \partial G/\partial y + \partial H/\partial z$, and ∇^2 as usual the operator $\partial^2/\partial x^2 + \partial^2/\partial y^2 + \partial^2/\partial z^2$. Similar equations are found in the same way for v and w. Thus we have the three equations of currents

$$\left. \begin{aligned} u &= \frac{1}{4\pi\mu}\left(\frac{\partial J}{\partial x} - \nabla^2 F\right) \\ v &= \frac{1}{4\pi\mu}\left(\frac{\partial J}{\partial y} - \nabla^2 G\right) \\ w &= \frac{1}{4\pi\mu}\left(\frac{\partial J}{\partial z} - \nabla^2 H\right) \end{aligned} \right\} \quad \ldots \ldots \quad (12)$$

Differential Equations of Propagation of an Electromagnetic Disturbance.

Equating these values of u, v, w, to those given in (10) we find

$$\left. \begin{aligned} 4\pi\mu k\frac{\partial F}{\partial t} + K\mu\frac{\partial^2 F}{\partial t^2} &= \nabla^2 F - \frac{\partial J}{\partial x} \\ 4\pi\mu k\frac{\partial G}{\partial t} + K\mu\frac{\partial^2 G}{\partial t^2} &= \nabla^2 G - \frac{\partial J}{\partial y} \\ 4\pi\mu k\frac{\partial H}{\partial t} + K\mu\frac{\partial^2 H}{\partial t^2} &= \nabla^2 H - \frac{\partial J}{\partial z} \end{aligned} \right\} \quad \ldots \quad (13)$$

These are Maxwell's differential equations of propagation of an electromagnetic disturbance of any type.*

Discussion as to Value of J.

It is easy to see that the quantity J which appears in them must be zero in the case of a periodic disturbance. For equations (13) differentiated with respect to x, y, z, and added give

$$4\pi\mu k\frac{dJ}{dt} + K\mu\frac{d^2J}{dt^2} = 0 \quad \ldots \ldots \quad (14)$$

of which the solution is

$$J = C + C' \exp.(-4\pi kt/K) \quad \ldots \ldots \quad (15)$$

---

* *Phil. Trans. R. S.* vol. clv. (1865), or *Reprint of Scientific Papers*, vol. i. p. 578; also *El. and Mag.* vol. ii. p. 395 (Sec. Ed.).

But if F, G, H, are periodic functions J must also be a periodic function, which is inconsistent with (15). J must therefore be zero. When $k = 0$, that is, when the medium is an insulator, the solution of (14) has the form

$$J = C + C't$$

and thus again if F, G, H, are periodic functions J must be zero. Thus in the propagation of periodic disturbances the terms involving J in equations (13) may be disregarded. We might of course by equation (57) of Chapter IV. have taken J to be zero at once, since we have assumed all the laws of induction established above to hold in wholly or partially insulating media.

If F, G, H be solutions of equations (13) their differential coefficients of any given order, and any linear function of these linear differential coefficients, are also solutions. It will be useful to verify this for the value of a given by (50) of Chapter I. We shall thereby obtain the differential equation for the propagation of magnetic induction. We have

$$4\pi\mu k \frac{\partial a}{\partial t} = 4\pi\mu k \frac{\partial}{\partial t}\left(\frac{\partial H}{\partial y} - \frac{\partial G}{\partial z}\right)$$

$$4\pi\mu k \frac{\partial^2 a}{\partial t^2} = K\mu \frac{\partial^2}{\partial t^2}\left(\frac{\partial H}{\partial y} - \frac{\partial G}{\partial z}\right).$$

Hence adding and rearranging we get

$$4\pi\mu k \frac{\partial a}{\partial t} + K\mu \frac{\partial^2 a}{\partial t^2}$$

$$= \frac{\partial}{\partial y}\left(4\pi\mu k \frac{\partial H}{\partial t} + \mu K \frac{\partial^2 H}{\partial t^2}\right) - \frac{\partial}{\partial z}\left(4\pi\mu k \frac{\partial G}{\partial t} + \mu K \frac{\partial^2 G}{\partial t^2}\right)$$

$$= \nabla^2\left(\frac{\partial H}{\partial y} - \frac{\partial G}{\partial z}\right), \text{ by (13)}$$

$$= \nabla^2 a.$$

Thus we have the three equations

$$\left.\begin{array}{l} 4\pi\mu k \dfrac{\partial a}{\partial t} + K\mu \dfrac{\partial^2 a}{\partial t^2} = \nabla^2 a \\[6pt] 4\pi\mu k \dfrac{\partial b}{\partial t} + K\mu \dfrac{\partial^2 b}{\partial t^2} = \nabla^2 b \\[6pt] 4\pi\mu k \dfrac{\partial c}{\partial t} + K\mu \dfrac{\partial^2 c}{\partial t^2} = \nabla^2 c \end{array}\right\} \quad \ldots \ldots \quad (16)$$

Differential Equations of Propagation of Magnetic Induction.

Again the first of equations (9) differentiated with respect to the time gives

$$\dot{u} = \frac{4\pi k}{K} \dot{f} + \ddot{f}$$

and the first of (12) gives likewise since $J = 0$,

$$\dot{u} = -\frac{1}{4\pi\mu} \nabla^2 \dot{F}$$

$$= \frac{1}{4\pi\mu} \nabla^2 P = \frac{1}{K\mu} \nabla^2 f,$$

Differential Equations of Propagation of Electric Displacement.

if we suppose that $\nabla^2(\partial\psi/\partial x) = 0$, at every point considered. Thus we have

and

$$\left.\begin{array}{l} 4\pi k\mu \dfrac{\partial f}{\partial t} + K\mu \dfrac{\partial^2 f}{\partial t^2} = \nabla^2 f \\[6pt] 4\pi k\mu \dfrac{\partial g}{\partial t} + K\mu \dfrac{\partial^2 g}{\partial t^2} = \nabla^2 g \\[6pt] 4\pi k\mu \dfrac{\partial h}{\partial t} + K\mu \dfrac{\partial^2 h}{\partial t^2} = \nabla^2 h \end{array}\right\} \quad \ldots \ldots (16')$$

which are the differential equations of propagation of electric displacement and therefore also of electromotive force.

We select equations (13) for solution and suppose a periodic disturbance produced in the field to be propagated along the axis of z, and the wave front at each instant to be a plane perpendicular to the axis of z. These conditions render F, G, H, independent of x and y, and since therefore

$$\frac{\partial^2 H}{\partial z^2} = \frac{\partial J}{\partial z} - \frac{\partial}{\partial z}\left(\frac{\partial F}{\partial x} + \frac{\partial G}{\partial y}\right) = 0$$

the equations become

Solution of Differential Equations of Propagation.

$$\left.\begin{array}{l} 4\pi k\mu \dfrac{\partial F}{\partial t} + K\mu \dfrac{\partial^2 F}{\partial t^2} = \dfrac{\partial^2 F}{\partial z^2} \\[6pt] 4\pi k\mu \dfrac{\partial G}{\partial t} + K\mu \dfrac{\partial^2 G}{\partial t^2} = \dfrac{\partial^2 G}{\partial z^2} \\[6pt] 4\pi k\mu \dfrac{\partial H}{\partial t} + K\mu \dfrac{\partial^2 H}{\partial t^2} = 0 \end{array}\right\} \quad \ldots \ldots (17)$$

PLANE WAVE IN NON-INSULATING MEDIUM

These equations have the periodic solution
$$(F, G) = (F_0, G_0) \exp. i(mz - nt) \quad . \quad . \quad . \quad (18)$$
where $i = \sqrt{-1}$, and $2\pi/n$ is the period of vibration. Substituting in the differential equation, we find the equation of condition
$$m^2 - n^2 K\mu - 4\pi k\mu n i = 0.$$
Thus since we suppose n to be real, m^2 is essentially a complex number, and therefore so also is m. If we write $m = q - pi$ we get by squaring and equating real and imaginary parts
$$q^2 - p^2 = n^2 K\mu, \quad pq = -2\pi k\mu n.$$
From these equations, remembering that p^2 must be a positive quantity since p is real, we obtain after reduction
$$\left. \begin{array}{l} p = -\dfrac{n\sqrt{\mu}}{\sqrt{2}}\{\sqrt{K^2 + 16\pi^2 k^2/n^2} - K\}^{\frac{1}{2}} \\[6pt] q = \dfrac{n\sqrt{\mu}}{\sqrt{2}}\{\sqrt{K^2 + 16\pi^2 k^2/n^2} + K\}^{\frac{1}{2}} \end{array} \right\} \quad . \quad . \quad (19)$$
where the positive sign is to be given to the radicals in each case, and p is made negative for the reason stated below. Thus the solution becomes
$$(F, G) = (F_0, G_0) \exp. i\{(q - pi)z - nt\}$$
$$= (F_0, G_0) \exp. pz \exp. i(qz - nt) \quad . \quad . \quad (20)$$
We are only concerned with the real part of (20) and hence
$$(F, G) = (F_0, G_0) \exp. pz \cos(pz - nt) \quad . \quad . \quad (20')$$
Thus the zero of time is so chosen that F_0, G_0 are the maximum values of F, G, at the point $z = 0$.

If the disturbance is not to increase in intensity as it travels out from the source p must be taken negative, and thus q is positive. Hence for the wave length λ, and velocity of propagation V the values are obtained

Wave-Length and Velocity of Propagation of Electromagnetic Waves.

$$\left. \begin{array}{l} \lambda = 2\pi/q = 2\pi / \left\{ \dfrac{n\sqrt{\mu}}{\sqrt{2}}(\sqrt{K^2 + 16\pi^2 k^2/n^2} + K)^{\frac{1}{2}} \right\} \\[6pt] V = n/q = 1 / \left\{ \dfrac{\sqrt{\mu}}{\sqrt{2}}(\sqrt{K^2 + 16\pi^2 k^2/n^2} + K)^{\frac{1}{2}} \right\} \end{array} \right\} \quad . \quad (21)$$

and for the ratio, at a given time, of the amplitude at any point $z + \lambda$ to the amplitude at any point z,

$$\exp. p\lambda = \exp. -\frac{2\pi(\sqrt{K^2 + 16\pi^2 k^2/n^2} - K)^{\frac{1}{2}}}{(\sqrt{K^2 + 16\pi^2 k^2/n^2} + K)^{\frac{1}{2}}} \quad . . \quad (22)$$

Perfectly Insulating Isotropic Medium. From these results we see that the wave length and velocity of propagation are less, and the rate of diminution of amplitude with distance from the source is greater, the greater the conductivity, other things remaining the same.

Further if $k = 0$, that is if the medium is an insulator, we have

$$p = 0, \quad q = n\sqrt{K\mu}, \quad V = \frac{1}{\sqrt{K\mu}} \quad . . . . \quad (23)$$

In this case the electric displacement and magnetic induction agree in phase, while the displacement current and magnetic induction differ in phase by a quarter period.

We obtain of course similar solutions for the propagation of electric displacement and magnetic induction. For the case supposed of a plane wave advancing parallel to z, we may, (if we put $t = 0$ for each quantity separately, so as to coincide with a maximum value of the quantity at $z = 0$), combine all the solutions for them in the single expression

Values of Electric Displacement and Magnetic Induction in Electromagnetic Wave.
$$(F, G, f, g, a, b) = (F_0, G_0, f_0, g_0, a_0, b_0) \exp. pz \exp. i(qz - nt) \quad (24)$$

where p, q have the values given in (19) and for which λ and V are given by (21). By the equations connecting f, g, with F, G, it is seen at once that the resultants of these pairs of components are in the same direction. But with regard to the magnetic induction we have, by (20)

$$\left. \begin{array}{l} a = \dfrac{\partial H}{\partial y} - \dfrac{\partial G}{\partial z} = -\dfrac{\partial G}{\partial z} = -(p + iq)G \\[6pt] b = \dfrac{\partial F}{\partial z} - \dfrac{\partial H}{\partial x} = \dfrac{\partial F}{\partial z} = (p + iq)F \end{array} \right\} \quad . . \quad (25)$$

Thus if θ, ϕ, be the angles which the resultants of F, and G, a and b respectively make with the axis of x, we have

$$\tan\theta = \frac{b}{a} = -\frac{F}{G} = -\cot\phi . \quad . . . . \quad (26)$$

Hence the direction of the magnetic induction is at right angles to that of the electric disturbance which is propagated along with it.

So far we have followed on the whole the theory as given by Maxwell himself, but a very symmetrical and instructive form of the equations of propagation which has been used by Heaviside and by Hertz in their investigations must not be passed over. By equations (9) above, and (5) of Chapter II. we have, when there is no electromotive force due to motion of the medium

Heaviside's and Hertz's Electromagnetic Equations.

$$\left(k + \frac{K}{4\pi}\frac{\partial}{\partial t}\right)P = \frac{1}{4\pi}\left(\frac{\partial \gamma}{\partial y} - \frac{\partial \beta}{\partial z}\right)$$
$$\left(k + \frac{K}{4\pi}\frac{\partial}{\partial t}\right)Q = \frac{1}{4\pi}\left(\frac{\partial a}{\partial z} - \frac{\partial \gamma}{\partial x}\right) \quad \ldots \quad (27)$$
$$\left(k + \frac{K}{4\pi}\frac{\partial}{\partial t}\right)R = \frac{1}{4\pi}\left(\frac{\partial \beta}{\partial x} - \frac{\partial a}{\partial y}\right)$$

But we find also by differentiating equations (50) of Chapter I., and using (9) above

$$\frac{\mu}{4\pi}\frac{\partial a}{\partial t} = -\frac{1}{4\pi}\left(\frac{\partial R}{\partial y} - \frac{\partial Q}{\partial z}\right)$$
$$\frac{\mu}{4\pi}\frac{\partial \beta}{\partial t} = -\frac{1}{4\pi}\left(\frac{\partial P}{\partial z} - \frac{\partial R}{\partial x}\right) \quad \ldots \quad (27')$$
$$\frac{\mu}{4\pi}\frac{\partial \gamma}{\partial t} = -\frac{1}{4\pi}\left(\frac{\partial Q}{\partial x} - \frac{\partial P}{\partial y}\right)$$

By differentiating the three equations of (9) with respect to x, y, z respectively, and using the relation [at once obtained from (5) of Chapter II.] $\partial u/\partial x + \partial v/\partial y + \partial w/\partial z = 0$ we find the condition

$$\frac{\partial P}{\partial x} + \frac{\partial Q}{\partial y} + \frac{\partial R}{\partial z} = 0 \quad \ldots \quad (28)$$

Conditions fulfilled in Homogeneous Isotropic Field by E and H.

which expresses that there is no electrification at the point (x, y, z).

A similar condition

$$\frac{\partial a}{\partial x} + \frac{\partial \beta}{\partial y} + \frac{\partial \gamma}{\partial z} = 0. \quad \ldots \quad (28')$$

is fulfilled by the magnetic force **H**, since that being purely inductive, must, like the magnetic induction **B**, and the electric force **E**, satisfy the solenoidal condition at every point except at the (vortex) origin of the disturbance.

There is thus a remarkable correspondence between the two quantities **E** and **H**, as shown by the foregoing equations involving their components $P, Q, R, a, \beta, \gamma$. This is brought out more distinctly by Oliver Heaviside (who first employed a similar set of relations for the investigation of electromagnetic waves) by the introduction of a non-existent factor g on the left of (27') to correspond to k in (27), so that (27') becomes

Derivation of Equations of Propagation.

$$\left(g + \frac{\mu}{4\pi}\frac{\partial}{\partial t}\right)a = -\frac{1}{4\pi}\left(\frac{\partial R}{\partial y} - \frac{\partial Q}{\partial z}\right) \\ \&c. \qquad \&c. \qquad \qquad \Bigg\} \quad \ldots \quad (27'')$$

Thus the quantities on the left of (27), which are the components of electric current, correspond to those on the left of (27''); and hence Heaviside has called the latter, from analogy, the magnetic currents, and g the magnetic conductivity.* By means of this analogy it is possible to obtain a set of electric theorems from the magnetic theorems given above, and we shall return to the subject later. As one result we are able to infer in the next section of this chapter from the value found for the electrokinetic energy, the existence of an exactly analogous quantity of electric energy.

Analogy between Electric and Magnetic Force.

By means of these equations we can at once investigate the propagation of electric and magnetic force. By differentiating the first of (27') with respect to the time, the second of (27) with respect to z, and the third of (27) with respect to y, subtracting the result of the last of these operations from that of the second, and having regard to the relation, $\partial a/\partial x + \partial \beta/\partial y + \partial \gamma/\partial z = 0$, we find

$$4\pi\mu k \frac{\partial a}{\partial t} + K\mu \frac{\partial^2 a}{\partial t^2} = \nabla^2 a \quad \ldots \quad (29)$$

and similarly we obtain two corresponding y and z equations. These are precisely equations (13)

In exactly the same way we obtain the equations

$$4\pi\mu k \frac{\partial P}{\partial t} + K\mu \frac{\partial^2 P}{\partial t^2} = \nabla^2 P \\ \&c. \qquad \&c. \qquad \Bigg\} \quad \ldots \quad (29')$$

---

* The reciprocal of this would be the specific magnetic resistance, or magnetic resistivity. This is the proper magnetic analogue of electric resistance, not that popularly regarded as such. See Heaviside "On the Self Induction of Wires," *Phil. Mag.* Aug. 1886. The reader should consult also, on the subject of this chapter, Heaviside's numerous papers in the *Phil. Mag.* and *Electrician passim*, during the last six or seven years.

which are the equations of propagation of electromotive force. The solutions and results given above (18) ... (24) are *mutatis mutandis* at once applicable.

Heaviside uses equations equivalent to (27) and (27′) above as the starting point of his very important and interesting researches regarding the propagation of electromagnetic waves. They have the advantage of giving the directly observable physical quantities E, H instead of the more abstract F, G, H, which are attended, as he puts it, by a parasitical ψ, the meaning of which is not clearly definable.

Equations (27) and (27′) are of the same form as those used by Hertz in his presentation of Maxwell's Theory, and, with $k = 0$, are those from which he deduced his solution of the problem of the radiation of electrical energy from a simple dumb-bell shaped vibrator.*

In the case of an æolotropic medium the resultant electric displacement is not generally in the direction of the resultant electromotive force. We shall suppose that the rectangular components of the displacement at any point are linear functions of the component electromotive forces, and write therefore

Perfectly Insulating Æolotropic Medium.

$$4\pi f = k_1 P + rQ + sR$$
$$4\pi g = r'P + k_2 Q + tR$$
$$4\pi h = s'P + t'Q + k_3 R$$

Components of Electric Displacement.

where the coefficients, apparently 9 in number, are constants. There are however in reality only 6 coefficients. For let f, g, h, be increased by small amounts df, dg, dh. The work done by the electromotive forces in effecting the displacements is $Pdf + Qdg + Rdh$, or, by (27), $(Pk_1 + Qr' + Rs')dP + (Pr + Qk_2 + Rt')dQ + (Ps + Qt + Rk_3)dR$, and this for conservation of energy must be a perfect differential, since the whole work in a complete cycle of changes must be zero. The conditions necessary for this are expressed by

Reduction of Number of Coefficients.

$$\frac{\partial(Pk_1 + Qr' + Rs')}{dQ} = \frac{d(Pr + Qk_2 + Rt')}{dP}$$

and two similar equations. These give $r' = r$, $s' = s$, $t' = t$, so that we have

$$\left.\begin{array}{l} 4\pi f = k_1 P + rQ + sR \\ 4\pi g = rP + k_2 Q + tR \\ 4\pi h = sP + tQ + k_3 R \end{array}\right\} \quad \ldots \quad (30)$$

Equations of Electric Displacement in Æolotropic Medium.

---

* Wiedemann's *Annalen*, No. 1, 1889. See also below Chap. XVI.

Now suppose the displacement to be taken in a direction in which it is parallel to the electromotive force, then since $4\pi f$, $4\pi g$, $4\pi h = KP, KQ, KR$, we obtain the equations

Determination of Principal Specific Inductive Capacities.

$$k_1 P + rQ + sR = KP$$
$$rP + K_2 Q + tR = KQ$$
$$sP + tQ + k_3 R = KR$$

which give the equation of condition

$$\begin{vmatrix} k_1 - K, & r, & s, \\ r, & k_2 - K, & t, \\ s, & t, & k_3 - K \end{vmatrix} = 0 \quad \ldots \ldots (31)$$

a cubic for K the three roots of which K_1, K_2, K_3, give three directions at right angles to one another having this property. Thus if l, m, n, be the cosines of one of these directions, equations just found may be written

$$\left. \begin{array}{c} k_1 l + rm + sn = Kl \\ \&c. \qquad \&c. \end{array} \right\} \quad \ldots \ldots (30')$$

from which by substitution of K_1, K_2, K_3 the corresponding values of l, m, n can be found, and it is easy to verify that the three directions specified by them are at right angles to one another (see p. 60 above). K_1, K_2, K_3 are called the principal specific inductive capacities of the medium.

We shall find the equations of wave propagation in such a medium on the supposition that the magnetic inductive capacity μ is constant. Taking as axes the directions of principal specific inductive capacity we obtain by the process used at p. 196 above

Equations of Wave Propagation in Æolotropic Medium.

$$\left. \begin{array}{c} \dfrac{\partial^2 F}{\partial t^2} = \dfrac{1}{K_1 \mu} \left(\nabla^2 F - \dfrac{\partial J}{\partial x} \right) \\ \dfrac{\partial^2 G}{\partial t^2} = \dfrac{1}{K_2 \mu} \left(\nabla^2 G - \dfrac{\partial J}{\partial y} \right) \\ \dfrac{\partial^2 H}{\partial t^2} = \dfrac{1}{K_3 \mu} \left(\nabla^2 H - \dfrac{\partial J}{\partial z} \right) \end{array} \right\} \quad \ldots \ldots (32)$$

Here the reasons assigned at p. 197 above for disregarding J do not hold, and it must be retained in the equations. The quantity ψ however is not a function of the time and does not appear.

DOUBLE REFRACTION IN ÆOLOTROPIC MEDIUM

Now consider a plane wave travelling at right angles to the plane $lx + my + nz = 0$, with velocity V. Then the disturbance may be expressed by the equation

Solution of Differential Equations.

$$(F, G, H) = (F_0, G_0, H_0) \exp. \frac{2i\pi}{\lambda}(lx + my + nz - Vt).$$

Writing $v_1^2, v_2^2, v_3^2 \equiv 1/K_1\mu, 1/K_2\mu, 1/K_3\mu$, we get, by substitution of these values of F, G, H in (31), equations which may be put in the form

$$\begin{rcases} (V^2 - v_1^2)F + v_1^2 l(lF + mG + nH) = 0 \\ (V^2 - v_2^2)G + v_2^2 m(lF + mG + nH) = 0 \\ (V^2 - v_3^2)H + v_3^2 n(lF + mG + nH) = 0 \end{rcases} \quad . \quad (33)$$

These divided by $V - v_1^2$, &c., multiplied by l, m, n, and added, give

$$\frac{v_1^2 l^2}{V^2 - v_1^2} + \frac{v_2^2 m^2}{V^2 - v_2^2} + \frac{v_3^2 n^2}{V^2 - v_3^2} = -1.$$

Relation of Velocity of Propagation to Direction of Ray.

This may be written, since $l^2 + m^2 + n^2 = 1$,

$$\frac{v_1^2 l^2}{V^2 - v_1^2} + l^2 + \frac{v_2^2 m^2}{V^2 - v_2^2} + m^2 + \frac{v_3^2 n^2}{V^2 - v_3^2} + n^2 = 0$$

or

$$\frac{l^2}{V^2 - v_1^2} + \frac{m^2}{V^2 - v_2^2} + \frac{n^2}{V^2 - v_3^2} = 0 \quad . \quad . \quad . \quad (33')$$

which is Fresnel's equation connecting the velocity of light in an æolotropic body with the direction of propagation, when v_1^2, v_2^2, v_3^2 are the squares of the principal velocities along the axes. These are the velocities of an *ordinary* wave in these directions, that is the velocities in isotropic media characterized by the constants K_1, K_2, K_3.

Double Refraction.

The direction of the electric displacement may be found as follows. Let λ, μ, ν be its direction cosines. The components of current are proportional to $K_1\dot{P}, K_2\dot{Q}, K_3\dot{R}$, that is to $\ddot{F}/v_1^2$, $\ddot{G}/v_2^2, \ddot{H}/v_3^2$, or by (33) to $l/(V^2 - v_1^2), m/(V^2 - v_2^2), n/(V^2 - v_3^2)$. Hence by (33') $l\lambda + m\mu + n\nu = 0$, or the displacement is in the plane of the wave front.

Let the direction of propagation be the axis of z, then in (33') above we must put $n = 1, l = m = 0$. If the direction of displacement be that of $y, \mu = 1, \lambda = \nu = 0$. But since $\mu \propto m/(V^2 - v_2^2)$, and $m = 0$, we must have $V = v_2$. Similarly

Ordinary and Extraordinary Waves.

when the displacement is in the direction of x, $V = v_1$. In the former case we have an *ordinary* wave travelling along z. But by optical experiment it is known that an *ordinary* ray of light travelling parallel to a principal axis of a crystal, for example at right angles to the optic axis of a uniaxal crystal, such as Iceland spar, is polarized in a plane parallel to the axis and to the direction of propagation. Hence the electromagnetic theory gives the direction of electric displacement as at right angles to the plane of polarization. This conclusion has been verified by the experiments of Hertz.

We have seen that the velocity of propagation of the disturbance through an insulating medium is $1/\sqrt{K\mu}$. The numerical value of this velocity for a medium such as air has been obtained in the following manner. If the units of measurement are electromagnetic the value of μ for air is unity (see Chap. XII. below). Now using electrostatic units we have for the force upon a point-charge of electricity q placed in the field due to another point-charge q', at a distance r from the former, is $qq'/K_s r^2$, if K_s be the value of the specific inductive capacity of the medium in electrostatic units. Now changing to electromagnetic units and putting Q, Q', K for the corresponding quantities, we obtain for the numerical expression of the same force as before QQ'/Kr^2. Hence

$$\frac{qq'}{K_s r^2} = \frac{QQ'}{Kr^2}$$

and therefore for air, since then $K_s = 1$,

$$\frac{1}{K} = \frac{qq'}{QQ'} = v^2 \quad \ldots \quad \ldots \quad (34)$$

Ratio of Electromagnetic Unit to Electrostatic Unit of Quantity = Velocity of Light.
if v denote the ratio of the numerical expression of a given quantity of electricity in electrostatic units, to the numerical expression of the same quantity in electromagnetic units. Thus v is numerically equal to the velocity of propagation of an electromagnetic disturbance in a medium, which has specific inductive capacity in electrostatic units, and magnetic inductive capacity in electromagnetic units, each equal to unity, or such that their product is unity. The value of v has been experimentally determined by a number of experimenters working in several different ways with strikingly concurrent results. A complete table of the principal results, with the names of the experimenters is given in Chapter XIII., where the methods of measurement are described. Here however

we have to note the remarkable fact that the velocity of propagation thus given for an electromagnetic disturbance, almost exactly coincides with the velocity of light! Maxwell, indeed, regarded light as an electromagnetic disturbance in the luminiferous ether, and put forward his theory as one of the propagation of light pure and simple; and his views have since been strikingly confirmed by experiment (see Chapter XVI.).

Section II

ELECTROSTATIC AND ELECTROKINETIC ENERGY OF THE MEDIUM. MOTION OF ENERGY IN THE ELECTRO-MAGNETIC FIELD

WE have now to consider the energy of the electromagnetic field. It consists of two parts, the electric energy, depending on the electromotive force, however produced, at each point of the field, and the magnetic or electrokinetic energy. The former which we denote by E_e is given (Vol. I. p. 33) by the equation

Energy of the Electromagnetic Field.

$$E_e = \iiint \frac{K}{8\pi}(P^2 + Q^2 + R^2)\,dx\,dy\,dz \quad . \quad . \quad (35)$$

the integral being taken throughout the whole field, that is wherever P, Q, R differ from zero. Here it is assumed that P, Q, R are the total component electric forces in the field, and that these used in the formula for the electrostatic energy give the electric energy. This point requires examination. We shall, however, infer the existence of the amount of electric energy presently from the analogy between electric and magnetic force.

Electrostatic Energy. Calculation of Electrokinetic Energy. First Method.

A similar expression can be found for the electromagnetic energy T. We have seen that

$$T = \tfrac{1}{2}\Sigma \dot{y}_k N_k$$

where N_k is the total magnetic induction through any circuit in which a current $\dot{y}_k$ is flowing. Now if F, G, H be the components of vector potential at the circuit, we have by (1) above

$$N_k = \int \left(F\frac{dx}{ds} + G\frac{dy}{ds} + H\frac{dz}{ds}\right)ds \quad . \quad . \quad (36)$$

where the integral is taken **round the circuit in** which $\dot{y}_k$ flows. If instead of any current $\dot{y}$ we **substitute its** components along the **axes** we get

$$T = \tfrac{1}{2}\Sigma \iiint (Fu + Gv + Hw)dxdydz$$

where the integral is extended throughout the whole substance of the current carrying conductor. If the integration is regarded as extending over the whole space in which currents exist we omit the symbol of summation, and write

$$T = \tfrac{1}{2}\iiint (Fu + Gv + Hw)dxdydz \quad \ldots \quad (37)$$

Now we have found that

$$u = \frac{1}{4\pi}\left(\frac{\partial \gamma}{\partial y} - \frac{\partial \beta}{\partial z}\right), \quad a = \frac{\partial H}{\partial y} - \frac{\partial G}{\partial z}$$
$$\&c. \qquad\qquad \&c.$$

Hence substituting in (37), rearranging, and integrating within a closed surface, we get

$$T = \frac{1}{8\pi}\iiint (a\alpha + b\beta + c\gamma)dxdydz$$
$$+ \frac{1}{8\pi}\iint \{l(H\beta - G\gamma) + m(F\gamma - Ha) + n(Ga - F\beta)\}dS \ . \ (38)$$

The surface integral vanishes when the surface is taken infinitely far from the system of currents, for, in the limit, if r denote the distance of dS from any point of the system of currents, a, β, γ are of the order of magnitude $1/r^3$, while dS is of the order r^2. Hence if, as supposed, the integral is taken throughout all space,

Value of Electrokinetic Energy.

or

$$\left.\begin{array}{l} T = \dfrac{1}{8\pi}\iiint (a\alpha + b\beta + c\gamma)dxdydz \\[6pt] T = \dfrac{\mu}{8\pi}\iiint (a^2 + \beta^2 + \gamma^2)dxdydz \end{array}\right\} \quad \ldots \quad (39)$$

If, according to Ampère's theory, we suppose all magnetic force to be due to systems of currents, T will include the

CALCULATION OF ELECTROKINETIC ENERGY

magnetic energy in the case of permanent steel magnets as well as that due to ordinary voltaic currents, or to displacement currents.

From this result we infer by the analogy between electric and magnetic force contained in equations (27), (27') above, the existence of the electric energy stated in (35).

This expression for T may also be found as follows. By the theorem, p. 113 above, as to the work done in carrying a unit pole in a closed path round a circuit, we have

Calculation of Electrokinetic Energy Second Method.

$$4\pi \dot{y}_k = \int \mathbf{H}_k \cos \theta_k ds$$

where the integration is taken round any closed path embracing the chosen circuit, $\mathbf{H}_k$ is the magnetic force *due to the current in that circuit*, and θ_k the angle which $\mathbf{H}_k$ makes with ds an element of the path. Hence drawing any surface in the field so as to form a cap with the circuit as bounding edge, and remembering that by the solenoidal condition fulfilled by the magnetic induction, the total induction through every such surface is the same, we have

$$N_k \dot{y}_k = \frac{1}{4\pi} \int N_k \mathbf{H}_k \cos \theta_k ds \quad \ldots \quad (40)$$

Now let the surface be taken at right angles to the lines of induction everywhere. These lines are closed curves round the conductors, and each threads through one or more of the circuits. It is possible to divide up the whole field by successive surfaces, each having for bounding edge *any given circuit*, so that every one of these surfaces shall be everywhere at right angles to the lines of induction. Every one, if it cut through a system of closed lines surrounding any one circuit, will pass through every point of that circuit. Of course no one of the closed tubes of induction, which the surface thus cuts through, contributes anything to the total induction through the surface.

Let then the direction of the closed curve, round which the integral of $\mathbf{H} \cos \theta ds$ is taken, be everywhere at right angles to these surfaces, and let $\mathbf{B}$ be the value of the induction at any point where this curve cuts one of these surfaces. Then, if dS be the area of an element of the surface at that point, the induction through it is $\mathbf{B}dS$. Thus we get

$$N_k \dot{y}_k = \frac{1}{4\pi} \iint \mathbf{B} \mathbf{H}_k \cos \theta_k ds dS \quad \ldots \quad (41)$$

where one integral is taken **over the surface** and the other round the closed curve. But this **is evidently the same** thing as

$$N_k \dot{y}_k = \frac{1}{4\pi} \int\!\!\int\!\!\int_{-\infty}^{+\infty} \mathbf{BH}_k \cos \theta_k \, dxdydz \quad \ldots \quad (42)$$

where x, y, z are the coordinates of the point at which the induction is $\mathbf{B}$, and $dxdydz$ is an element of volume.

Finally, take any point on one of these surfaces and let $\mathbf{B}$ be the induction there. For *every* such point a surface can be drawn having any one of the circuits as its boundary, and hence by (30) of Chapter IV. and (42) above, we have the equation

$$T = \frac{1}{8\pi} \int\!\!\int\!\!\int_{-\infty}^{+\infty} \mathbf{B}(\mathbf{H}_1 \cos \theta_1 + \mathbf{H}_2 \cos \theta_2 + \&c.) \, dxdydz \ . \quad (43)$$

But if $\mathbf{H}$ be the total magnetic force at the point, and θ the angle which it makes with the normal to the surface, we have

$$\mathbf{H} \cos \theta = \mathbf{H}_1 \cos \theta_1 + \mathbf{H}_2 \cos \theta_2 + \&c.$$

Hence (43) becomes

$$T = \frac{1}{8\pi} \int\!\!\int\!\!\int_{-\infty}^{+\infty} \mathbf{BH} \cos \theta \, dxdydz \quad \ldots \quad (44)$$

which is identical with (39).

Electro-kinetic Energy in Medium Isotropic as to Magnetic Quality.

If we suppose, what is always the case in an isotropic medium, that $\mathbf{B}$ and $\mathbf{H}$ have **the same** direction, we have

$$T = \frac{1}{8\pi} \int\!\!\int\!\!\int_{-\infty}^{+\infty} \mathbf{BH} \, dxdydz$$

$$= \frac{1}{8\pi} \int\!\!\int\!\!\int_{-\infty}^{+\infty} \mu \mathbf{H}^2 \, dxdydz = \frac{1}{8\pi} \int\!\!\int\!\!\int_{-\infty}^{+\infty} \frac{1}{\mu} \mathbf{B}^2 \, dxdydz \ . \quad (44')$$

since $\mu = \mathbf{B}/\mathbf{H}$.

The expressions given in (35) and (39) for the electric and magnetic energies suggest at once the idea that energy may be

DISSIPATION OF ENERGY IN MAGNETIC CYCLE 211

distributed throughout space, so that at any given point where the electric force is **E**, the magnetic induction **B**, and the electric and magnetic inductive capacities K and μ, there is in the medium there per unit of volume $K\mathbf{E}^2/8\pi$ of electric energy and $\mathbf{B}^2/8\pi\mu$ of magnetic (or electrokinetic) energy. It is to be observed, however, that this localization is a matter of assumption.

Localization of Energy in the Medium.

We may now prove an important proposition as to the energy dissipated in putting a piece of magnetizable matter through a closed cycle of magnetization. It is found by experiment (see Chap. XIV.) that if the magnetization of a piece of iron is carried through a closed cycle beginning and ending with the same values of **B** and **H**, the curve formed by plotting the successive values of **H** as abscissæ, and the corresponding values of **B** as ordinates, forms a loop; that is to say the values of **B**, corresponding to the same values of **H**, are not the same in the forward and backward portions of the curve. This indicates as we shall see dissipation of energy in the iron to an amount equal to the area of the loop divided by 4π.

Energy Dissipated in Closed Cycle of Magnetization.

Let the magnetism be produced by a current flowing in a given circuit, and let the total induction through that circuit be increased by an amount dN, produced by increasing the current in that circuit. The energy drawn from the battery over and above that dissipated is (small quantities of the second order neglected) $\dot{y}dN$, by (33) of Chap. IV.; and clearly by the process used above to find (44) we may prove that

$$\dot{y}dN = \frac{1}{4\pi}\int\!\!\int\!\!\int_{-\infty}^{+\infty} \mathbf{H}d\mathbf{B}\,dxdydz . \quad . \quad . \quad . \quad (45)$$

The change of electrokinetic energy dT is given by

$$dT = \frac{1}{8\pi}\int\!\!\int\!\!\int_{-\infty}^{+\infty} (\mathbf{H}d\mathbf{B} + \mathbf{B}d\mathbf{H})\,dxdydz . \quad . \quad . \quad (46)$$

Hence for the energy spent *otherwise* than in increasing the electrokinetic energy in the medium we have by (45) and (46)

$$\dot{y}dN - dT = \frac{1}{4\pi}\int\!\!\int\!\!\int_{-\infty}^{+\infty} \{\mathbf{H}d\mathbf{B} - \tfrac{1}{2}(\mathbf{H}d\mathbf{B} + \mathbf{B}d\mathbf{H})\}dxdydz \quad . \quad (47)$$

P 2

Expression for Total Energy Dissipated in Closed Cycle.

If the magnetization be carried through a closed cycle, so that the medium is brought back to the same state as at first, the electrokinetic energy returns to the same value, and the integral of the quantity within the inner brackets in (47) which is $d(\mathbf{BH})$ is zero. Thus the energy furnished to the medium in the closed cycle is

$$\frac{1}{4\pi}\int\int\int_{-\infty}^{+\infty}\left\{\int \mathbf{H}d\mathbf{B}\right\}dx\,ly\,dz\quad\ldots\quad(48)$$

where the inner integral is taken with respect to $\mathbf{B}$ round the cycle.

If we take the changes per unit of volume at a place where the induction is $\mathbf{B}$ and the magnetic force $\mathbf{H}$, we have for the energy given to the medium, the value $\mathbf{H}d\mathbf{B}/4\pi$, and for the

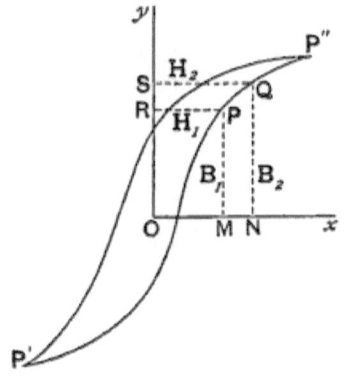

Fig. 40.

increase of electrokinetic energy $d(\mathbf{BH})/8\pi$. Therefore the energy dT' spent otherwise than in increasing the electrokinetic energy is given by the equation

Energy Dissipated per Unit of Volume in Closed Cycle.

$$dT' = \frac{1}{4\pi}\mathbf{H}d\mathbf{B} - \frac{1}{8\pi}(\mathbf{H}d\mathbf{B} + \mathbf{B}d\mathbf{H})\quad\ldots\quad(49)$$

Hence if PQ (Fig. 39) be two points in a curve of which the abscissæ are values of $\mathbf{H}$, and the ordinates values of $\mathbf{B}$, we have for the whole energy T' spent otherwise than in increasing T, in this part of the curve the value

$$T' = \frac{1}{4\pi} \left\{ \int_{\mathbf{B}_1}^{\mathbf{B}_2} \mathbf{H} d\mathbf{B} - \tfrac{1}{2} \int_{\mathbf{B}_1}^{\mathbf{B}_2} (\mathbf{H} d\mathbf{B} + \mathbf{B} d\mathbf{H}) \right\} \quad . \quad (50)$$

that is by Fig. 39

$$T' = \frac{1}{4\pi} (\text{area } PQSR - \tfrac{1}{2} \text{ area } NQSRPM).$$

The second area vanishes when Q coincides with P as it does when the curve forms a closed cycle, for example the loop $P'P''$.

If the magnetic permeability is a constant there is no loop, and thus the energy dissipated in any cycle of changes is zero.

On the assumption made above that the electric and magnetic energies may be regarded as stored up in the medium, so that at any point the amount of electric energy is $K\mathbf{E}^2/8\pi$, and of magnetic energy $\mathbf{HB}/8\pi$, per unit of volume in each case, we may inquire how the energy moves in the electromagnetic field when any change takes place. We have, supposing that K and μ do not vary with the time

Motion of Energy in Electromagnetic Field.

$$\frac{d(\mathbf{E}_e + T)}{dt} = \frac{1}{4\pi} \iiint \left\{ K\left(P\frac{\partial P}{\partial t} + Q\frac{\partial Q}{\partial t} + R\frac{\partial R}{\partial t}\right) \right. $$
$$\left. + \left(a\frac{\partial a}{\partial t} + \beta\frac{\partial b}{\partial t} + \gamma\frac{\partial c}{\partial t}\right) \right\} dx\,dy\,dz \quad . \quad (51)$$

But if u, v, w, p, q, r, be the components of the total current and the conduction current respectively (the latter, in the general case, not merely generating heat in the conductor), we have for those of the displacement current

$$\frac{K}{4\pi}\frac{\partial P}{\partial t} = u - p = \frac{1}{4\pi}\left(\frac{\partial \gamma}{\partial y} - \frac{\partial \beta}{\partial z}\right) - p$$

$$\frac{K}{4\pi}\frac{\partial Q}{\partial t} = v - q = \frac{1}{4\pi}\left(\frac{\partial a}{\partial z} - \frac{\partial \gamma}{\partial x}\right) - q$$

$$\frac{K}{4\pi}\frac{\partial R}{\partial t} = w - r = \frac{1}{4\pi}\left(\frac{\partial \beta}{\partial x} - \frac{\partial a}{\partial y}\right) - r.$$

Also if we write equations (5) in the form

$$P, Q, R = P' + c\dot{y} - b\dot{z},\ Q' + a\dot{z} - c\dot{x},\ R' + b\dot{x} - a\dot{y}. \quad . \quad (52)$$

we obtain by differentiation, remembering that $\partial a/\partial t$, $\partial b/\partial t$, $\partial c/\partial t$, denote time-rates of variation at a given point of space the coordinates of which are not functions of the time,

$$\frac{\partial a}{\partial t} = \frac{\partial}{\partial t}\left(\frac{\partial H}{\partial y} - \frac{\partial G}{\partial z}\right) = \frac{\partial Q'}{\partial z} - \frac{\partial R'}{\partial y}$$

$$\frac{\partial b}{\partial t} = \frac{\partial}{\partial t}\left(\frac{\partial F}{\partial z} - \frac{\partial H}{\partial x}\right) = \frac{\partial R'}{\partial x} - \frac{\partial P'}{\partial z}$$

$$\frac{\partial c}{\partial t} = \frac{\partial}{\partial t}\left(\frac{\partial G}{\partial x} - \frac{\partial F}{\partial y}\right) = \frac{\partial P'}{\partial y} - \frac{\partial Q'}{\partial x}$$

Substituting in (51) the values of $K(\partial P/\partial t)/4\pi$, &c., $\partial a/\partial t$, &c., just found, and using (52) we find after rearrangement

$$\frac{d(\mathsf{E}_e + T)}{dt} = \frac{1}{4\pi}\iiint \left\{ \frac{\partial}{\partial x}(R'\beta - Q'\gamma) + \frac{\partial}{\partial y}(P'\gamma - R'a) \right.$$
$$\left. + \frac{\partial}{\partial z}(Q'a - P'\beta) \right\} dxdydz$$

$$- \iiint \{(cv - bw)\dot{x} + (aw - cu)\dot{y} + (bu - av)\dot{z}\} dxdydz$$

$$- \iiint (Pp + Qq + Rr) dxdydz \quad . \quad . \quad . \quad (53)$$

Time-rate of Change of Energy within Closed Space

After transposition and substitution of the components (p. 119) X, Y, Z, of electromagnetic force for $cv - bw$, $aw - cu$, $bu - av$, this gives by integration over the closed surface

$$\frac{d(\mathsf{E}_e + T)}{dt} + \iiint (X\dot{x} + Y\dot{y} + Z\dot{z}) dxdydz$$
$$+ \iiint (Pp + Qq + Rr) dxdydz$$
$$= \frac{1}{4\pi}\iint \{l(R'\beta - Q'\gamma) + m(P'\gamma - R'a)$$
$$+ n(Q'a - P'\beta)\} dS \quad . \quad . \quad (54)$$

where l, m, n are the direction cosines of the normal to the surface element, drawn outwards.

If l', m', n' be the direction cosines of a line at right angles to the plane defined by the resultant $\mathbf{E}'$ of the component electric forces P', Q', R', and the resultant magnetic force $\mathbf{H}$, and drawn in the direction in which a right-handed screw would move if the handle were turned round in the plane of **E**′ and **H**, from the direction of **H** to that of **E**′, the element of the surface integral on the right has the value $HE' \sin\theta(ll' + mm' + nn')/4\pi$, which shows that the rate of flow of energy per unit of area is represented in magnitude and direction by the vector product of **H** and **E**′ (that is the **vector** $HE' \sin\theta$ at right angles to the plane of **H** and **E**′), divided by 4π. The component of flow across unit of area at right angles to the direction (l', m', n') is thus $HE' \sin\theta(ll' + mm' + nn')/4\pi$. The direction of flow is

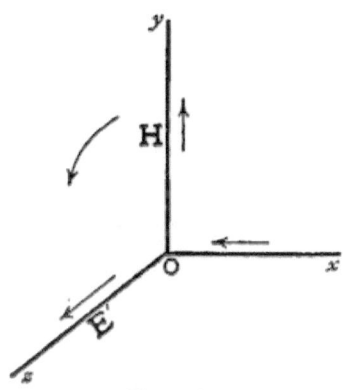

Fig. 41.

opposite to that in which the line (l', m', n') is drawn as specified above. Thus in the diagram, if the resultant **H** be along Oy, and **E**′ along Oz, the flow of energy is in the direction xO.

It is to be remarked that the first term on the left of (54) is the rate of increase of the electric and magnetic energies within the closed space, the second term the rate at which work is done by electromagnetic forces, and the third the rate at which energy is dissipated, and the rate at which it is expended in effecting chemical changes, and the equation asserts that these rates combined are equal to that of the flow of energy across the surface given by the integral on the right-hand side. This very important theorem is due to Professor J. H. Poynting.*

Margin notes: Expressed as of Energy Flow across Bounding Surface. Interpretation of Result.

---

* *Phil. Trans. R.S.* Part VI. 1885. The following examples of the theory and Figs. 42—47 are taken from this paper.

It must be noticed that the addition of any term $(l\phi + m\chi + n\psi)dS$, of proper dimensions, to the element of the integral would not alter the value of the integral over the closed surface provided ϕ, χ, ψ are functions of the coordinates fulfilling the condition $\partial\phi/\partial x + \partial\chi/\partial y + \partial\psi/\partial z = 0$. That the actual flow across a given element of the closed surface is that stated above is not strictly demonstrated, but the assumption that it is, is here made. The following examples, agreeing as they do with known facts, so far bear out the assumption.

Examples of Flow of Energy.
1. Long Straight Wire carrying a current.

As an example consider the flow of energy at the surface of a long straight wire of circular section of radius r, in which a

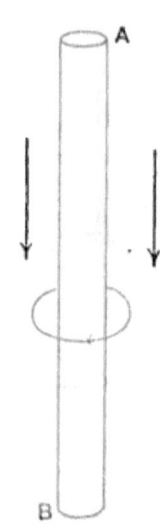

Fig. 42.

steady current of strength γ is flowing. Since the displacement is maintained constant there is no displacement current, and the direction of the magnetic force (shown in Fig. 42) is tangential to a normal cross-section of the conductor and of amount $2\gamma/r$; that of the electric force is parallel to the conductor, and of amount equal to the current per unit of area divided by the specific conducting power of the conductor, that is $\gamma/\pi r^2 k$. The rate of flow of energy across unit of area is thus $\{(2\gamma/r) \times (\gamma/\pi r^2 k)\}/4\pi = \gamma^2/2\pi^2 r^3 k$, and the direction of flow is, by

ILLUSTRATIONS OF FLOW OF ENERGY

the rule found above, inwards from the surrounding medium to the wire. The rate of flow of energy, inwards upon unit length of the wire, is thus $\gamma^2/2\pi r^2 k$, and across l units of length is $\gamma^2 l/\pi r^2 k$, or $\gamma^2 R$ if R is the resistance of the length l of the wire. This is the well-known amount of energy transformed into heat in the wire per unit of time.

According to this theory our conception of a current in a conductor must not involve any notion of the transport of energy actually along the conductor. The manner of arrangement of the electric and magnetic equipotential surfaces in the field is conditioned by the existence and position of the conducting wire, which therefore also controls the flow of energy. Thus in a metallic conductor there is dissipation of energy received from the medium; and further, if at any place electrical energy is utilised in doing work, this, by the theory, does not come along the conductor but from the surrounding medium, being guided to the required place by the conformation of the equipotential surfaces produced by the conductor.

New Conception of Current in a Wire.

All Energy used moves from Source through Medium and not through Conductor.

Consider as another example the case of a charged condenser the plates of which are connected by wires to another pair of plates, so that the capacity of the condenser is increased. The tubes of force, formerly existing for the most part in the portion of the medium between the conducting plates of the original condenser, and entirely depending for their arrangement on these plates, which thus localised the energy of strain in the medium, have moved out sideways with their ends on the connecting conductors until the state of strain has been set up between the other pair of plates. Along with this motion is produced magnetic action in the medium which dies away as the motion of the tubes of force ceases.

Examples. 2. One Condenser discharging into another.

As however the tubes of force move out guided by the conductors, points on the wire connecting the insulated plates must be at potentials intermediate between the potentials of these plates, and so some of the equipotential surfaces which pass between the plates of the original condenser must cut the conducting wire. Thus there is, by the theory given above, a flow of energy into the conducting wire, which, dissipated in its substance, gives the heat commonly said to be produced by the current.

This view of the matter seems exceedingly reasonable. The medium was, we know, in a state of strain between the plates of the original condenser, and must therefore have contained a corresponding quantity of energy; and this state of strain has been communicated to the medium between the plates of the other condenser. The result has been a diminution in the

218 GENERAL ELECTROMAGNETIC THEORY

Energy moves laterally into Guiding Conductor where Strain spreads in Medium.

intensity of the strain between the plates of the original condenser, or, as we say, the electric force at each point of the medium, or, otherwise, the difference of potential between the plates, has been diminished. Now it is very difficult to suppose that the energy has been transferred along the conductor to the other condenser, and there inserted as energy of strain between the plates. In fact we know that, *pari passu* with the process of charging the other condenser, has gone on the growth of electric strain in the medium between the conductors; and since

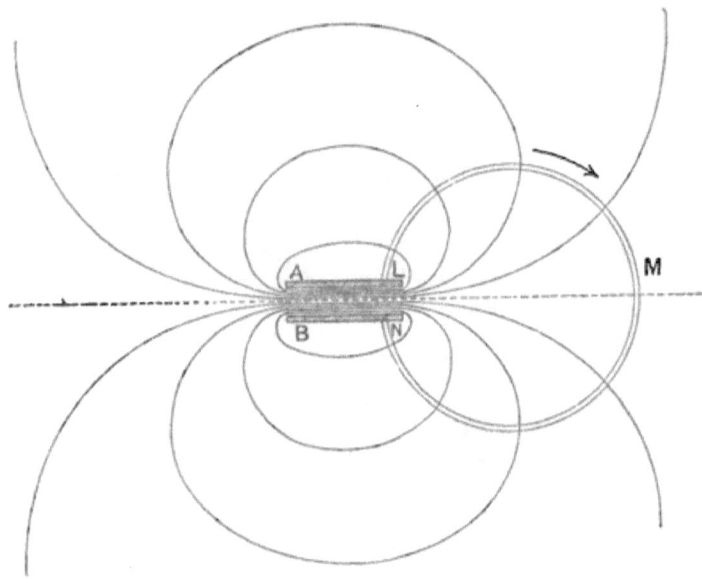

FIG. 43.

the energy is undoubtedly stored up in the strain of the medium, it is rational to regard the strain as being propagated through the medium under the guidance of the conductors which as we have seen localise it.

3. *Condenser Discharging through Wire.*

If the condenser have, as shown in Fig. 43, its plates A, B connected by a wire LMN, the process of discharge will consist in the passage outwards of the tubes of force with their ends on the wire at each instant and everywhere at right angles to the equipotential surfaces for the time being. The flow of energy

ILLUSTRATIONS OF FLOW OF ENERGY

is parallel to these surfaces, or rather parallel alike to the electric and the magnetic equipotential surfaces. The tubes of electric force sink into the conductor, and thus shorten as they advance, until at a point on the conductor midway between the plates as regards potential, the last remnant disappears laterally into the conductor. The magnetic tubes of force which encircle the conductor contract down upon it and disappear within it, giving up also their energy as they do so.

Next consider a single voltaic cell, for simplicity a cell consisting of plates of copper and zinc immersed in a dilute acid and connected by a wire. The arrangement of the equipotential surfaces is shown in Fig. 44. All these surfaces are there shown

4. Voltaic cell with Plates connected by Wire.

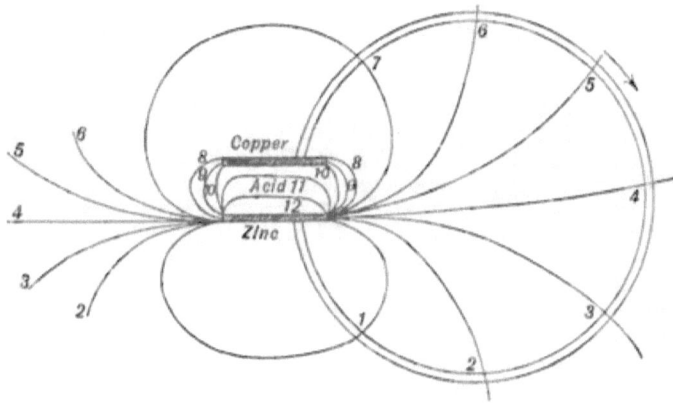

Fig. 44.

passing between the zinc plate and the acid, and closed by returning, some (of which specimens are numbered 1—7) so as to intersect the conducting wire, others (8—10) between the copper and the acid, and the remainder through the acid. Now according to the contact theory of voltaic action there exists a certain difference of potential on the two sides of each of the separating surfaces of dissimilar substances. According to the theory of the motion of energy there ought to be at every such surface a gain or loss of energy, from or to the surrounding medium. If going round the circuit in the direction in which the current (in the ordinary sense) flows, an electric force is encountered at any such surface opposed in direction to the current, the impressed electromotive force, whatever its origin,

Seat, according to this Theory, of Electromotive Force in Voltaic Circuit.

does work there and sends energy out from that part of the circuit to the medium. Such a surface we have between the zinc and the acid in the cell. The line integral of the electric force across the surface is the difference of potential V on its two sides and the line integral of the magnetic force round the conductor is $4\pi\gamma$. Hence by the rule (p. 215 above) the energy sent out per unit of time from this part of the circuit is $V \times 4\pi\gamma/4\pi = V\gamma$, which is the well-known rate of working against the opposing difference of potential. Part of this passes *into* the connecting wire which forms the external resistance, since there the electric force is *with* the current and is converted into heat; part is sent into the surface separating the copper from the acid, and goes to supply the potential energy required for the separated hydrogen given off at the copper plate; the remainder passes to the acid to be dissipated in heat or used in effecting any chemical change which may there take place. It is to be observed that, as before, the quantity dissipated in heat in the conductor is equal to the line integral of the electric force along it (that is the difference of potential between its extremities) multiplied by $1/4\pi$ of the line integral of the magnetic force round the conductor, that is by γ.

Bearing of this Theory on Contact Difference of Potential between Metals.

It must be noticed that the theory appears to negative the view that there exists on the two sides of the surface of separation of the zinc and copper a finite difference of potential. For if the theory be true there is no real transference of energy along the conductor, and the energy, if any, received at the junction between the zinc and copper ought to produce there some effect. The drop in potential from zinc to copper shown by ordinary contact methods is so considerable, that, on the supposition of its reality, the energy there received should form a large proportion of the total energy given out elsewhere. No effect is however observed. This result does not, in view of the observed contact difference of potentials, necessarily decide against the theory of flow of energy, as the apparent contact difference between zinc and copper involves zinc-air and air-copper contact differences of potential which as yet we have no means of directly measuring.

5. Generator and Motor in Same Circuit.

If work is done at any part of the circuit otherwise than in generating heat, as for example in driving a motor, or in effecting chemical decomposition, there is a flow of energy to that part of the circuit corresponding to the difference of potential which exists between its terminals. Such a case may be taken as illustrated by the equipotential surfaces 4, 5, 6 shown in Fig. 45, proceeding to one part of the circuit from that part through which pass all the equipotential surfaces, and which is the source of the energy. Along these surfaces energy moves from the source and flows into the conductor in amount equal to that locally used up.

In reality Fig. 44 illustrates a thermoelectric circuit of two metals A (say copper) and B (iron). One junction is hot, the other cold. At the hot junction where the current flows from copper to iron heat is absorbed, and at the cold junction heat is evolved. In the rest of the circuit heat is generated according to the ordinary law, and it is supposed that the Thomson effect is zero, that is, that there is neither absorption nor evolution of heat where the current flows from cold to hot or from hot to cold in the metals. Thus the equipotential surfaces 1, 2, &c.,

6. Thermoelectric Circuit of Two Metals.

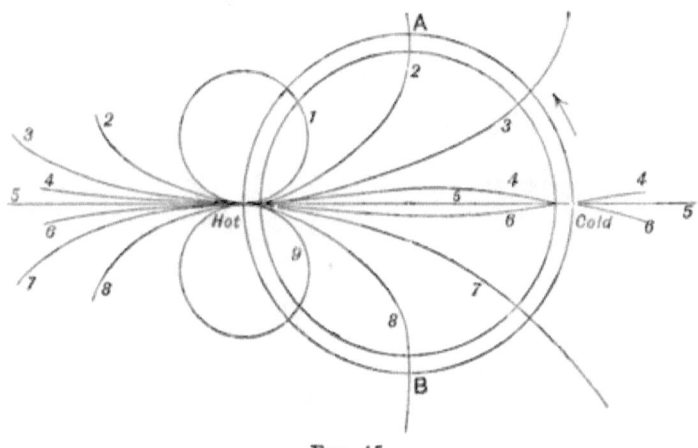

FIG. 45.

pass out from the hot junction and intersect the conductor, a number of them concentrating on the cold junction where there is a finite step of potential downward from the iron to the copper. The Peltier effect is illustrated in Fig. 46 which shows a circuit of two metals A (copper) and B (iron), as before, but with a voltaic cell placed in the latter part of the circuit. The junctions apart from the effects of the current are supposed to be at the same temperature. At the junction D, which corresponds to the former cold junction, heat is evolved, at the other, C, heat is absorbed. Thus the junction D is heated, and the junction C is cooled when the current passes, provided, of course, the temperatures be below the neutral temperature for

7. Circuit of Two Metals, one of which includes Voltaic Cell.

222 GENERAL ELECTROMAGNETIC THEORY

Bearing of this Theory on Peltier effect.

8. Thermoelectric Circuit of Iron, and Two Metals of Lead Type.

the metals in question (see Chap. X.). There is a finite step down of potential from B to A at D, and a finite step up of potential from A to B at C. Thus the equipotential surfaces which converge on D intersect the circuit only once after leaving the source, while those which pass through C intersect it three times, once at C, and once on each of the two sides of C. Energy flows out from the cell and from C along these surfaces, and is

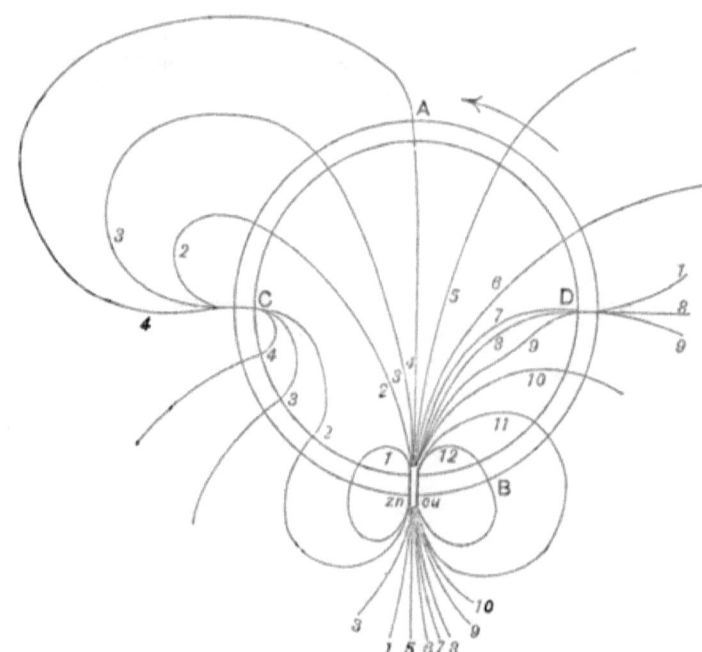

Fig. 46.

received by the conductor at each of the other points of intersection.

Lastly, a very interesting case is shown in Fig. 47. AB, CD are supposed to be two metals in which no Thomson effect occurs, that is, two metals of the lead type, while BC joining them is a metal of the iron type in which a current flowing from hot to cold absorbs heat. B is a hot junction, C a cold junction, each supposed to be at the neutral temperature for the pair of metals in contact. **Thus if a current** flow from a battery in the circuit there is no convergence of equipotential surfaces upon *B* or *C*, and neither absorption nor evolution **of heat** at those places. If the resistance of *BC* be sufficiently small there will be a gradual rise of potential in the metal from *B* to *C*, and the gradient of electromotive force being opposed to the current, heat will be absorbed there, transformed into electric energy, and carried out through the medium to the conductor where it is dissipated. Thus the equipotential surfaces which pass through *BC* will cut the circuit twice elsewhere (besides at the battery), once beyond *B*, and once beyond *C*. In consequence however of the resistance of the conductor, there is a fall of potential from *B* to *C* superimposed on the rise produced by the Thomson effect and a corresponding evolution of heat; **and it is** possible

Bearing of this Theory on Thomson effect.

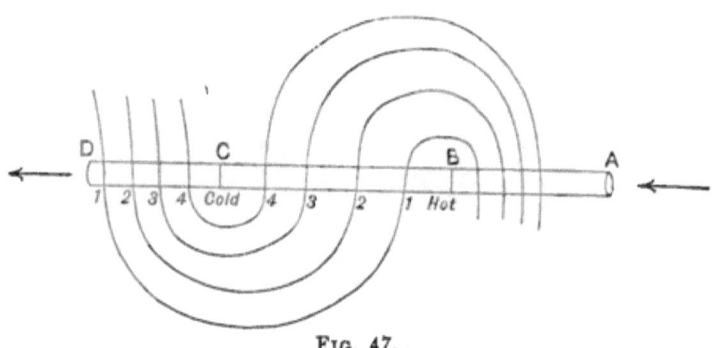

Fig. 47.

so to adjust matters **that the rise shall just balance the fall of** potential; and the evolution of heat just balance the absorption at every point. In that case we should have the curious case of a homogeneous conductor throughout at one potential with a current flowing through it. It is to be observed however that there is a gradient of temperature along *BC*, so that there is no real paradox in the result.

Apparently Paradoxical Result.

We may apply the theory of the flow of energy to the determination of the rate of transmission of electromagnetic waves through an insulating medium. Consider a **plane wave** in which the electric and magnetic forces are **at right angles to one another in the** plane of the wave front. **Let** P **be the electric force,** β the magnetic. Then the total energy in unit volume of the medium is $(KP^2 + \mu\beta^2)/8\pi$. **If** V be the velocity of wave

Velocity of Electromagnetic wave deduced from Theory of Motion of Energy.

transmission the energy will travel forward a distance Vdt in time dt, and an amount of energy $(KP^2 + \mu\beta^2)Vdt/8\pi$ will pass across unit of area of a fixed plane in that time. But by Poynting's theorem the time-rate of flow of energy across unit of area is $P\beta/4\pi$. Hence equating these two expressions we find

$$V = \frac{2P\beta}{KP^2 + \mu\beta^2} \quad \ldots \ldots (55)$$

But the current, which is one of displacement, is $K(\partial P/\partial t)/4\pi$, where $\partial P/\partial t$ denotes the rate of variation of P with time at a given point in space, and if we take a distance dz in the direction of propagation, and dy in that of the magnetic force, the current through the rectangle $dydz$ is $K(\partial P/\partial t)dydz/4\pi$. But the line-integral of magnetic force round the rectangle, taken with due regard to the relation of the directions of the current and magnetic force, is

$$-\beta dy + (\beta - \frac{\partial \beta}{\partial z}dz)dy = -\frac{\partial \beta}{\partial z}dydz.$$

Hence we find

$$K\frac{\partial P}{\partial t} = -\frac{\partial \beta}{\partial z}. \quad \ldots \ldots (56)$$

By the nature of a wave the value of any quantity characteristic of it at a given point and instant, is found after an interval dt at a distance Vdt in advance of the former point in the direction of propagation. Hence if at any point z, the value of the electric force be P at a given instant, its value at the same instant at a point distant Vdt behind the former point is $P - (\partial P/\partial z)Vdt$. Hence the change which P undergoes at z in time dt is $-(\partial P/\partial z)Vdt$. But this is also $\partial P/\partial t \cdot dt$. Thus we have

$$-\frac{\partial P}{\partial t} = V\frac{\partial P}{\partial z}$$

and therefore by (56)

$$KV\frac{\partial P}{\partial z} = \frac{\partial \beta}{\partial z}$$

or

$$KVP = \beta$$

since the arbitrary function of t and constant which enter with the integration must be zero, as both P and β are zero in regions which the wave has not yet reached. Substituting this value of β in (55) we get finally

$$V^2 = \frac{1}{K\mu}. \qquad \ldots \ldots \ldots (57)$$

as before.

It is to be carefully noted that the result expressed in (57) which was obtained by a method independent of the theory of flow of energy, taken with the value of the velocity of propagation found above leads to the theorem that the energy per unit of volume at any point in the wave is half electric and half magnetic. For we have

Energy per Unit of Volume of Medium is half Electric, half Magnetic.

$$\frac{\mu\beta^2}{8\pi} = \frac{\mu K^2 V^2 P^2}{8\pi} = \frac{KP^2}{8\pi} \quad \ldots \ldots (58)$$

which proves the theorem.

SECTION III.

MAGNETO-OPTIC ROTATION.

WE give here a short account of the theory of the rotation of the plane of polarized light by passage along the lines of force in a magnetic field. It was discovered by Faraday that a beam of plane polarized light sent through perforated pole pieces so as to pass from end to end along a prism of a certain kind of heavy glass, placed parallel to the lines of force of a powerful electromagnet, had its plane of polarization rotated through an angle, which observation showed to be dependent on the wave length of the light and the intensity of the magnetic field. The usual method adopted was to polarize the light by means of a Nicol's prism, and adjust a second prism so as to receive the light and produce extinction of the beam after its passage through the glass with the magnet unexcited, then to excite the magnet and rotate the analysing prism until extinction was again produced, noting the angle of turning.

Turning of Plane of Polarization of Light in Magnetic Field.

It was found that the effect was produced in a large number of substances each of which gave an amount of rotation depending in the same way on the wave length of the light employed and the intensity of the magnetic field, but differing for different

VOL. II. Q

Law of the Phenomenon. substances. Measurements made by **Verdet** gave results agreeing fairly well with the formula

$$\theta = ml\mathbf{H}\frac{r^2}{\lambda_a^2}\left(r - \lambda_a \frac{dr}{d\lambda_a}\right). \quad \ldots \ldots \quad (59)$$

where θ is the angle of rotation, l the length of the path in the medium, $\mathbf{H}$ the intensity of the magnetic field, λ_a the wave length of the light in air, r the index of refraction of the light in the medium, and m a constant depending on the nature of the substance.

The following table gives a comparison of the results with the formula:—

Bisulphide of Carbon at 24°·9 C.

Line of the spectrum	C	D	E	F	G
Observed rotation	592	768	1000	1234	1704
Calculated	589	760	1000	1234	1713

Rotation of ray E   25° 88′.

Creosote at 24°·3 C.

Line of the spectrum	C	D	E	F	G
Observed rotation	573	758	1000	1241	1723
Calculated	617	780	1000	1210	1603

Rotation of ray E   21° 58′.

Other two formulas which have been proposed may be obtained by replacing the factor r^2/λ_a^2 of (59) by $1/\lambda_a^2$, or by 1. These do not agree nearly so well with the results of experiment.

Verdet's Constant. The amount of rotation per unit of $l\mathbf{H}$ for light of given wave length and for a given substance is called Verdet's constant for that light and substance. It is thus the amount of turning produced in unit length of beam when unit difference of magnetic potential exists between its extremities.

Non-magnetic Turning of Plane of Polarization. Comparison of Phenomena. Turning of the plane of polarization of light is shown also by quartz, solutions of sugar, tartaric acid, &c., and the amount of turning follows apparently a similar law to that stated above for the magnetic effect. But between the phenomenon in such cases and that in a magnetic field there is an essential difference. The turning of the plane in space is in the same direction in a magnetic field whether the ray be travelling with or against the direction of the magnetic force. In the other cases the direction of turning is reversed by reversing the ray. Thus if a ray after passage through the field be reversed by reflection, and sent back again, the effect of the second passage in turning the plane of polarization will be the same as before, and the plane of polarization of the emergent beam will have been turned by the passage through a double angle. Thus by repeated reflection from the silvered ends of a piece of heavy glass as indicated in Fig. 48, a considerable effect can be obtained by the addition of the turnings produced by successive passages.

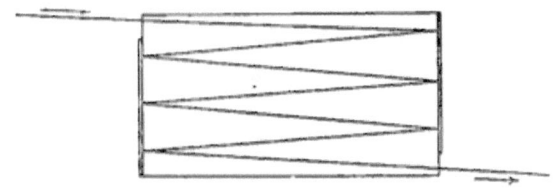

Fig. 48.

This difference between the two kinds of turning of the plane points to an essential difference between the molecular condition of the substances concerned. In the case of quartz, sugar solutions, &c., the phenomenon can be explained by supposing the substances to have some kind of helical structure, so that the restitutional forces, for displacements of the nature of shears in parallel planes, are greater when the displacements are in one direction than when they are in the opposite direction. *Physical Interpretation of their Points of Difference.*

Thus, in a medium possessing this property, particles displaced from a straight line, so as to be situated on a helix round the line as axis, will be acted on by greater restitutional forces when the helix is a right-handed one than when it is a left-handed one, or *vice versâ*, according to the structure.

For example, in diagram (2) of Fig. 49, particles displaced from the axis, so as to occupy the positions there shown, will, according to our supposition as to the structure of the medium, be subjected to (say) greater elastic forces than if the displacements were as shown in diagram (1). *Optical Behaviour of Sugar Solutions, &c., agrees with that of Medium of Helical Structure.*

In consequence of these forces the particles in diagram (2) moving under elastic forces in the circular orbits indicated by the closed curves move, we suppose, more quickly than those in diagram (1), whether the motion be clockwise or counter-clockwise. To fix the ideas let the motion of the particles be, as shown in Fig. 49, clockwise in (1) and counter-clockwise in (2). The configurations will after a given time have advanced in the direction from A to C and from B to D, but (because of the difference of configuration, not of motion) through a greater *Motions of Particles in Circularly Polarized Wave.*

MAGNETO-OPTIC ROTATION

distance in the latter case than in the former. Each of these motions will represent precisely the propagation of a wave of

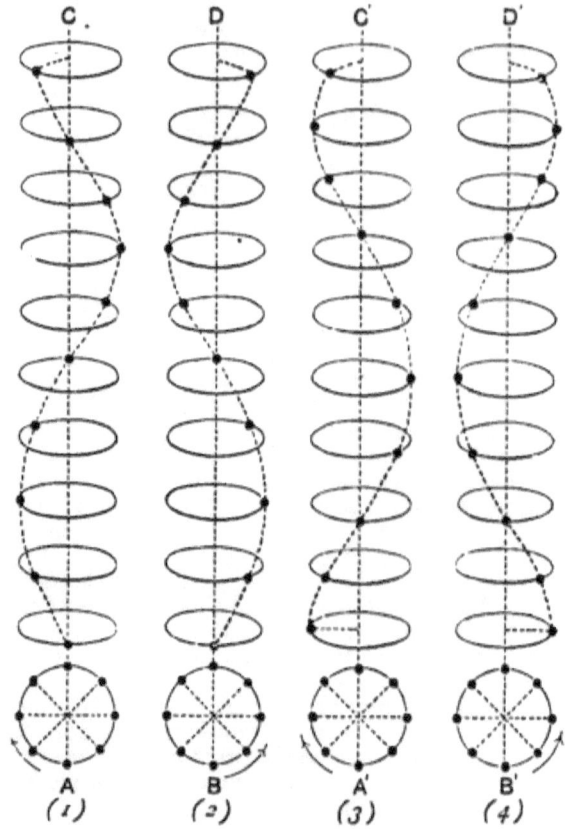

This Figure represents particles moving in parallel circular orbits as in a ray of circularly polarized light. The curved line shows the helical configuration of the corresponding particles in the wave. The circles below show the relative phases of the particles in each case, as seen by an eye looking from the lower to the upper end.

FIG. 49.

circularly polarized light, and the wave represented in (2) travels more quickly than that represented by (1).

But now let the waves be reflected from the ends C, D, as from the surface of a rarer medium, each after travelling the same distance. The reflected rays are shown in diagrams (3) and (4) respectively of Fig. 49. The right-handed and left-handed configurations (viewed in the same direction) have become left-handed and right-handed, with the same directions of motion as before.

Thus the configuration in diagram (4) travels from D' to B' more slowly than that in (3) does from C' to A', so that the gain by (2) of distance traversed in a given time in one direction is just balanced by the loss in (4) of distance traversed in the opposite direction; and there is the same relation of phase between the terminal particles at A' and B' in the return waves as there was in the advancing waves at A and B.

In the magnetic phenomenon on the other hand the velocity of the particles depends only on the direction of motion, for the relative speeds of propagation of a right-handed circularly polarized ray and a left-handed * one are the same whether the ray is direct or reflected. Thus, referring to Fig. 49, the ray represented in (2) and (4) travels faster or slower than the other in both cases, and by the same amount. If then two such rays travel simultaneously from one end of the medium to the other, and back again after reflection, the difference in phase produced by the forward will be doubled by the backward passage. The same final result will be found on examination to hold for reflection from a denser medium.

<small>Dynamical Inference from Magnetic Phenomenon</small>

In a diamagnetic medium such as **Faraday's heavy glass**, carbon-disulphide, &c., that ray travels faster, the direction of motion of the particles in which is round in the direction of the current producing the field. The reverse is the case for a magnetic substance.

These results, as was originally pointed out by Sir William Thomson,† indicate the existence of a motion in the magnetic field capable of being compounded with that motion of the medium which constitutes light. This remark was of extreme importance as forming the first step towards a dynamical theory of magnetic action, and since its publication considerable progress has been made with the application of dynamical principles to the explanation of the observed phenomena. We conclude the present chapter with a short account of this theory.

<small>that Medium contains similarly Oriented Rotating Particles.</small>

---

* That is as specified by the direction of the orbital motion of each particle, as seen by an observer imagined looking always from the same side of the plane of the orbit.

† *Proc. R. S.* June 1856. See also Rep. *El. and Mag.* p. 423 (2nd edn.), footnote; and *Baltimore Lectures on Molecular Dynamics* (Papyrograph Report), p. 241 *et seq*.

The fundamental idea of the theory is contained in the following extract from Thomson's paper: "The magnetic influence on light discovered by Faraday depends on the direction of motion of moving particles. For instance, in a medium possessing it, particles in a straight line parallel to the lines of magnetic force displaced to a helix round this line as axis, and then projected tangentially with such velocities as to describe circles, will have different velocities according as their motions are round in one direction (the same as the nominal direction of the galvanic current in the magnetizing coil) or in the contrary direction. But the elastic reaction of the medium must be the same for the same displacements whatever be the velocities and directions of the particles, that is to say, the forces which are balanced by centrifugal forces of the circular motions are equal, while the luminiferous motions are unequal. The absolute circular motions being therefore either equal or such as to transmit equal centrifugal forces to the particles initially considered, it follows that the luminiferous motions are only components of the whole motion; and that a less luminiferous component in one direction, compounded with a motion existing in the medium when transmitting no light, gives an equal resultant to that of a greater luminiferous motion in the contrary direction compounded with the same non-luminous motion."

Diamagnetism probably not a Differential Effect. It is worthy of special note that this dynamical explanation of the phenomenon is inconsistent with the explanation of the diamagnetism of a substance as a differential effect due to the greater magnetization of the surrounding medium. If admitted it decides that the difference between paramagnetic and diamagnetic substances is due to an absolute difference between the states of motion of their particles.

Dynamical Consideration of Luminiferous Motions. We have now to consider these luminiferous motions. The displacement of each particle of the medium in a ray of plane polarized light may be regarded as compounded of displacements corresponding to two circularly polarized rays of equal amplitude and period (see p. 228), but one right-handed and the other left-handed. To prove this it is only necessary to note that if the positions of a given particle in its orbit in obeying the two motions be as shown in the diagram, the component displacements (from the centre) at right angles to the diameter which bisects the arc of the orbit between the two positions cancel one another, while the components along that diameter are equal and in the same direction, and therefore give by their addition a double displacement.

The positions and motions of the particles of the medium on a ray of circularly polarized light travelling along the axis of z,

KINEMATICS OF PLANE POLARIZED RAY

may, if ξ, η denote the displacements parallel to the axis of x and y, be represented by the equations

$$\left.\begin{aligned}\xi &= a\cos\left\{\frac{2\pi}{\lambda}(z-vt)+e\right\} \\ \eta &= a\sin\left\{\frac{2\pi}{\lambda}(z-vt)+e\right\}\end{aligned}\right\} \quad \ldots \ldots (60)$$

Displacements of Particle of Medium in Circularly Polarized Ray.
1. Right-handed Ray.

as this gives for each particle, at a given distance z from the zero plane, motion in a circular orbit with periodic time λ/v, and radial displacement a. This represents, to a person looking in the direction in which the wave is travelling, a right-handed circular motion.

A left-handed wave travelling in the same direction, and $2e$ in phase-angle behind the former wave at the plane z, is represented by

$$\left.\begin{aligned}\xi' &= a\cos\left\{\frac{2\pi}{\lambda}(z-vt)-e\right\} \\ \eta' &= -a\sin\left\{\frac{2\pi}{\lambda}(z-vt)-e\right\}\end{aligned}\right\} \quad \ldots \ldots (61)$$

2. Left-handed Ray.

Compounding these two motions we get for the components of the resultant

$$\left.\begin{aligned}\xi+\xi' &= 2a\cos\frac{2\pi}{\lambda}(z-vt).\cos e \\ \eta+\eta' &= 2a\cos\frac{2\pi}{\lambda}(z-vt).\sin e\end{aligned}\right\} \quad \ldots \ldots (62)$$

Resultant of two Opposite Circular Motions is Rectilinear.

which show that the resultant displacement is along a diameter making an angle e with the axis of x. That is, the compounded ray is plane polarized.

Now considering two opposite circularly polarized rays, of the same period, travelling at different speeds v, v', through the medium, we get instead of (62) for the resultant displacements

$$\left.\begin{aligned}\xi+\xi' &= a\left[\cos\left\{\frac{2\pi}{\lambda}(z-vt)+e\right\}+\cos\left\{\frac{2\pi}{\lambda'}(z-v't)-e\right\}\right] \\ \eta+\eta' &= a\left[\sin\left\{\frac{2\pi}{\lambda}(z-vt)+e\right\}-\sin\left\{\frac{2\pi}{\lambda'}(z-v't)-e\right\}\right]\end{aligned}\right\} \quad (63)$$

Opposite Circularly Polarized Rays of Different Velocities

or (with the condition expressed that $\lambda/v = \lambda'/v' = T$)

$$\xi + \xi' = 2a \cos \pi \left\{ \left(\frac{1}{\lambda} + \frac{1}{\lambda'} \right) z - \frac{2t}{T} \right\} \cos \left\{ \pi \left(\frac{1}{\lambda} - \frac{1}{\lambda'} \right) z + e \right\}$$
$$\eta + \eta' = 2a \cos \pi \left\{ \left(\frac{1}{\lambda} + \frac{1}{\lambda'} \right) z - \frac{2t}{T} \right\} \sin \left\{ \pi \left(\frac{1}{\lambda} - \frac{1}{\lambda'} \right) z + e \right\} \quad (64)$$

produce Rectilinear Motion of Changing Direction.

This represents as before a plane polarized ray the direction of vibration in which makes an angle $\pi (1/\lambda - 1/\lambda') z + e$ with the axis. This angle increases with z, that is the directions of vibration for successive points along the ray lie on a screw surface of constant twist round the axis of z.

Propagation of Light in Medium containing similarly Oriented Rotating Particles.

To obtain a dynamical explanation of this action, we shall consider the transmission of waves of transverse displacement through an elastic medium in which are imbedded small molecules rapidly rotating round the axis of x. We shall suppose that these molecules are very small in volume, as compared with the volume of the rest of the medium, and of so slight mass as not appreciably to *load* the medium, and are uniformly distributed; so that we may regard each element of the medium as containing a large number of them, and as homogeneous in quality. The presence of these rotating particles will modify the equations of motion in a peculiar manner, in consequence of the existence of "gyrostatic domination" produced by their rotation.*

"Gyrostat," or Top, within Closed Non-rotating Case.

A gyrostat (Fig. 50) is composed of a fly-wheel of great moment of inertia pivoted within a case so that the instrument can be handled and moved about at pleasure. When the fly-wheel is spun so as to rotate very rapidly very curious dynamical results are obtained. It is found to rest in stable equilibrium in positions which without rotation would be essentially unstable; for example when supported, as shown in Fig. 50, on a glass plate by the thin edge passing symmetrically round the case, or placed with its axis of rotation vertical and supported on a universal joint in the line of the axis and *below* the gyrostat.

Again, and this is most important for our present purpose, if to the rapidly rotating gyrostat a couple be applied, so as to

---

* See Sir W. Thomson's *Baltimore Lectures on Molecular Dynamics* p. 241 *et seq.* : also Thomson and Tait's *Nat. Phil.* 2nd ed. Part I. § 345. See also an important paper by Dr. Larmor, *Proc. London Math. Soc.* No. 396 (1890) on *Rotatory Polarization, illustrated by the Vibrations of a Gyrostatically Loaded Chain.*

generate angular momentum round an axis making an angle ϕ with the axis of rotation of the fly-wheel, the axis of angular momentum will, in a small interval of time, change its position

Motion of Gyrostat.

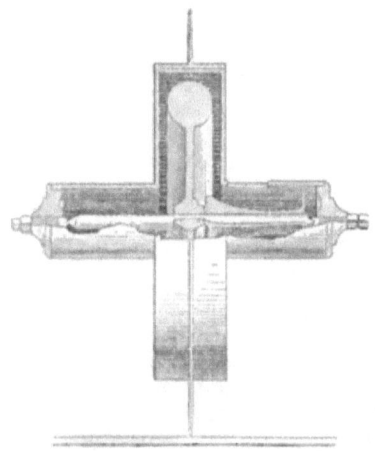

This Figure shows a gyrostat resting on a thin edge on a glass plate. The case is represented as cut open to show the fly-wheel, which is pivoted on a spindle turning in bearings attached to the case. As the section indicates, the fly-wheel is a thin disk with a massive rim. [This cut is reduced from Thomson and Tait's *Natural Philosophy* (Vol. I. Part 1, p. 397), to which the reader may refer for further information regarding gyrostatic action.]

FIG. 50.

through an angle which is found by compounding the initial angular momentum with that generated by the couple in the interval. Thus if G be the moment of the couple, the angular

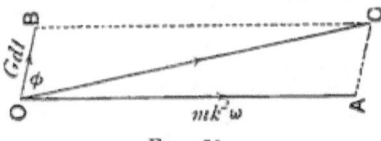

FIG. 51.

momentum generated by it in the fly-wheel in time dt is Gdt. Let this be represented by OB, Fig. 51, drawn in the usual way in the direction of the axis of the couple so as to represent the direction of rotation, and the magnitude of the angular **momentum**. Let OA in the **same** way represent the angular momentum of the fly-wheel, which, if m be the **mass** of the wheel, k its radius of gyration round the **axle**, and ω its angular velocity, is numerically $mk^2\omega$. Then OA and OB can be compounded according to the ordinary parallelogram law, **and give** a resultant OC which represents the **new angular** momentum in magnitude and direction. By the diagram we have, approximately,

$$\sin COA = \frac{G dt}{mk^2\omega} \sin\phi$$

or denoting COA by $d\theta$, and proceeding to the limit when dt is infinitely small, we have accurately

$$\frac{d\theta}{dt} = \frac{G}{mk^2\omega}\sin\phi \quad \dots \dots \dots \quad (65)$$

This is the angular velocity with which the axis of angular momentum turns in the plane of itself and the axis of the couple.

Examples of Precessional Motion.
For example, if the gyrostat referred to above be suspended by a cord attached to the case at some point in the prolongation of the axis, and the axis be placed in a horizontal position, the weight of the instrument acting downwards will give a couple round a horizontal axis at right angles to that of rotation. If the rotation be very rapid, the axis of the fly-wheel will remain approximately horizontal, while the whole instrument turns slowly round in azimuth about a vertical axis with angular velocity $\Omega = G/mk^2\omega$. This azimuthal motion corresponds to the precessional motion (of period nearly 26,000 years) which the earth has in consequence of the differential couple, due to attractional forces, tending to bring the planes of the equator and ecliptic into coincidence.

Reactional Couple due to Motion of the Axis.
Conversely, if a motion of the axis of rotation take place, the gyrostat will react on the constraining system with a couple equal and opposite to that given by the construction in Fig. 51 above. Thus if the angular velocity with which the direction of the axis of rotation changes be Ω, the magnitude of the couple round a moving axis at right angles to the axis of rotation and the instantaneous axis, with which the gyrostat will react on its supports, will be given by

$$G = mk^2\omega\,\Omega \quad \dots \dots \dots \quad (66)$$

We shall now apply this result to form the equations of motion for a **wave of transverse** displacement through the medium referred to above. As before let the direction of propagation be along the axis of z, and consider a transverse displacement of the medium of amount ξ parallel to the axis of x, at a distance z from the zero plane of x, y. The strain at that point is $\partial\xi/\partial z$. At a distance dz in advance along the ray the strain is $\partial\xi/\partial z + \partial^2\xi/\partial z^2 . dz$. Thus by the suppositions made on p. 232 above, the elastic force producing acceleration, in the positive direction of ξ, of the matter between the sections is $n\partial^2\xi/\partial z^2 . dz$, per unit of area of the cross sections, if n denote the proper elastic modulus.

Wave of Transverse Displacement in Gyrostatic Medium. Ordinary Elastic Return Force.

But besides this elastic force there is a gyrostatic reaction due to the angular motion produced in the portion of the medium by the strain in the direction of η. The velocity of displacement at z is $\partial\eta/\partial t$, and at $z + dz$ is $\partial\eta/\partial t + \partial^2\eta/\partial z\partial t . dz$. Hence the rate at which the element is changing direction is $\partial^2\eta/\partial z\partial t$. The reacting gyrostatic couple per unit of volume is by (66) $c\rho\partial^2\eta/\partial z\partial t$, if $c\rho$ be the angular momentum of the rotating matter in unit of volume. Each force of this couple is parallel to ξ, and it must be balanced by the action of the surrounding medium exerted across the ends of the element considered. The resulting gyrostatic couple per unit of volume of the next element is, if $c\rho$ be constant, $c\rho(\partial^2\eta/\partial z\partial t + \partial^3\eta/\partial z^2\partial t . dz)$. Thus the mutual force between the successive elements is $c\rho\partial^3\eta/\partial z^2\partial t$. If the direction of rotation of the molecules be right-handed, as seen by an observer looking in the direction in which the wave is travelling, this force will be in the positive direction along x. Thus we have finally if ρ be the density of the medium

Gyrostatic Reaction.

$$\rho\frac{\partial^2\xi}{\partial t^2} = n\frac{\partial^2\xi}{\partial z^2} + c\rho\frac{\partial^3\eta}{\partial z^2\partial t} \quad \ldots \quad (67)$$

In the same way we should find for the η equation

Differential Equations of Motion.

$$\rho\frac{\partial^2\eta}{\partial t^2} = n\frac{\partial^2\eta}{\partial z^2} - c\rho\frac{\partial^3\xi}{\partial z^2\partial t} \quad \ldots \quad (68)$$

These equations are satisfied by the values of ξ, η given by (60) above. Substituting we get from each the equation of condition

Solution of Differential Equation.

$$v^2 - \frac{2\pi c}{T}\frac{n}{\rho} = 0 . \quad \ldots \quad (69)$$

The equation of condition for a left-handed vibration is by (68) evidently to be obtained by simply changing the sign of c, and is therefore

Velocities of Right-handed and Left-handed Circularly Polarized Rays.

$$v^2 + \frac{2\pi c}{T} - \frac{n}{\rho} = 0. \quad \ldots \quad (70)$$

Thus for the right-handed ray we have

$$v = \pm \frac{\lambda}{T} = \pm \sqrt{\frac{n}{\rho} + \frac{2\pi c}{T}}. \quad \ldots \quad (71)$$

and for the left-handed ray,

$$v' = \pm \frac{\lambda'}{T} = \pm \sqrt{\frac{n}{\rho} - \frac{2\pi c}{T}}. \quad \ldots \quad (72)$$

Rate of Turning of Plane of Polarization.

Now by the result on p. 232 the alteration θ, of direction of vibration, per unit of distance travelled through the medium, is $\pi(1/\lambda' - 1/\lambda)$. Hence equations (71) and (72) give

$$\theta = \frac{1}{T}\left\{ \frac{1}{\sqrt{\frac{n}{\rho} - \frac{2\pi c}{T}}} - \frac{1}{\sqrt{\frac{n}{\rho} + \frac{2\pi c}{T}}} \right\}.$$

If we suppose $2\pi c/T$ to be small in comparison with n/ρ we may write the equation in the form

$$\theta = \frac{2\pi c}{T^2} \frac{\left(\frac{\rho}{n}\right)^{\frac{3}{2}}}{1 - 2\frac{\pi^2 c^2 \rho^2}{T^2 n^2}}. \quad \ldots \quad (73)$$

Relation of Rate of Turning to Wave Length.

Now if r be the index of refraction of the substance with respect to air and λ the length of a wave in the medium which has length λ_a in air we have $\lambda_a = r\lambda$. Moreover $T = \lambda/v = \lambda_a/rv$, in which we may put $\sqrt{n/\rho}$ for v, since the velocities in opposite directions differ only very slightly. Also if V be the velocity of light in air, $v = V/r$. Hence we get finally

$$\theta = \frac{r^3}{\lambda_a^2} \frac{2\pi c}{V\left(1 - 2\frac{\pi^2 c^2 r^4}{T^2 V^4}\right)} \quad \ldots \quad (74)$$

which agrees with the formula (59) above, in giving rotation of the direction of vibration inversely as the square of the wave length. The rotation also depends on the direction of vibration, not on that of propagation. It is to be observed however that the medium as we have supposed it constituted does not produce ordinary dispersion, and therefore the remaining factor of the expression for θ in (74) does not here find any explanation. The effect is very small in all ordinary transparent **substances** for which it has been investigated; and this of course is in accordance with what we should expect, taking into account the smallness **of the** magnetization (or diamagnetization) of these substances. But it ought **to be** great in a highly magnetized substance **such** as iron, **and this** has been found by Kundt * to be **the case for a** thin film of metallic iron magnetized transversely.

Magnitude of Effect in Diamagnetic and Paramagnetic Substances.

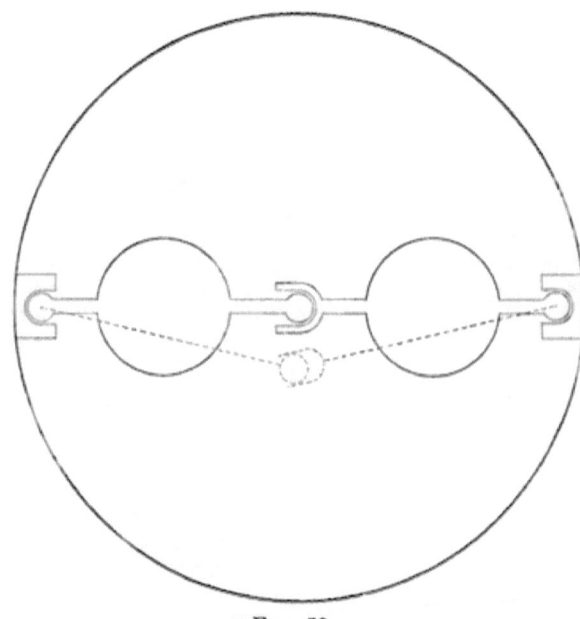

Fig. 52.

When the substance is magnetized the already existing magnetic molecules turn round towards coincidence of direction of their axes, and this must involve some freedom of motion of the particles relatively to the medium in which they are imbedded. Sir William Thomson † has suggested a form of gyrostatic

Gyrostatic Molecule.

---

* *Phil. Mag.* Oct. 1884.
† *Baltimore Lectures*, p. 320.

Gyrostatic Molecule. molecule, consisting (Fig. 52) of a spherical sheath enclosing two equal gyrostats. Each gyrostat is attached at one extremity of its axis of rotation to the case by a ball-and-socket joint, and the two are connected together by another ball-and-socket joint uniting their other extremities. The attachments to the case are at opposite extremities of a diameter, so that the gyrostats when their axes are in line have the centres of the fly-wheels on the diameter.

Supposing a couple applied to the case so as to turn it round any axis through its centre, no disalignment of the gyrostats would take place, and it would turn simply as a rigid body. If however the case were made to undergo translation in any direction, except along the line of the axis, the gyrostats would lag behind, and the two-link chain which they form would bend at the centre. This bending, however, would be resisted by the quasi-rigidity of the chain produced by the rotation, and the gyrostats would react on the sheath at the ball-and-socket joints, with forces at right angles to the plane in which the bending takes place.

The result when worked out is that if the centre of the gyrostatic molecule be carried round with uniform velocity in a circle in a plane at right angles to the diameter joining the ball-and-socket joints, the necessary centreward force is greater or less according as the direction in which the molecule is carried round is with or against the direction of rotation of the gyrostats. In other words the effect of the rotation is to virtually increase the inertia of the molecule in the one case and diminish it in the other, by a certain amount. If the molecule be small enough the effect of any rotation on the sheath is infinitesimal in comparison with that of translation ; which is found to produce the required magneto-optic effect. The sheath might therefore be smooth.

This modification of the molecule seems important, as it accounts for the magneto-optic effect, while rendering it possible for the molecules to turn round in obedience to magnetizing force towards coincidence of direction so that the medium becomes magnetized.

Magneto-Optic Rotation on Electro-magnetic Theory of Light. It is obviously suggested by the gyrostatic investigation that it ought to be possible to explain the magneto-optic rotation on the electromagnetic theory of light as a consequence of the existence of the small magnets imbedded in the medium with their axes in the direction of propagation of the ray, and therefore producing a component of magnetization in that direction.*

---

* A theory of this kind has, I learn from M. Poincaré's *Théories de Maxwell*, been proposed by M. Potier, and published in the

In consequence of the motions of the ether the direction of the chains of these molecules which we suppose to exist along the axis of z in the undisturbed state of the medium are continually undergoing change at every point, and thus the direction of the axial magnetic force along each chain also undergoes alteration. It is obvious that if the displacements be everywhere small the actual magnitude of this force will not sustain any sensible change, but that each small change of direction will produce a component magnetic force in each of the two directions at right angles to the axis. The calling into existence of these components will produce corresponding electromotive forces tending to increase the electric displacements. We shall *assume* that the electric displacements are in the same direction as, and proportional to the displacements of the medium in which the magnetized molecules are supposed imbedded.

Displacement of Magnetized Medium in Wave

The electromotive force in the direction of y is given, (5) above, by

$$Q = -\frac{dG}{dt} - \frac{\partial \psi}{\partial y},$$

where dG/dt stands for the *total* time rate of change of G. Also since H does not perceptibly vary along x, if the direction of propagation be, as taken here, along z, $-\partial G/\partial z$ denotes magnetic induction through unit of area in the plane of yz. Hence any part of the total time rate of variation of $-\partial G/\partial z$ will denote the space rate of variation in the direction of z of an electromotive force parallel to z, provided the time and space differentiations of the part are commutative.

produces Transverse Electromotive Forces.

Now if the displacements of the ether particles from their undisturbed positions be taken as parallel and proportional to the electric displacement, and C be the component of magnetization of the substance in the direction of z due to the existence of the molecular magnets, the component magnetic force in the direction of x will be approximately $eC\partial f/\partial z$, and thus the magnetic induction through unit of area in the plane of yz is $\mu e C \partial f/\partial z$. The time rate of variation of this is

Calculation of Transverse Electromotive Force.

$$\mu e C \frac{\partial}{\partial t}\frac{\partial f}{\partial z}.$$

Comptes Rendus. This theory itself is not given in M. Poincaré's book, and I have not seen the paper; but the treatment of the matter given above occurred to me after reading M. Poincaré's statement of the results.

But we have by (5), Chapter II.

$$\frac{\partial f}{\partial t} = \frac{1}{4\pi}\left(\frac{\partial \gamma}{\partial y} - \frac{\partial \beta}{\partial z}\right) = -\frac{1}{4\pi}\frac{\partial \beta}{\partial z} \quad . \qquad (75)$$

since there is no conduction current.

Further, by (50) p. 33, $\beta = (\partial F/\partial z)/\mu$, and therefore (75) becomes

$$\frac{\partial f}{\partial t} = -\frac{1}{4\pi\mu}\frac{\partial^2 F}{\partial z^2} . \quad . \quad . \quad . \quad . \quad (76)$$

Now since the differentiation of f with respect to t is partial only, we may use the substitution,

$$\frac{\partial}{\partial z}\frac{\partial f}{\partial t} \equiv \frac{\partial}{\partial t}\frac{\partial f}{\partial z},$$

and therefore (76) gives

$$\mu e C \frac{\partial}{\partial t}\frac{\partial f}{\partial z} = -\frac{eC}{4\pi}\frac{\partial^3 F}{\partial z^3},$$

which gives an electromotive force in the direction of y, of amount

$$-\frac{eC}{4\pi}\frac{\partial^2 F}{\partial z^2},$$

since there can be no constant or arbitrary time function concerned. Hence we have finally

$$Q = -\frac{\partial G}{\partial t} - \frac{eC}{4\pi}\frac{\partial^2 F}{\partial z^2} - \frac{\partial \psi}{\partial y},$$

and therefore

$$\frac{\partial Q}{\partial t} = -\frac{\partial^2 G}{\partial t^2} - \frac{eC}{4\pi}\frac{\partial^3 F}{\partial t \partial z^2} . \qquad . \quad (77)$$

But by (9) and (10),

$$\frac{K}{4\pi}\frac{\partial Q}{\partial t} = \frac{\partial g}{\partial t} = -\frac{1}{4\pi\mu}\frac{\partial^2 G}{\partial t^2},$$

which used in (77) gives,

$$\frac{\partial^2 G}{\partial t^2} = \frac{1}{K\mu}\frac{\partial^2 G}{\partial z^2} - \frac{eC}{4\pi}\frac{\partial^3 F}{\partial t \partial z^2} . \quad . \quad . \quad (78)$$

HALL'S PHENOMENON

Similarly for the other component in the case of circularly polarized light we find the equation

$$\frac{\partial^2 F}{\partial t^2} = \frac{1}{K\mu} \frac{\partial^2 F}{\partial z^2} + \frac{eC}{4\pi} \frac{\partial^3 G}{\partial t \partial z^2} \quad . \quad . \quad . \quad (79)$$

These two equations are identical in form with those already found, and of course lead to the same results, namely, magneto-optic rotation independent of whether the ray is direct or reflected.

This is the proper place in which to refer to the phenomenon discovered by Hall (as the result of a research undertaken at the suggestion of Professor Rowland), that in a conductor carrying a current in a magnetic field a component electromotive force is produced in the direction of the electromagnetic force. The experimental arrangement used by Hall is shown in Fig. 53. A

Hall's Phenomenon.

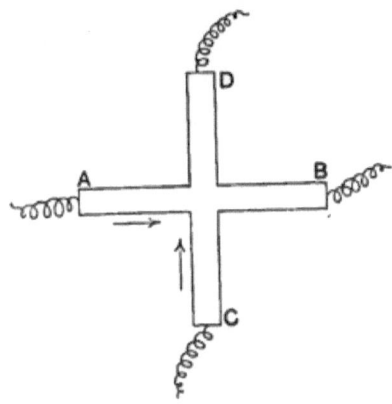

FIG. 53.

cross consisting of a thin film of the conductor, deposited, or fixed in some other manner, on glass, is placed with its plane at right angles to the lines of force in the field, which is produced by an electromagnet, and a current is sent along the arm AB. The terminals of a sensitive galvanometer are placed at two such points, at the extremities of the arm CD, that no current flows through it when the magnet is not excited. The magnet is now excited, and a current is found to flow through the galvanometer, proving that its points of attachment are no longer at the same potential; that is, a component of electromotive force across the direction of the current in the plane of the film has been produced. When the current flows from A to B and the direction of the lines of magnetic force is upwards through the paper, the direction of the electromotive force produced is from C to D.

With this arrangement the field was not uniform but was concentrated on the central portion of the film. The arrangement of the conductor in a very thin film is necessary for the production of an observable effect.

This result has been attributed by several physicists to inequality of strain in the material set up by the unequal action of the electromagnetic force on the matter of the film, and Shelford Bidwell has suggested thermoelectric action due to the current flowing through the no longer homogeneous substance as the immediate cause. This view however does not appear to be shared by Rowland and Hall. It would seem to be possible to test its accuracy by properly arranging a film so that it should be wholly within a *uniform* field of sufficient intensity and observing whether or not the effect is produced.

Electromotive Force in Hall Effect.

If an electromotive force proportional to the electromagnetic force (X, Y, Z) exists at each point in a homogeneous medium, its components are

$$\left.\begin{aligned} e'X &= e'(vc - wb) \\ e'Y &= e'(wa - uc) \\ e'Z &= e'(ub - va) \end{aligned}\right\} \quad \ldots \ldots \quad (80)$$

where e' is a constant. Thus, Ψ being used to denote electrostatic potential, produced by a battery or otherwise, independently of ordinary magnetic induction, the equations of electromotive force [(5) p. 191] are now

$$\left.\begin{aligned} P &= -\frac{\partial F}{\partial t} + e'(vc - wb) - \frac{\partial \Psi}{\partial x} \\ Q &= -\frac{\partial G}{\partial t} + e'(wa - uc) - \frac{\partial \Psi}{\partial y} \\ R &= -\frac{\partial H}{\partial t} + e'(ub - va) - \frac{\partial \Psi}{\partial z} \end{aligned}\right\} \quad \ldots \quad (81)$$

and if a, b, c be the components of total magnetic induction, a', b', c' those of the part due to the constant field,

WAVE IN MEDIUM SHOWING HALL EFFECT

$$\left.\begin{aligned}a &= \frac{\partial H}{\partial y} - \frac{\partial G}{\partial z} + a' \\ b &= \frac{\partial F}{\partial z} - \frac{\partial H}{\partial x} + b' \\ c &= \frac{\partial G}{\partial x} - \frac{\partial F}{\partial y} + c'\end{aligned}\right\} \quad \ldots \quad (82)$$

For a plane wave travelling in the direction of z through an insulating medium showing the Hall phenomenon, (82) becomes

Rowland's Theory of Plane Wave through Medium showing Hall Effect.

$$a = -\frac{\partial G}{\partial z} + a'$$

$$b = \frac{\partial F}{\partial z} + b'$$

$$c = c'.$$

Hence the equations of electromotive force (81) become, since $w = 0$ and $R = 0$,

$$P = -\frac{\partial F}{\partial t} - \frac{e'c'}{4\pi\mu}\frac{\partial^2 G}{\partial z^2} - \frac{\partial \Psi}{\partial x}$$

$$Q = -\frac{\partial G}{\partial t} + \frac{e'c'}{4\pi\mu}\frac{\partial^2 F}{\partial z^2} - \frac{\partial \Psi}{\partial y}$$

and therefore

$$\left.\begin{aligned}\frac{\partial P}{\partial t} &= -\frac{\partial^2 F}{\partial t^2} - \frac{e'c'}{4\pi\mu}\frac{\partial^3 G}{\partial t \partial z^2} \\ \frac{\partial Q}{\partial t} &= -\frac{\partial^2 G}{\partial t^2} + \frac{e'c'}{4\pi\mu}\frac{\partial^3 F}{\partial t \partial z^2}\end{aligned}\right\} \quad \ldots \quad (83)$$

But we have by (9) p. 195

$$\frac{\partial P}{\partial t} = \frac{4\pi u}{K} = -\frac{1}{K\mu}\frac{\partial b}{\partial z} = -\frac{1}{K\mu}\frac{\partial^2 F}{\partial z^2}$$

$$\frac{\partial Q}{\partial t} = \frac{4\pi v}{K} = +\frac{1}{K\mu}\frac{\partial a}{\partial z} = -\frac{1}{K\mu}\frac{\partial^2 G}{\partial z^2}.$$

Hence substituting from these equations in (83) and rearranging we find

Equations of Propagation indicate Magneto-Optic Effect.

$$\frac{\partial^2 F}{\partial t^2} = \frac{1}{K\mu}\frac{\partial^2 F}{\partial z^2} - \frac{e'c'}{4\pi\mu}\frac{\partial^3 G}{\partial t \partial z^2}$$
$$\frac{\partial^2 G}{\partial t^2} = \frac{1}{K\mu}\frac{\partial^3 G}{\partial z^2} + \frac{e'c'}{4\pi\mu}\frac{\partial^3 F}{\partial t \partial z^2}\Bigg\} \quad \ldots \quad (84)$$

which are equations of the same form as before obtained, and show that in such a medium a magneto-optic effect would be produced.

Kerr's Magneto-Optic Effect.

Dr. John Kerr, of Glasgow, discovered about fifteen years ago * that when plane polarized light is incident on the polished pole of a magnet, so that the wavefront is not parallel to the direction of magnetization, the reflected beam is elliptically polarized. This result has been explained by Professor G. F. Fitzgerald † in accordance with Maxwell's Electromagnetic Theory of Light; but want of space prevents our giving here a sketch of his investigation. The effect has been used by Dr. H. Dubois as the foundation of a method of measuring magnetic permeability described below in the Chapter on Magnetic Measurements.

Kerr's Electro-Optic Effect.

Dr. Kerr had previously discovered ‡ that when a beam of polarized light is incident upon a transparent medium subjected to electrostatic strain the transmitted beam is is general elliptically polarized. The maximum effect is produced when the wave front is parallel to the direction of the lines of force, and the plane of polarization inclined to them at an angle of 45° No effect is produced when the plane of polarization is either parallel or at right angles to the lines of force.

---

* *Phil. Mag.* May 1877, and March 1878.
† *Phil. Trans. R. S.* Pt. II. 1880.
‡ *Phil. Mag.* Nov. 1875.

CHAPTER VI.

CALCULATION OF CONSTANTS OF COILS AND CO-EFFICIENTS OF INDUCTION.

SECTION I.

MAGNETIC ACTION OF CIRCUITS AND COILS.

It has been proved above [(81) p. 47] that the solid angle subtended by a circle at any point is given by the equation *

Solid Angle subtended by Circle.

$$\omega = 2\pi \left\{ 1 - \cos\psi + \sin^2\psi \sum \frac{1}{i} \cdot {_\psi}Z'_i \cdot {_\phi}Z_i \left(\frac{\rho}{r}\right)^i \right\} . \quad (1)$$

where (Fig. 54) ϕ is the angle between CP and the axis, ψ the angle between CA and the axis, ${_\phi}Z_i$ the zonal spherical harmonic of order i, ${_\psi}Z'_i$ the differential coefficient of ${_\psi}Z_i$ with respect to $\cos\psi$, ρ the distance $P'P$, and r the distance $P'A$. If as indicated in the figure we take $\phi = 90°$, then all the zonal harmonics of odd order vanish,† and the general expression of the zonal harmonic of even order $2i$ is

$$(-1)^i \frac{1 \cdot 3 \ldots (2i-1)}{2 \cdot 4 \ldots 2i}.$$

---

* It is to be noted that (81), p. 47, is given for the case (Fig. 9) of P' on the opposite side of the shell from the point C, and so (p. 46) the angle β at P' is used. Here the solid angle is taken from P' coincident with C of Fig. 9, and is $2\pi(1 - \cos\psi)$. It would have been better to have taken P' in Fig. 9, on the same side of the circle as C. Then in (78) and (81) we should have had $-\cos\alpha$ instead of $+\cos\alpha$, corresponding to $-\cos\psi$ above, and $+\sin^2\alpha$ for $-\sin^2\alpha$ in the third term.

† See Appendix on Spherical Harmonics.

Potential at any Point due to Circular Current.

Now the solid angle subtended at any point by a closed curve is equal to the potential which a unit current flowing in the curve would produce at that point. Hence if a current γ flow in the circle, and Ω be the potential which the current produces, we have, writing y for $P'P$ in the particular case in which it is at right angles to the axis, and x for OP',

$$\Omega = 2\pi\gamma\{1 - \frac{x}{r} - \frac{3}{2^2}\frac{y^2}{r^5}a^2x - \frac{3 \cdot 5}{2^2 \cdot 4^2}\frac{y^4}{r^9}a^2x(3a^2 - 4x^2)$$
$$- \frac{3^2 \cdot 5}{2^2 \cdot 4^2 \cdot 6^2}\frac{y^6}{r^{13}}a^2x(35a^4 - 140a^2x^2 + 56x^4) - \ldots\} \quad (2)$$

a series which is convergent if $y < r$.

This equation can be found as follows without the use of zonal harmonics, and a comparison of the processes gives results which will be of great service in some more complex applications of zonal harmonics later in the present chapter.

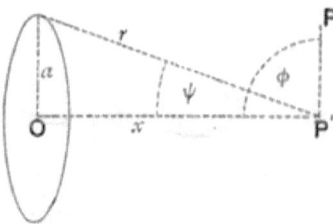

FIG. 54.

Potential due to Circular Surface Distribution of Magnetism.

Consider the potential produced at any point P (Fig. 54) at a distance y from the axis of the circle by a circular plane distribution of magnetism. Let the dimensions be as in Fig. 54, and denote by σ the surface density of the magnetic distribution, and by dV the potential at P' produced by a narrow concentric ring of the magnetism of radius p and breadth dp. Then

$$dV = 2\pi\sigma\frac{pdp}{\sqrt{p^2 + x^2}}.$$

Hence integrating from $p = 0$, to $p = a$, we find

$$V = 2\pi\sigma(\sqrt{a^2 + x^2} - a)$$

for the potential at P' due to the whole distribution.

POTENTIAL DUE TO THIN CIRCULAR COIL

Now assume for the potential at P

$$V = 2\pi\sigma(A_0 + A_1 y^2 + A_2 y^4 + \ldots) \quad \ldots \quad (3)$$

where $A_0, A_1, \ldots$ are functions of x. No odd powers of y can enter, since the potential is not altered by reversing the sign of y; and since when $y = 0$, the value of V reduces to that for P' we have

$$A_0 = \sqrt{a^2 + x^2} - x \quad \ldots \quad \ldots \quad (4)$$

At all points external to the distribution V must satisfy Laplace's equation, which, for the case of symmetry round the axis of x, takes the form

Potential due to Disk formed by solving Laplace's Equation by Expansion.

$$\frac{\partial^2 V}{\partial x^2} + \frac{\partial^2 V}{\partial y^2} + \frac{1}{y}\frac{\partial V}{\partial y} = 0 \quad \ldots \quad \ldots \quad (5)$$

Differentiating (3) and substituting in (4) we find

$$\frac{\partial^2 A_0}{\partial x^2} + \frac{\partial^2 A_1}{\partial x^2}y^2 + \frac{\partial^2 A_2}{\partial x^2}y^4 + \ldots$$

$$+ 2A_1 + 3.4 A_2 y^2 + 5.6 A_3 y^4 + \ldots$$

$$+ 2A_1 + \quad 4A_2 y^2 + \quad 6 A_3 y^4 + \ldots = 0.$$

The coefficients of the different powers of y in this series equated separately to zero give

$$A_1 = -\frac{1}{2^2}\frac{\partial^2 A_0}{\partial x^2}, \quad A_2 = \frac{1}{2^2 \cdot 4^2}\frac{\partial^4 A_0}{\partial x^4}, \quad A_3 = -\frac{1}{2^2 \cdot 4^2 \cdot 6^2}\frac{\partial^6 A_0}{\partial x^6}, \ldots$$

Hence finally

$$V = 2\pi\sigma\left(A_0 - \frac{y^2}{2^2}\frac{\partial^2 A_0}{\partial x^2} + \frac{y^4}{2^2 \cdot 4^2}\frac{\partial^4 A_0}{\partial x^4} - \ldots\right) \quad \ldots \quad (6)$$

where $A_0 = \sqrt{a^2 + x^2} - x$.

From (6) of course by differentiation with respect to x and y respectively, the axial and radial component forces at the point x, y, can be obtained for the given distribution.

If now another circular plane distribution, of equal density but opposite sign, be supposed placed coaxial with and at a distance $-dx$ from the former, its potential at the point (x, y) will be the same as that produced at the point $(x + dx, y)$ by the former distribution, except that the sign will be changed. Thus it is $-(V + \partial V/\partial x \cdot dx)$. The potential at the point (x, y), due to the two plane distributions together, is thus $-\partial V/\partial x \cdot dx$. Calling this Ω we have

Potential of Circular Magnetic Shell deduced from Potential of Disk.

$$\Omega = 2\pi\sigma dx \left(-\frac{\partial A_0}{\partial x} + \frac{y^2}{2^2}\frac{\partial^3 A_0}{\partial x^3} - \frac{y^4}{2^2 \cdot 4^2}\frac{\partial^5 A_0}{\partial x^5} + \ldots \right) \quad (7)$$

This is the potential at x, y of a magnetic **shell** of strength σdx. If the shell be replaced by a current of strength γ flowing in a circle coinciding with the edge of the shell, we have $\sigma dx = \gamma$. Performing then the differentiations of A_0, (writing for brevity r for $\sqrt{a^2 + x^2}$, and replacing σdx by γ) we find again (2).

By comparison of (1) and (7) remembering that ϕ is now $90°$, and $\rho = y$, we see that

$$\left. \begin{array}{l} 1 - \dfrac{x}{r} = -\dfrac{\partial A_0}{\partial x} \\[6pt] (2i-1)! \dfrac{a^2}{r^{2i+2}} \psi Z'_{2i} = -\dfrac{\partial^{2i+1} A_0}{\partial x^{2i+1}} \end{array} \right\} \quad (8)$$

a general result which enables $\int \{\psi Z_{2i}/r^{2i+2}\}dx$ to be calculated for any value of i by successive differentiation of A_0. We shall show when we come to use this result that a similar theorem is true for zonal harmonics of odd order.

Magnetic Forces due to Circular Magnetic Shell. The axial and radial component forces F, R, are $-\partial\Omega/\partial x$, $-\partial\Omega/\partial y$ respectively. These could be obtained directly from (2), but it is easier to differentiate (7) with respect to x and y, and insert the differential coefficients of A_0 in the result. Thus

$$F = -\frac{\partial\Omega}{\partial x} = 2\pi\gamma\frac{a^2}{r^3}\left\{1 + \frac{3}{2^2}\frac{y^2}{r^4}(a^2-4x^2) + \frac{3^2 \cdot 5}{2^2 \cdot 4^2}\frac{y^4}{r^8}(a^4-12a^2x^2+8x^4)\right.$$
$$\left. + \frac{3^2 \cdot 5}{2^2 \cdot 4^2 \cdot 6^2}\frac{y^6}{r^{12}}(35a^6 - 840a^4x^2 + 1680x^4 - 448x^6) + \ldots\right\}. \quad (9)$$

$$R = -\frac{\partial\Omega}{\partial y} = 3\pi\gamma\frac{a^2xy}{r^5}\left\{1 + \frac{5}{2 \cdot 4}\frac{y^2}{r^4}(3a^2-4x^2)\right.$$
$$\left. + \frac{3 \cdot 5}{2 \cdot 4^2 \cdot 6}\frac{y^4}{r^8}(35a^4 - 140a^2x^2 + 56x^4) + \ldots\right\} \quad (10)$$

The field due to a circular conductor is shown in section by Fig. 55.*

---

* This cut is taken from Maxwell's *Electricity and Magnetism*, Vol. II., with an extension to show the field symmetrically about the centre of the circular conductor.

FIELD OF THIN CIRCULAR COIL

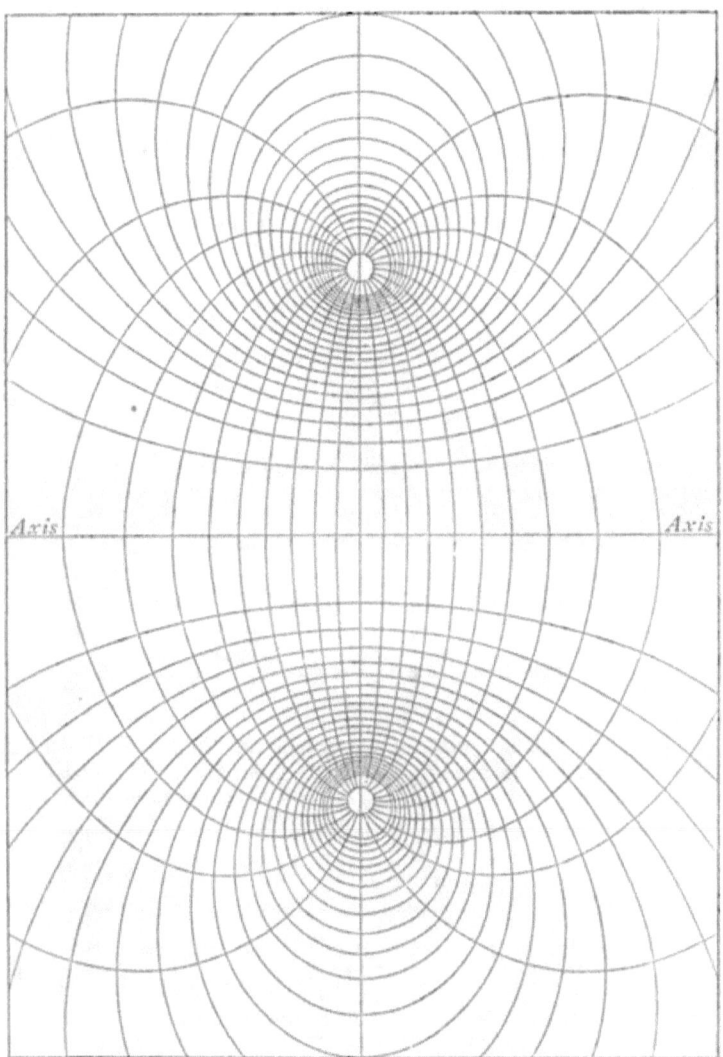

Fig. 55.

Couple on Magnetic Needle produced by Circular Current.

From these results we could calculate the couple on a thin uniformly magnetized needle A, B (Fig. 56) placed with its centre on the axis, and deflected into any given position; but the following method is preferable. Let $2l$ be the length of the

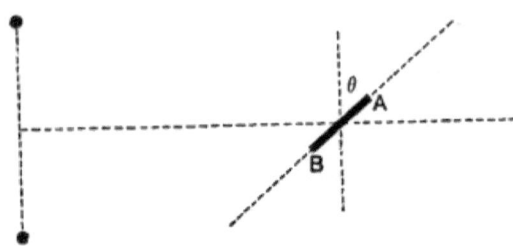

FIG. 56.

needle, and θ the angle which its axis makes with the plane of the circuit. The coordinates of its ends are $x + l\sin\theta$, $l\cos\theta$, for A, and $x - l\sin\theta$, $-l\cos\theta$ for B. Now if Ω_1, Ω_2 be the potentials at A and B respectively we have by Taylor's theorem

$$\left.\begin{aligned}\Omega_1 &= \Omega + l\sin\theta\frac{\partial\Omega}{\partial x} + \frac{l^2\sin^2\theta}{1 \cdot 2}\frac{\partial^2\Omega}{\partial x^2} + \cdots \\ \Omega_2 &= \Omega - l\sin\theta\frac{\partial\Omega}{\partial x} + \frac{l^2\sin^2\theta}{1 \cdot 2}\frac{\partial^2\Omega}{\partial x^2} - \cdots\end{aligned}\right\} \quad (11)$$

Thus if the strength of each pole of the needle be m, the energy of the needle in the given position is $m(\Omega_2 - \Omega_1)$, supposing the positive end at B. By (11) we have, writing M, the magnetic moment of the needle, for $2ml$

$$m(\Omega_2 - \Omega_1) = -M\sin\theta\left\{\frac{\partial\Omega}{\partial x} + \frac{l^2\sin^2\theta}{3!}\frac{\partial^3\Omega}{\partial x^3} + \frac{l^4\sin^4\theta}{5!}\frac{\partial^5\Omega}{\partial x^5} + \cdots\right\}$$

$$= 2\pi\gamma M\left\{\sin\theta\left|\frac{\partial^2 A_0}{\partial x^2} - \frac{\cos^2\theta}{2^2}\right|l^2\frac{\partial^4 A_0}{\partial x^4} + \frac{\cos^4\theta}{2^2 \cdot 4^2}\right|l^4\frac{\partial^6 A_0}{\partial x^6} - \cdots\right\} \quad (12)$$
$$+ \frac{\sin^2\theta}{3!} \quad -\frac{\sin^2\theta\cos^2\theta}{3! \, 2^2}$$
$$+ \frac{\sin^4\theta}{5!}$$

CORRECTION FOR DIMENSIONS OF SECTION

where y is replaced by $l\cos\theta$. If instead of a single turn of wire there be N turns, which may be taken as coincident, we must write $N\gamma$ instead of γ in this equation.

The couple Θ acting on the needle is thus given numerically by

$$\Theta = m\frac{\partial(\Omega_2 - \Omega_1)}{\partial\theta} = 2\pi N\gamma M \cos\theta \frac{a^2}{r^3}\left\{1 + \frac{3}{2^2}\frac{l^2}{r^4}(a^2 - 4x^2)(1 - 5\sin^2\theta)\right.$$

$$\left. + \frac{3^2 \cdot 5}{2^2 \cdot 4^2}\frac{l^4}{r^3}(a^4 - 12a^2x^2 + 8x^4)(1 - 14\sin^2\theta + 21\sin^4\theta) + \ldots\right\} \quad (13)$$

a formula of great importance in galvanometry.

If the needle be not uniformly magnetized the value of l is not definite. It is easy to see however that M, the magnetic moment of the magnet, should be used in the first term. In the other terms l^2, l^4, &c. should be replaced by quantities depending on the distribution of magnetism on the needle. This however it is in general impossible to determine for a small needle.

If l be very small the expression on the right of (13) reduces to the first term approximately; and if also $x = 0$, that is, if the centre of the needle is at the centre of the circle, we have

$$\Theta = 2\pi N\gamma M \cos\theta/a \quad \ldots \quad (13')$$

The principal term in the expression on the right of (13) is the first $2\pi N\gamma M \cos\theta a^2/r^3$, which, by (9) and (10), is the value of the couple when l is so small that the component R of magnetic force is negligible, and the value which F has at the centre of the needle is taken as the force at each pole. Now we have for the couple in that case

Modification of Formulæ to allow for Dimensions of Coil-Section.

$$\Theta = 2\pi N\gamma M \cos\theta a^2/r^3 = FM\cos\theta$$

so that $F = 2\pi N\gamma a^2/r^3$. We have to find what takes the place of $2\pi N\gamma a^2/r^3$, or F, in (13) when the coil cannot be treated as a simple circular conductor. For the other terms unless the dimensions of the bobbin are larger than usual the coil may be taken as a single circular conductor coinciding with the mean circle of the bobbin, and carrying the whole current. The case of a long bobbin we shall consider specially.

Let the breadth in the direction of the axis of the cross-section of the coil by a plane through the axis be $2b$, and the radial depth of the section $2d$. Let BC (Fig. 56) be a radius drawn from the centre C of the coil in that plane which cuts the coil into two equal and similar coils, and taking $DE(=h)$, $CD(=k)$ at right

MAGNETIC ACTION OF COILS

angles to one another, we have $dhdk$ for the area of the element E of the cross-section of the coil by a plane passing through the axis and through BC. Also $PE^2 = (x-h)^2 + k^2$. Let, further, n be the number of turns crossing unit of area of cross-section, and γ the current in each. The current crossing the element E is $n\gamma dhdk$, for we here suppose the wire so fine that we may suppose the current everywhere crossing any area of cross-section proportional to that area.* Hence by the law (p. 143 above) which we may assume as to the magnetic action of the

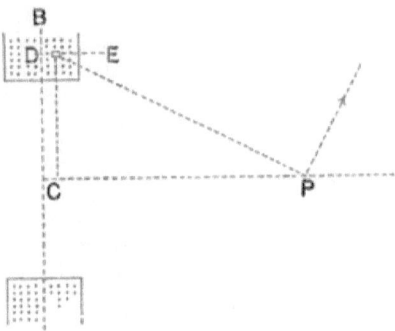

Fig. 57.

elements of a circuit, the force exerted on a unit magnetic pole at P, by an element, of area $dhdk$ and length ds, at right angles to the plane of the paper is $n\gamma ds dhdk/\{(x-h)^2 + k^2\}^{\frac{3}{2}}$. Hence if dF be the component in this direction due to the whole ring, of which the element E is the cross-section

$$dF = 2\pi n\gamma \frac{k^2 dhdk}{\{(x-h)^2 + k^2\}^{\frac{3}{2}}}.$$

The whole magnetic force parallel to the axis is therefore

$$F = 2\pi n\gamma \int_{-b}^{+b}\int_{a-d}^{a+d} \frac{k^2 dhdk}{\{(x-h)^2 + k^2\}^{\frac{3}{2}}},$$

---

* When the layers of wire form each a helix we here neglect the axial component of flow. How this may be compensated will be explained in the next chapter.

or after integration

$$F = 2\pi n\gamma \left\{ (x+b) \log \frac{a+d+\sqrt{(x+b)^2+(a+d)^2}}{a-d+\sqrt{(x+b)^2+(a-d)^2}} \right.$$
$$\left. - (x-b) \log \frac{a+d+\sqrt{(x-b)^2+(a+d)^2}}{a-d+\sqrt{(x-b)^2+(a-d)^2}} \right\} \quad . \quad (14)$$

which reduces when $x = 0$, to

$$F = \pi N\gamma \frac{1}{d} \log \frac{a+d+\sqrt{(a+d)^2+b^2}}{a-d+\sqrt{(a-d)^2+b^2}} \quad . \quad . \quad (15)$$

and when b and d are small enough to

$$F = 2\pi N\gamma/a \quad . \quad . \quad . \quad . \quad . \quad . \quad (16)$$

where N is the total number of turns in the coil.

The value of F in (14) is to be used, when required, instead of $2\pi N\gamma a^2/r^3$, so far as the first term of the series for F is concerned; the remainder of the series is to be retained without alteration as sufficiently accurate for practical purposes.

The second term of the series in (13), involving the product $(a^2 - 4x^2)(1 - 5\sin^2\theta)$, may be made to vanish by arranging so that one or both of the factors may vanish. The value of the second factor is 1 when $\theta = 0$, and diminishes as θ (whether positive or negative) increases in numerical value, until, when $\theta = \pm \sin^{-1}(1/\sqrt{5}) = \pm 26° 34'$, it is zero. Thereafter it becomes negative, and approaches -4 as θ approaches $90°$. At $45°$ its value is $-3/2$. *Removal of Second Term in Series for F.*

The first factor may be made to vanish by placing the needle so that $x = a/2$. This was done by Gaugain in his galvanometer, which consisted of a vertical coil with a needle so suspended that its centre was as nearly as possible on the axis of the coil, at a distance equal to half its radius. The uncertainty as to the proper distance, caused by the dimensions of the cross-section of the coil itself, was got over by winding the wire on a conical surface of semi-vertical angle $\tan^{-1} 2$, so that the distance of the needle, suspended with its centre as nearly as possible at the vertex, might be in the proper position relatively to each spire. *Gaugain's Galvanometer.*

With proper arrangements this winding of the coil, though more difficult than that of an ordinary bobbin, might be carried out with sufficient exactness; but any inaccuracy in the placing

Objections to Gaugain's Arrangement.

of the needle is serious. For by (13) the value of $\partial\Theta/\partial x$ is $-2\pi N\gamma M \cos\theta \, 3a^2/r^4 \cdot \partial r/\partial x$, and therefore Θ requires correction for an error dx in placing the needle, by multiplication by the factor $1 + 1/\Theta \cdot \partial\Theta/\partial x \cdot dx$, or $1 - 3xdx/r^2$, or since $x = a/2$, by the factor $1 - 6dx/5a$. Thus if dx is sensible, this factor, depending as it does on $1/a$, seriously affects the value of Θ.

Helmholtz's Arrangement of Galvanometer Coils.

Gaugain's galvanometer has been improved upon by von Helmholtz, in whose arrangement two equal parallel coils are placed with their medial planes at a distance apart equal to their mean radius. The needle is suspended with its centre as nearly as may be on the axis, at a point about which the arrangement of coils is symmetrical; and the coils are so joined that the current flows in the same direction round both. This makes $a^2 - 4x^2 = 0$, very approximately, in Θ, and further obviates the uncertainty just referred to. For any displacement of the needle towards the coil is attended by a diminution of the couple due to the other coil, and a very nearly equal increase of the couple due to that which is approached.

Uniformity of Field about Centre of Helmholtz's Arrangement.

The field due to the arrangement is shown in Fig. 58,* and may be contrasted with that for a simple coil shown in Fig. 55. It will be seen from the diagram of lines of force, and the same thing is obvious from (10) (since the values of x for the two coils are equal and opposite), that R is zero at every point in the plane midway between the coils, and passing therefore (approximately) through the centre of the needle, and also very nearly zero at points even at some distance on either side of this plane. Thus over quite a considerable space surrounding the centre of the needle, the field due to the coils is practically uniform and parallel to the axis, and the couple practically independent of θ, and unaffected by an error in centring the needle which would have a serious effect on the couple in the case of a single coil.

Couple on Needle in Helmholtz's Arrangement.

It is clear that the energy of the needle in the field of the double coil is that given in (12). For the energy of the positive pole, supposed nearer to the coil from which it is repelled, is $m\Omega_2$ in the field of that coil, and $-m\Omega_1$ in the field of the other coil. The energy of the other pole has evidently the same value, so that the whole energy is $2m(\Omega_2 - \Omega_1)$. The couple is thus 2Θ, where Θ is given by (13), subject to the condition that $a^2 = 4x^2$. It may be written therefore to terms of the fourth order inclusive

$$\Theta = 4\pi N\gamma M \cos\theta \frac{a^2}{r^3}\left\{1 - \frac{3}{2}\frac{45}{64}\frac{l^4 a^4}{r^8}(1 - 14\sin^2\theta + 21\sin^4\theta)\right\} \quad (17)$$

---

* This cut is taken from Maxwell's *Electricity and Magnetism*, vol. ii.

FIELD OF HELMHOLTZ GALVANOMETER

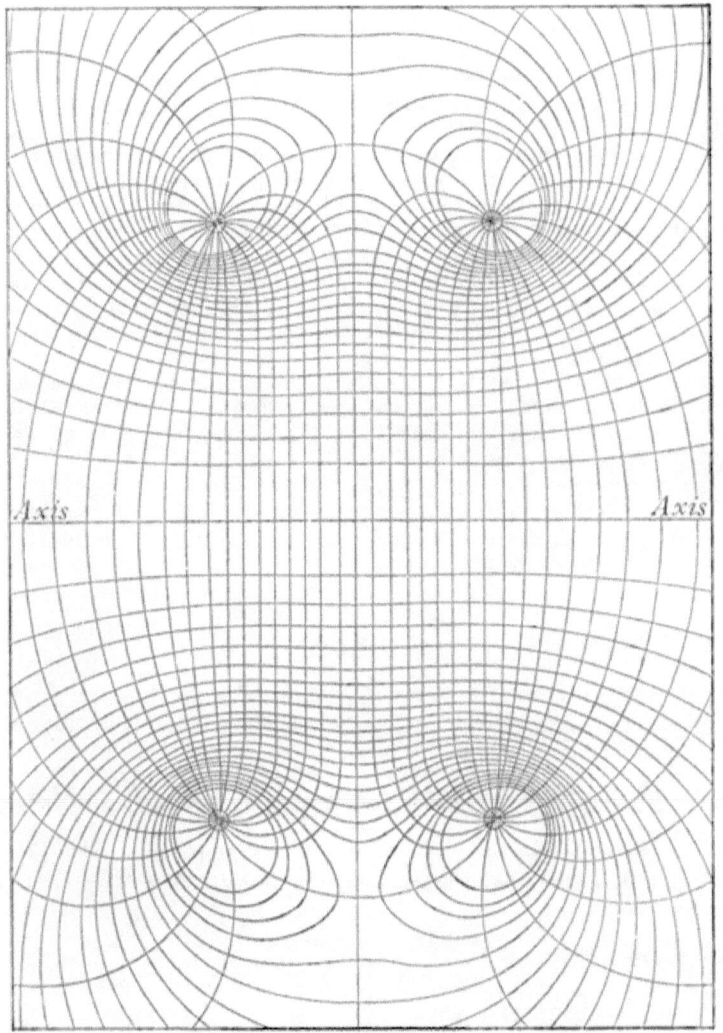

Fig. 58.

where since $a^2 = 4x^2$, $r^2 = 5x^2$, and N is the number of turns in *each* coil.

Values of θ which satisfy the equation $1 - 14\sin^2\theta + 21\sin^4\theta = 0$, render the factor of the second term in brackets zero. These values are $16° 34'$ and $49° 55'$. The factor in brackets has two maximum numerical values, viz. 8 for $\theta = \pm 90°$, and $-4/3$ for $\theta = \pm 35° 16'$.

Approximate Calculation of Effect of Finite Cross-section of Coil.

To take into account the distribution of the wire over the finite cross-section of the bobbin, we may take the coil just considered as an elementary ring of the real coil, and, regarding the distance x and radius a of this ring as subject to variation, find from each term in the expression of any effect produced on the needle by the central ring, the corresponding term of the effect produced by any other parallel ring of the coil. From this we can find an expression for the average value of the term for the whole coil.

Method of Calculation of Average Effect.

Thus let P_0 denote any term of the expression for the action, whatever its nature, on the needle produced by the central circular filament. If then P be the corresponding term for a filament the coordinates of which reckoned from the centre of the cross-section coil are h, k, and the area of cross-section of which is $dhdk$, $\overline{P}$ the average term for the action of the whole coil, and $2b$, $2d$ be the axial breadth and radial depth of the coil, we have by definition

$$4bd\overline{P} = \int_{-b}^{b}\int_{-d}^{d} P\,dh\,dk \quad \ldots \ldots \quad (18)$$

But, since the value of P for this term is obtained by substituting in the expression $x - h$ for x and $a + k$ for k, by Taylor's theorem

$$P = P_0 - h\frac{\partial P_0}{\partial x} + k\frac{\partial P_0}{\partial a} + \frac{h^2}{1.2}\frac{\partial^2 P_0}{\partial x^2} + \frac{k^2}{1.2}\frac{\partial^2 P_0}{\partial a^2} + \cdots$$

Multiplying this value of P by $dhdk$, and integrating as indicated in (18) between the limits $-b, +b$ for h, and $-d, +d$ for k we find

$$4bd\overline{P} = 4bd P_0 + \frac{4b^3 d}{6}\frac{\partial^2 P_0}{\partial x^2} + \frac{4bd^3}{6}\frac{\partial^2 P_0}{\partial a^2}$$
$$+ \frac{4b^5 d}{5!}\frac{\partial^4 P_0}{\partial x^4} + \frac{8b^3 d^3}{5!}\frac{\partial^4 P_0}{\partial x^2 \partial a^2} + \frac{4bd^5}{5!}\frac{\partial^4 P_0}{\partial a^4} \quad . \quad . \quad (19)$$

since the terms of odd order vanish in the integration.

CORRECTION FOR CROSS-SECTION OF COIL.

We apply this result to the correction of the values of F and L given in (9) and (13) by treating the terms separately as follows. It will suffice to take F, as the results obtained will apply at once to Θ also.

First Correction. Terms for Simple Coil.

A first approximation to F' for the whole coil is obtained by writing $4ab\kappa\gamma$ (or $N\gamma$ if N is the whole number of turns) for γ, since this is the whole current flowing across each section. To correct for the distribution of the turns, we take first the factor a^2/r^3, and call it P_0. Differentiating we find $\partial^2 P_0/\partial x^2 = 3a^2(4x^2 - a^2)/r^7$, $\partial^2 P_0/da^2 = (2x^4 - 11x^2 a^2 + 2a^4)/r^7$, so that taking the first three terms of (19)

$$P = \frac{a^2}{r^3} + \frac{b^2}{6}\frac{3a^2}{r^7}(4x^2 - a^2) + \frac{d^2}{6}\frac{1}{r^7}(2x^4 - 11x^2 a^2 + 2a^4). \quad (20)$$

and this takes the place of a^2/r^3 in (9) and (13).

If the coil is a Helmholtz arrangement, in which $4x^2 = a^2$, the second term disappears, and we have after reduction

Application to Helmholtz Double-Coil.

$$\bar{F} = \frac{8}{5\sqrt{5}a}\left(1 - \frac{1}{15}\frac{d^2}{a^2}\right),$$

and the first term of F takes the corrected form

$$\frac{32\pi N\gamma}{5\sqrt{5}a}\left(1 - \frac{1}{15}\frac{d^2}{a^2}\right).$$

where N is the number of turns in each of the two coils.

The second term of F may be corrected in the same way by taking $a^2(a^2 - 4x^2)/r^7$ for P_0. We have $\partial^2 P_0/\partial x^2 = -3 \cdot 5a^2(8x^4 - 12x^2 a^2 + a^4)/r^{11}$, and $\partial^2 P_0/\partial a^2 = -(8x^6 - 136x^4 a^2 + 159x^2 a^4 - 12a^6)/r^{11}$, so that to three terms

Second Correction. Terms for Single Coil.

$$\bar{P} = \frac{a^2}{r^7}(a^2 - 4x^2) - \frac{b^2}{6}\frac{3\cdot 5}{r^{11}}a^2(8x^4 - 12x^2 a^2 + a^4)$$

$$- \frac{d^2}{6}\frac{1}{r^{11}}(8x^6 - 136x^4 a^2 + 159x^2 a^4 - 12a^6) \quad (21)$$

which takes the place of $a^2(a^2 - 4x^2)/r^7$ wherever the latter occurs.

Again, if the coil is a Helmholtz arrangement, this value of $\bar{P}$ is simplified. Its first term disappears altogether on account of

258   MAGNETIC ACTION OF COILS

Application to Helmholtz Double-Coil. the relation $4x^2 = a^2$, which also reduces the remaining two terms so that

$$\bar{P} = \frac{5}{6 \cdot 2^3} \frac{a^6}{r^{11}} (36b^2 - 31d^2) \quad \ldots \quad (22)$$

where $r^2 = 5/4 \cdot a$.

Hence, taking in only second powers of b and d, and the first three terms of (9), we have for the Helmholtz arrangement

$$F = \frac{32\pi N\gamma}{5\sqrt{5}a} \left\{ (1 - \frac{1}{15}\frac{d^2}{a^2}) + \frac{2^4 y^2}{5^3 a^4}(36b^2 - 31d^2) - \frac{2 \cdot 3^3}{5^3} \frac{y^4}{a^4} \right\} . \quad (23)$$

Further Correction by Properly Proportioned Cross-Section. The value of $\bar{P}$ in (22), and therefore also the second term of F for any arrangement can be made to vanish by constructing the coil so that $b^2 = 31/36 \cdot d^2$. If this is done for a Helmholtz galvanometer, the value is, for that instrument, given to a very high degree of approximation by

$$\Theta = \frac{32\pi N\gamma M \cos\theta}{5\sqrt{5}a} \left\{ (1 - \frac{1}{15}\frac{d^2}{a^2}) \right.$$

$$\left. - \frac{2 \cdot 3^3}{5^3}\frac{l^4}{a^4}(1 - 14\sin^2\theta + 21\sin^4\theta) \right\} \quad \ldots \quad (24)$$

If the length of the needle, as it ought always to be, is small in comparison with a, the value of Θ for the Helmholtz arrangement may, within the limits of errors of observation, be taken as given by the formula obtained by omitting the term involving l^4/a^4 on the right in (24).

Galvanometer with Four Coaxial Coils. If four coaxial coils be arranged so that the current flows through them all in the same direction, the values of F at the same point due to the separate coils will have the same sign. Consider then the component magnetic force at a point O symmetrically situated with reference to the coils, which are arranged in pairs, those of each pair having equal radii, and being at equal distances along the axis on opposite sides of the point at which F is taken. Let a, a, be the radii of the coils, x, ξ, the distances of their planes from O, N, N', the number of turns in each, and $r^2 = x^2 + a^2, \rho^2 = \xi^2 + a^2$. Then to three terms

$$F = 4\pi\gamma \left[N\frac{a^2}{r^3} + N'\frac{a^2}{\rho^3} + \frac{3}{2^2}y^2 \left\{ \frac{N}{r^7}a^2(a^2 - 4x^2) + \frac{N'}{\rho^7}a^2(a^2 - 4\xi^2) \right\} \right.$$

$$\left. + \frac{3^2 \cdot 5}{2^2 \cdot 4^2}y^4 \left\{ \frac{N}{r^{11}}a^2(a^4 - 12a^2x^2 + 8x^4) + \frac{N'}{\rho^{11}}a^2(a^4 - 12a^2\xi^2 + 8\xi^4) \right\} \right] \quad (25)$$

Now we can impose the condition that $r = \rho$, that is, that the coils should lie on a sphere having its centre at O, and so choose a, a, x, ξ, that the coefficients of y^2, y^4, may vanish identically. We thus have fulfilled by these four quantities the equations

Conditions that Terms in y^2 and y^4 should vanish.

$$Na^2(a^2 - 4x^2) + N'a^2(a^2 - 4\xi^2) = 0$$
$$Na^2(a^4 - 12a^2x^2 + 8x^4) + N'a^2(a^4 - 12a^2\xi^2 + 8\xi^4) = 0.$$

We may write $a^2 - 4x^2 = 5a^2 - 4r^2$, and $a^4 - 12a^2x^2 + 8x^4 = 21a^4 - 28a^2r^2 + 8r^4$, so that calling ϕ, ϕ', the angles which the radii of the coils subtend at O, and putting m for N/N', we may write the equations in the form

$$\left. \begin{array}{l} m\sin^2\phi(4 - 5\sin^2\phi) + \sin^2\phi'(4 - 5\sin^2\phi') = 0 \\ m\sin^2\phi(21\sin^4\phi - 28\sin^2\phi + 8) \\ \qquad + \sin^2\phi'(21\sin^4\phi' - 28\sin^2\phi' + 8) = 0 \end{array} \right\} \quad (26)$$

Since $\sin\phi$, $\sin\phi'$ can never exceed 1, these equations necessitate the fulfilment of certain conditions by m, $\sin\phi$, $\sin\phi'$, and, subject to these, any number of arrangements can be found to carry out the object stated. If however $\sin\phi = 1$, so that one pair of circles coincide in the equatorial plane through O, we have from (26)

Galvanometer with Three Coaxial Coils.

$$21\sin^4\phi' - 33\sin^2\phi' + 12 = 0$$

which is satisfied by $\sin^2\phi' = 4/7$, or by $\sin^2\phi' = 1$.

The second solution, in which all the coils are round the equator of the sphere, is not relevant, inasmuch as it would make $m = -1$, which may be interpreted to mean that the number of turns on each coil should be the same, and that the currents should flow in opposite directions, that is that there should be no current at all on the whole, and therefore no magnetic effect.

The solution $\sin^2\phi' = 4/7$, gives $m = 32/49$, that is, the circles surrounding the centre should each contain 32 turns for every 49 turns contained in each of the others, and the latter should be placed on the two sides of the great circle of the sphere bisecting the axis, at a distance in each case of $\sqrt{3/7}$ of the radius, and have a corresponding radius of $2/\sqrt{7}$ of that of the sphere.

Conditions for Uniformity of Field about Axis.

We now consider a long right cylindrical solenoid. Such a solenoid can be very approximately constructed by winding a close single layer of fine wire, so that the mean radius of the single layer may be taken with sufficient accuracy as the radius

Long Coil of Single Layer of Fine Wire.

s 2

of the wire, and the wire may be regarded as everywhere at right angles to the axis. Such a single layer coil is very convenient for accurate work since there can be no uncertainty as to the winding.

The value of F for a single turn of such a coil is given by (9), which, taking here for convenience the origin of coordinates at the centre of the coil, and x, ξ, as the axial distances of the point P considered, and the turn in question, we may write

$$F = 2\pi\gamma \frac{a^2}{r^3}\left[1 + \frac{3}{2^2}\frac{y^2}{r^4}\{a^2 - (x-\xi)^2\}\right.$$
$$\left. + \frac{3^2}{2^2 \cdot 4^2}\frac{5}{r^8}\{a^4 - 12a^2(x-\xi)^2 + 8(x-\xi)^4\} + \ldots\right] \quad . \quad (27)$$

where $r^2 = a^2 + (x-\xi)^2$.

Similarly the value of the radial component, R, may be written down. The value of the couple Θ exerted on the needle could easily be found also; but it will be given later [see equation (54) below].

If then n be the number of turns per unit of length γ we have to replace γ by $n\gamma d\xi$. Hence if $2l$ be the axial length of the coil we have for the total force

$$F = 2\pi n\gamma \int_{-l}^{+l} d\xi \frac{a^2}{r^3}\left[1 + \frac{3}{2^2}\frac{y^2}{r^4}\{a^2 - (x-\xi)^2\} + \ldots\right] \quad (27')$$

But clearly the expansion in (27) is $-\partial\Omega/\partial x$. if Ω be given by (2) with x replaced by $x - \xi$. But $-\partial\Omega/\partial x$, is $+\partial\Omega/\partial\xi$, so that (2) gives at once the integral $+\Omega$ for (27'). Hence taking the integral between the limits $-l$ and $+l$ for ξ, and writing $r_1 = \sqrt{a^2 + (x-l)^2}$, $r_2 = \sqrt{a^2 + (x+l)^2}$, we find

$$F = 2\pi n\gamma\left[\frac{x+l}{r_1} - \frac{x-l}{r_2} + \frac{y^2}{2^2}3a^2\left\{\frac{x+l}{r_1^5} - \frac{x-l}{r_2^5}\right\}\right.$$
$$+ \frac{y^4}{2^2 \cdot 4^2}3 \cdot 5a\left\{\frac{x+l}{r_1^9}(3a^2 - 4(x+l)^2)\right.$$
$$\left.\left. - \frac{x-l}{r_2^9}(3a^2 - 4(x-l)^2)\right\} + \ldots\right] \quad . \quad . \quad (28)$$

which holds for all points whether inside or outside the solenoid.

POTENTIAL AND FORCE AT CENTRE OF COIL

If $y = 0$ this gives

$$F = 2\pi n\gamma \left(\frac{x+l}{r_1} - \frac{x-l}{r_2} \right) \quad \ldots \quad (28')$$

that is if the point at which F is taken be on the axis, and ψ_1, ψ_2 be the angles which r_1, r_2 make with the axis, as shown in Fig. 59.

$$F = 2\pi n\gamma (\cos\psi_2 - \cos\psi_1) \quad \ldots \quad (28'')$$

If the coil be very long r_1, r_2, approximate for internal points not near the ends, more and more nearly to $x + l$, $l - x$, so that all terms vanish in (28) except the first two. For such points $x - l$ is negative, and approximately $(x + l)/r_1 - (x - l)/r_2 = 2$. Thus the field within a long coil is uniform except near the ends and its intensity is given by

$$F = 4\pi n\gamma \quad \ldots \ldots \quad (29)$$

To take into account different layers if there are more than one, the best course in any practical case is (since only a limited number of layers would be employed) to calculate F, by (27) above, for each, and add the results together.

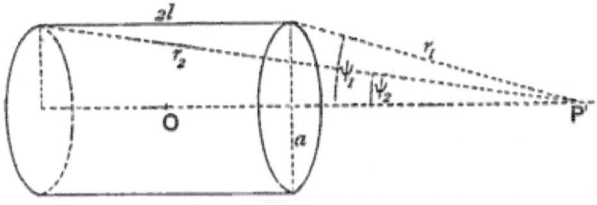

FIG. 59.

The result expressed in (28'') can of course be obtained at once by direct calculation. The potential due to a circular current of strength $n\gamma dx$, at a point P' (Fig. 59) on the axis at numerical distance x from the plane of the circle is $n\gamma\omega dx$, where ω is the solid angle subtended at the point by the circle. But if ψ be the angle subtended by the radius of the circle $\omega = 2\pi(1 - \cos\psi)$. Thus if Ω be the potential of magnetic induction due to the whole solenoid

Direct Calculation for Potential and Force at Centre of Coil.

$$\Omega' = 2\pi n\gamma \int_{\psi=\psi_1}^{\psi=\psi_2} (1 - \cos\psi)\,dx$$

Ω is also the mutual energy of the solenoid and a unit pole placed at P'. Reckoning then x as the distance of any turn from P', the force in the direction of x on the solenoid (that is from the pole towards the solenoid) is $-\partial\Omega/\partial x$, and this is the force F on the pole at P' in the opposite direction. Thus

$$F = -2\pi n\gamma \int_{\psi=\psi_1}^{\psi=\psi_2} \frac{\partial}{\partial x}(1 - \cos\psi)\,dx = 2\pi n\gamma(\cos\psi_2 - \cos\psi_1).$$

Or Ω and F may be found thus. The potential produced by a circular disk of positive magnetism of surface density $n\gamma$ and radius a, at a point on the axis distant x from the disk is $2\pi n\gamma(\sqrt{a^2 + x^2} - x)$. The repulsion due to the disk on unit pole at P', is therefore $2\pi n\gamma(1 - x/\sqrt{a^2 + x^2}) = 2\pi n\gamma(1 - \cos\psi)$. Hence for two equal positive and negative coaxial disks subtending angles ψ_1, ψ_2 respectively at P', and at distances x_1, x_2, the potential and force are $2\pi n\gamma\{\sqrt{a^2 + x_2^2} - x_2 - (\sqrt{a^2 + x_1^2} - x_1)\}$ and $2\pi n\gamma(\cos\psi_2 - \cos\psi_1)$.

The magnetic potential and force at a point at distance y from the axis can also be found as follows. It has been shown (p. 43 above), that the energy of a magnetic shell in a magnetic field is equal to the total induction through the shell multiplied by the strength of the shell. Hence in order to find the force on a pole placed in the field of the solenoid we have to calculate the magnetic induction at the point.

Solenoid regarded as Lamellar Magnet. Now we may regard the solenoid as a lamellar distribution of magnetism, the direction of magnetization of which is everywhere parallel to the axis. Hence by (68) and (70) of Chap. I. above, if Ω be the potential of magnetic induction in the interior of the solenoid

$$\Omega = V - 4\pi\phi$$

where $\phi\,(= \int \partial\phi/\partial x\,.\,dx)$ is the sum of the strengths of the shells traversed by a point imagined to move parallel to the axis from an adopted zero, to the point where the potential is to be found. But if we suppose x to increase from the negative towards the positive end of the solenoid, we have $\partial\phi/\partial x = n\gamma$, and hence reckoning from the zero of x, $4\pi\phi = 4\pi n\gamma x$.

Potential at Internal Point.

V in the present case is simply the potential due to the ends of the solenoid, which may be regarded as two uniform parallel circular disks of magnetism having densities σ, $-\sigma$, respectively. Hence if V_1 be the potential due to the positive disk, V_2 that due to the negative $V = V_1 - V_2$, and

$$\Omega = V_1 - V_2 - 4\pi n\gamma x \quad \ldots \quad \ldots \quad (30)$$

For an external point $\Omega = V_1 - V_2$ simply. The values of V_1, V_2 can be found from (6) above and the value of F then found by differentiation of (30). The result, as the reader may verify, agrees with (28).

Long Coil of Several Layers.

The method described above (p. 256) may also, if desired, be employed to take into account the radial depth of the coil. Supposing the number of layers per unit of depth to be n', the number in unit area of cross-section is nn'. Thus if $2d$ be the depth of the coil, the number of turns in unit of length is $2nn'd$, and this must replace n in (27). Taking then as P_0 any term of the expression for the effect of the mean coaxial current sheet, the average value, $\overline{P}$, of the term, for all the coaxial sheets into which the coil may be supposed divided, is given by the equation

$$\overline{P} = P_0 + \frac{d^2}{6}\frac{\partial^2 P_0}{\partial a^2} + \frac{d^4}{5!}\frac{\partial^4 P_0}{\partial a^4}.$$

Taking first from (28)

$$\overline{P}_0 = \frac{x+l}{r_1} - \frac{x-l}{r_2},$$

we have

$$\frac{\partial^2 P_0}{\partial a^2} = (x+l)\frac{2a^2 - (x+l)^2}{r_1^5} - (x-l)\frac{2a^2 - (x-l)^2}{r_2^5}.$$

Therefore to the second power of d

$$P = \frac{x+l}{r_1} - \frac{x-l}{r_2} + \frac{d^2}{6}\bigg((x+l)\frac{2a^2 - (x+l)^2}{r_1^5}$$

$$- (x-l)\frac{2a^2 - (x-l)^2}{r_2^5}\bigg).$$

Next taking the second term of (28),

$$P_0 = \frac{a^2(x+l)}{r_1^5} - \frac{a^2(x-l)}{r_2^5}$$

and therefore

$$\overline{P} = \frac{a^2(x+l)}{r_1^5} - \frac{a^2(x-l)}{r_2^5} + \frac{d^2}{6}\Big\{\frac{x+l}{r_1^5}(12a^4 - 21a^2(x+l)^2 + 2(x+l)^4)$$

$$- \frac{x-l}{r_2^5}(12a^4 - 21a^2(x-l)^2 + 2(x-l)^4)\Big\}.$$

Hence to terms in d^2 and y^2, (27) becomes (x_1 being put for $2nn'd$, the number of terms per unit length)

$$F = 2\pi n_1 \gamma \Big[\frac{x+l}{r_1} - \frac{x-l}{r_2} + \frac{d^2}{6}\Big((x+l)\frac{2a^2 - (x+l)^2}{r_1^5}$$

$$- (x-l)\frac{2a^2 - (x-l)^2}{r_2^5}\Big)$$

$$+ \frac{3}{2^2} y^2 \Big\{\frac{a^2(x+l)}{r_1^5} - \frac{a^2(x-l)}{r_2^5} + \frac{d^2}{6}\frac{x+l}{r_1^5}(12a^4 - 21a^2(x+l)^2 + 2(x+l)^4)$$

$$- \frac{d^2}{6}\frac{x-l}{r_2^5}(12a^4 - 21a^2(x-l)^2 + 2(x-l)^4)\Big\} \quad . \quad . \quad . \quad (31)$$

Potential, &c. of Circular Current. Equations equivalent to (2), (9), (10), (12) may be obtained by first expanding A_0 in ascending powers of x or a according as $x <$ or $> a$. These equations are convenient only when the point considered is near to or far from the plane of the circular current, as only then are the series sufficiently convergent. We have

Expansions available for Near or Distant Points.

$$A_0 = \sqrt{a^2 + x^2} - x$$

$$= a\Big\{1 + \tfrac{1}{2}\frac{x^2}{a^2} - \frac{1.1}{2.4}\frac{x^4}{a^4} + \frac{1.1.3}{2.4.6}\frac{x^6}{a^6} - \ldots\Big\} - x$$

$$(x < a)$$

$$= x\Big\{\tfrac{1}{2}\frac{a^2}{x^2} - \frac{1.1}{2.4}\frac{a^4}{x^4} + \frac{1.1.3}{2.4.6}\frac{a^6}{x^6} - \ldots\Big\}$$

$$x > a$$

$$\quad . \quad (32)$$

EXPANSIONS FOR CIRCULAR CURRENT

Calculating $\partial A_0/\partial x$, $\partial^3 A_0/\partial x^3$, &c. from these and substituting in (6), we find if $x < a$

Expansions for Potential.

$$\Omega = 2\pi\gamma \Big\{ 1 - \frac{x}{a} + \frac{1}{2}\frac{x^3}{a^3} - \frac{1.3}{2.4}\frac{x^5}{a^5} + \dots$$

$$+ \frac{1}{2^2}\frac{y^2}{a^3}\Big(-1.3x + \frac{1.3.5}{2}\frac{x^3}{a^2} - \frac{1.3.5.7}{2.4}\frac{x^5}{a^4} + \dots\Big)$$

$$+ \frac{1}{2^2.4^2}\frac{y^4}{a^5}\Big(-1.3^2.5x + \frac{1.3.5^2.7}{2}\frac{x^3}{a^2} - \frac{1.3.5.7^2.9}{2.4}\frac{x^5}{a^4} + \dots\Big)$$

$$+ \dots \dots \dots \dots \dots \dots \dots \dots \Big\} \quad (33)$$

Or if $x > a$

$$\Omega = 2\pi\gamma \Big\{ \tfrac{1}{2}\frac{a^2}{x^2} - \frac{1.3}{2.4}\frac{a^4}{x^4} + \frac{1.3.5}{2.4.6}\frac{a^6}{x^6} - \dots$$

$$+ \frac{1}{2^2}\frac{y^2}{x^2}\Big(-1.3\frac{a^2}{x^2} + \frac{1.3.5}{2}\frac{a^4}{x^4} - \frac{1.3.5.7}{2.4}\frac{a^6}{x^6} - \dots\Big)$$

$$+ \frac{1}{2^2.4^2}\frac{y^4}{x^4}\Big(3.5.4\frac{a^2}{x^2} - \frac{1.3.5.7}{2}6\frac{a^4}{x^4} + \frac{1.3.5.7.9}{2.4}8\frac{a^6}{x^6} - \dots\Big)$$

$$+ \dots \dots \dots \dots \dots \dots \dots \Big\} \quad (33')$$

Hence since $F = -\partial\Omega/\partial x$

Expansions for Axial Component Force.

$$F = 2\pi\gamma\frac{1}{a}\Big\{ 1 - \frac{1.3}{2}\frac{x^2}{a^2} + \frac{1.3.5}{2.4}\frac{x^4}{a^4} - \dots$$

$$+ \frac{1}{2^2}\frac{y^2}{a^2}\Big(1.3 - \frac{1.3^2.5}{2}\frac{x^2}{a^2} + \frac{1.3.5^2.7}{2.4}\frac{x^4}{a^4} - \dots\Big)$$

$$+ \frac{1}{2^2.4^2}\frac{y^4}{a^4}\Big(1.3^2.5 - \frac{1.3^2.5^2.7}{2}\frac{x^2}{a^2} + \frac{1.3.5^2.7^2.9}{2.4}\frac{x^4}{a^4} - \dots\Big)$$

$$+ \dots \dots \dots \dots \dots \dots \dots \Big\} \quad (34)$$

if $x < a$; or if $x > a$

$$F = 2\pi\gamma\frac{a^2}{x^3}\Big\{1 - \frac{1.3}{2}\frac{a^2}{x^2} + \frac{1.3.5}{2.4}\frac{a}{x^4} - \cdots$$

$$+ \frac{1}{2^2}\frac{y^2}{x^2}\Big(-3.4 + \frac{1.3.5}{2}6\frac{a^2}{x^2} - \frac{1.3.5.7}{2.4}8\frac{a^4}{x^4} + \cdots\Big)$$

$$+ \frac{1}{2^2.4^2}\frac{y^4}{x^4}\Big(3.5.4.6 - \frac{1.3.5.7}{2}6.8\frac{a^2}{x^2} + \frac{1.3.5.7.9}{2.4}8.10\frac{a^4}{x^4} - \cdots\Big)$$

$$+ \cdots\cdots\cdots\cdots\cdots\cdots\cdots\cdots\Big\} \quad (34')$$

Expansions for Couple on Small Magnet with Centre on Axis.

If instead of a single turn of wire, the circle consist of N turns, each carrying a current γ, the above expressions must of course be multiplied by N.

Finally, multiplying these values of F by $M\cos\theta$ the second term in each by $(1 - 5\sin^2\theta)$, the third terms by $1 - 14\sin^2\theta + 21\sin^4\theta$, and changing y into l, we get the values of Θ for the respective cases $x < a$, $x > a$.

It is to be carefully observed that in all these expressions it is necessary for convergence that $y < a$ when $x < a$, and $y < x$ when $a < x$.

It will be seen that the formulas just obtained are simply those previously found for the different cases, with the finite expressions which constitute the different terms in the latter replaced by infinite series. In the majority of practical cases it is much more convenient to calculate numerically the values of the finite expressions. The series are in fact only useful for points very near the plane of the circle, or very far from it. In the former case equations (33), (34) are applicable, in the latter (33'), (34').

When the coil has a finite cross-section the last found expressions may be readily corrected by direct integration; or the process explained at p. 256 above may be used. We cannot here afford space for the corrected expressions, which would seldom be needed; but the reader will have no difficulty in writing them down for himself.

Mutual Action of Two Circular Conductors.

It has been shown (p. 48 above) that the mutual potential energy of two circular magnetic shells is given by the equation*

---

* As noted in an erratum the sign *minus* should be prefixed to the quantity on the right-hand side of the expressions for E on pp. 48, 49.

ENERGY OF TWO CIRCULAR CURRENTS

$$E = -4\pi^2 \Phi \Phi' \sin^2\psi \sin^2\psi' \rho \sum \frac{1}{i(i+1)} \cdot {}_\psi Z'_i \cdot {}_{\psi'} Z'_i \cdot {}_\phi Z_i \left(\frac{\rho}{r}\right)^i \quad (35)$$

$$(r > \rho)$$

if ψ, ψ', Fig. 60, denote the angles which the radii of the shells subtend at the intersection of their axes, r, ρ the distances of the circular arcs from the origin, ϕ the angle between the axes of the shells (denoted by θ in Fig. 10 above), and ${}_\phi Z_i$ the zonal surface harmonic of the ith order taken for the angle ϕ, and similarly for the others as explained at p. 47. This value of E with its sign changed, and γ, γ' written for Φ, Φ', is the mutual electrokinetic energy T of two circular currents and is at once available for the calculation of their mutual action. The result enables the mutual action of two coils to be found, and is therefore the foundation of the theory of absolute electrodynamometers and current balances, which measure currents in absolute units by the forces exerted on a movable coil by a fixed coil, through both of which the current to be measured is flowing, or in which the currents flowing have a certain known ratio.

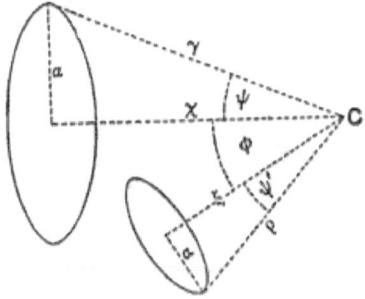

Fig. 60.

Putting then (Fig. 60) a, a, for the radii of the larger and smaller circles respectively, and x, ξ, for the distances of their planes from the origin, we have $\sin\psi = a/r$, and $\sin\psi' = a/\rho$, and substituting in the zonal harmonic expressions, as given in the Appendix on Spherical Harmonics, their values in terms of a, x, a, ξ, we have

Electrokinetic Energy of Two Circular Currents.

MAGNETIC ACTION OF COILS

$$T = \pi^2 \gamma \gamma' \frac{a^2 a'^2}{r^3} \Big\{ 1 \cdot 2 \cos \phi + 2 \cdot 3 \frac{x}{r^2} \xi (\cos^2 \phi - \tfrac{1}{2} \sin^2 \phi)$$

$$+ 3 \cdot 4 \frac{x^2 - \tfrac{1}{4}a^2}{r^4} (\xi^2 - \tfrac{1}{4}a'^2)(\cos^3 \phi - \tfrac{3}{2} \sin^2 \phi \cos \phi)$$

$$+ 4 \cdot 5 \frac{x(x^2 - \tfrac{3}{4}a^2)}{r^6} \xi(\xi^2 - \tfrac{3}{4}a'^2)(\cos^4 \phi - 3 \cos^2 \phi \sin^2 \phi + \tfrac{3}{8} \sin^4 \phi)$$

$$+ 5 \cdot 6 \frac{x^4 - \tfrac{3}{2}x^2 a^2 + \tfrac{1}{8}a^4}{r^8} (\xi^4 - \tfrac{3}{2}\xi^2 a'^2 + a'^4)(\cos^5 \phi - 5 \cos^3 \phi \sin^2 \phi$$

$$+ \frac{3 \cdot 5}{8} \sin^4 \phi) + \quad \ldots \ldots \quad (36)$$

As explained at p. 160 the couple Θ due to the mutual action of the two circuits tending to increase ϕ is $\partial T/\partial \phi$. Hence for this couple we have

$$\Theta = -\pi^2 \gamma \gamma' \sin \phi \Big\{ 1 \cdot 2 \frac{a^2}{r^3} a'^2 + 2 \cdot 3 \frac{x}{r^2} \xi \cdot 3 \cos \phi$$

$$+ 3 \cdot 4 \frac{x^2 - \tfrac{1}{4}a^2}{r^4} (\xi^2 - \tfrac{1}{4}a'^2) \cdot 2 \cdot 3 (\cos^2 \phi - \tfrac{1}{4} \sin^2 \phi)$$

$$+ 4 \cdot 5 \frac{x(x^2 - \tfrac{3}{4}a^2)}{r^4} \xi(\xi^2 - \tfrac{3}{4}a'^2) \cdot 2 \cdot 5 \cos \phi (\cos^2 \phi - \tfrac{3}{4} \sin^2 \phi)$$

$$+ \ldots \ldots \ldots \ldots \ldots \ldots \Big\} \quad (37)$$

Attraction between Two Parallel Circular Currents. The attraction between the circuits when they are coaxial, that is when $\phi = 0$, may be found by putting $\phi = 0$ in (35), and calculating $\partial T/\partial \xi$. We have

$$T = \pi^2 \gamma \gamma' \frac{a^2 a'^2}{r^3} \Big\{ 1 \cdot 2 + 2 \cdot 3 \frac{x}{r^2} \xi + 3 \cdot 4 \frac{x^2 - \tfrac{1}{4}a^2}{r^4} (\xi^2 - \tfrac{1}{4}a'^2)$$

$$+ 4 \cdot 5 \frac{x(x^2 - \tfrac{3}{4}a^2)}{r^6} \xi(\xi^2 - \tfrac{3}{4}a'^2) + \ldots \quad (36')$$

$$\frac{\partial T}{\partial \xi} = \pi^2 \gamma \gamma' \frac{a^2 a'^2}{r^4} \Big\{ 1 \cdot 2 \cdot 3 \frac{x}{r} + 2 \cdot 3 \cdot 4 \frac{x^2 - \tfrac{1}{4}a^2}{r^3} \xi$$

$$+ 3 \cdot 4 \cdot 5 \frac{x(x^2 - \tfrac{3}{4}a^2)}{r^5} (\xi^2 - \tfrac{1}{4}a'^2) + \ldots \quad (38)$$

ATTRACTION OF ONE COIL ON THE OTHER

We may now proceed from two simple circles to two mutually influencing coils. This may be done by direct integration with respect to x, a, and ξ, a, in the two cases, or by the method explained above and already used for a coil and a magnet.

Action between Two Coils of Finite Cross-Section.

Proceding first according to the latter method and dealing with the terms of (36) separately, putting for the axial breadth and radial depth $2b$, $2d$ in the case of the larger coil, 2β, 2δ in the case of the smaller (both being supposed of rectangular cross-section), while x, a, ξ, a, are retained for the mean filaments in the two cases, we find if N, n, be the numbers of turns in the two coils, larger and smaller respectively

$$T = Nn\gamma\gamma' \{G_1 g_1 . \phi Z_1 + G_2 g_2 . \phi Z_2 + G_3 g_3 . \phi Z_3 + \ldots\} \quad . \quad (39)$$

where

Coil of Finite Section. Value of Electrokinetic Energy.

$$G_1 = 2\pi \frac{a^2}{r^3} \left\{ 1 + \frac{3b^2}{6r^2}(4x^2 - a^2) + \frac{d^2}{6a^2 r^4}(2x^4 - 11x^2 a^2 + 2a^4) + \ldots \right\}$$

$$G_2 = 3\pi \frac{a^2 x}{r^5} \left\{ 1 + \frac{5b^2}{6r^4}(4x^2 - 3a^2) + \frac{5d^2}{6a^2 r^4}(2x^4 - 21x^2 a^2 + 12a^4) + \ldots \right\}$$

$$G_3 = \frac{\pi}{r^7} \left\{ a^2(4x^2 - a^2) + \frac{3 \cdot 5 \, a^2 b^2}{6 r^4}(8x^4 - 12x^2 a^2 + a^4) \right.$$

$$\left. + \frac{d^2}{6r^4}(8x^6 - 136x^4 a^2 + 159x^2 a^4 - 12a^6) + \ldots \right\}$$

. . . .

$$g_1 = \pi(a^2 + \tfrac{1}{3}\beta^2 + \ldots)$$
$$g_2 = 2\pi\xi(a^2 + \tfrac{1}{3}\beta^2 + \ldots)$$
$$g_3 = \pi\{\tfrac{1}{4}3a^2(4\xi^2 - a^2) + \tfrac{1}{2}\delta^2(2\xi^2 - 3a^2) + \beta^2 a^2 + \ldots\}$$

. . . .

Hence we have from (35)

Value of the Turning Couple and Attraction of one Coil on the other.

$$\Theta = -Nn\gamma\gamma' \sin\phi \{G_1 g_1 . \phi Z'_1 + G_2 g_2 . \phi Z'_2 + G_3 g_3 . \phi Z'_3$$
$$+ \ldots\} \quad . \quad . \quad . \quad . \quad . \quad . \quad (40)$$

which is the corrected form of (37). Similarly we could write down from (38) the corrected value of the attraction between the coils.

Turning Couple of Coil in Magnetic Needle deduced from Theory of Two Coils.

Equation (37) is applicable to the determination of the couple due to the action of a coil on a uniformly magnetized thin magnet the centre of which is on the origin. We have only to suppose another coil equal in all respects to the smaller placed coaxial with the latter on the other side of the origin at a mean distance ξ from that point, and further suppose a current of the same strength to flow in like directions round both. The couple acting on the second coil will be got from that on the first by merely supposing the angle ϕ to be increased by 180°, and the current in it to be reversed. But changing ϕ into $\phi + 180°$ changes the signs of $_\phi Z'_1$, $_\phi Z'_2$, &c., and taking into account the change of sign of the current and of $\sin \phi$, we have for the couple, Θ_1 say, on the second coil

$$\Theta_1 = - Nn\gamma\gamma' \sin\phi \left\{ G_1 g_1 . {}_\phi Z'_1 - G_2 g_2 . {}_\phi Z'_2 + G_3 g_3 . {}_\phi Z'_3 - \&c. \right\}$$

Hence for the total couple we get

$$\Theta + \Theta_1 = -2Nn\gamma\gamma' \sin\phi \left\{ G_1 g_1 . {}_\phi Z'_1 + G_3 g_3 . {}_\phi Z'_3 + \&c \right\} \quad (41)$$

But the double coil here supposed to exist is equivalent to a needle with its centre at the origin, and of moment $M = 2\pi a^2 n \gamma'$. Also if we make the section of each coil very small, and the radius a very small, but preserve $2\pi a^2 n\gamma'$ a finite quantity, we may regard the pair of coils as equivalent to a uniformly magnetized magnet of moment $2\pi a^2 n\gamma'$, and of length 2ξ, and put, in the values of g_1, g_2, &c., $\beta = 0$, $\delta = 0$, $Ma^2 = 0$, &c. In this way we shall obtain from (41) a formula equivalent to that given in (12) when the latter is corrected for the finite cross-section of the large coil.

Electro-Dynamometer with Double-Coil Arrangement.

If instead of two single coils, one fixed and the other movable, the Helmholtz double arrangement is adopted for both the fixed and movable parts of the dynamometer, so that the centres of both are made coincident with the origin,* the expressions for their mutual action are much simplified.

Let A, B Fig. 61 denote the large coils A', B', the small coils. Then the mutual energy of A and A', and the couple on A' due to the action of A, are equal in numerical amount and sign to those of B and B'. These are given by (39) and (40). Hence for these two pairs of coils the energy is

$$2T = 2Nn\gamma\gamma' \left\{ G_1 g_1 . {}_\phi Z_1 + G_2 g_2 . {}_\phi Z_2 + G_3 g_3 . {}_\phi Z_3 + \ldots \right\} \quad (42)$$

---

* This was the arrangement adopted for the Absolute Electro-dynamometer made by Mr. Latimer Clark for the British Association Committee on Electrical Standards. See Chapter VI. below.

where ϕ is the angle ACA' indicated in Fig. 60, and $a^2 = 4a^2$, $a^2 = 4\xi^2$. Now the mutual energy of the coils B', A, is that which the value of T would become for A' and A if θ were increased by 180° and the current in A' were then reversed. The mutual electrokinetic energy of B and A' has evidently the same value. But $\cos(\phi + 180°) = -\cos\phi$, so that the zonal harmonics of odd

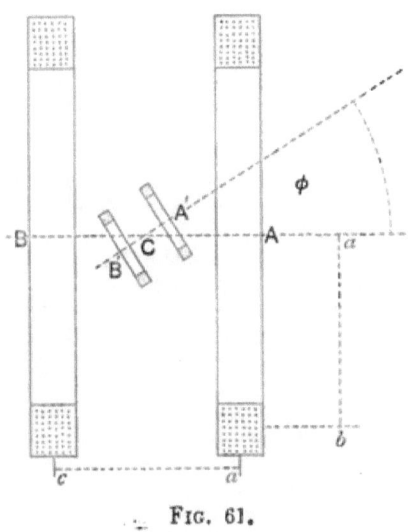

FIG. 61.

order change sign. Hence taking into account the change of sign of current, we have for the electrokinetic energy of the other two pairs of coils A, B' and A', B the value

$$2T_1 = 2Nn\gamma\gamma'\{G_1g_1 \cdot \phi Z_1 - G_2g_2 \cdot \phi Z_2 + G_3g_3 \cdot \phi Z_3 - \&c.\} \quad (43)$$

Mutual Energy of Coil Systems.

where all the quantities have the same values as before. Hence for the total energy of the arrangement we have

$$2(T+T_1) = 4Nn\gamma\gamma'\{G_1g_1 \cdot \phi Z_1 + G_3g_3 \cdot \phi Z_3 + G_5g_5 \cdot \phi Z_5 + \&c.\} \quad (44)$$

and the turning couple on the pair of small coils is

$$\Theta = -4Nn\gamma\gamma'\sin\phi\{G_1g_1 \cdot \phi Z'_1 + G_3g_3 \cdot \phi Z'_3 + G_5g_5 \cdot \phi Z'_5\} \quad (45)$$

Turning Couple on Movable Coil. Effect of Arrangement in Simplifying Calculation of Action.

The values of G_1, G_3, G_5, ..., g_1, g_3, g_5, ..., are given on p. 269 above, and it is to be noticed that in these $4x^2 - a^2 = 0$, and $4\xi^2 - a^2 = 0$, so that to a considerable degree of approximation G_3 and g_3 vanish, and the couple reduces to

$$\Theta = -4Nn\gamma\gamma' \sin\phi \; G_1 g_1 \cdot {}_\phi Z'_1 \quad \ldots \ldots \quad (46)$$

or neglecting the correction terms in b^2, a^2, &c., to

$$\Theta = -64Nn\gamma\gamma' \frac{a^2}{5\sqrt{5}a} \sin\phi \quad \ldots \quad (46')$$

Considering the movable coil-system as equivalent to a needle of moment $2n\gamma'\pi a^2$, this agrees with (24) above.

In the same manner as at p. 270 we could deduce the action of a Helmholtz double-coil on a magnetic needle, with its centre at the centre of symmetry, from the theory of the double electrodynamometer just given.

Mutual Energy of Two Long Single-Layer Coils.

From (36) we can find an expression for the mutual energy of two long cylindrical coils, consisting each of a single layer of fine wire, carrying currents γ, γ', and so placed that their axes intersect at an angle ϕ, as shown in Fig. 60. Such coils are capable of being constructed with very great accuracy, and the expression of the electrokinetic energy of the arrangement enables the coefficients of mutual and self-induction to be obtained for a number of important cases.

Expressed by Zonal Harmonics.

Let x_1, x_2, ξ_1, ξ_2, be the distances of the nearer and farther ends of the coil from the intersection of their axes, x, ξ, those of two circular elements of lengths dx, $d\xi$. If n, n', γ, γ', be the numbers of turns per unit length and the currents in the two coils, the currents in the elements are $n\gamma dx$, $n'\gamma' d\xi$. Writing down then by (35) the expression for the energy of the two elements, and integrating from $x = x_1$ to $x = x_2$ in the one case, and from $\xi = \xi_1$ to $\xi = \xi_2$ in the other, we get for the mutual electrokinetic energy of the two coils of lengths $x_1 - x_2$, $\xi_1 - \xi_2$, the expression

$$T = 4\pi^2 nn' a^2 a^2 \sum \frac{1}{i(i+1)} \cdot {}_\phi Z_i \left\{ \int_{x_1}^{x_2} \frac{{}_\psi Z'_i}{r^{i+2}} dx \right\}$$
$$\left\{ \int_{\xi_1}^{\xi_2} \rho^{i-1} \cdot {}_{\psi'} Z'_i dx \right\} \quad \ldots \ldots \quad (47)$$

${}_\psi Z'_i$, ${}_{\psi'} Z'_i$ can be found by differentiation with respect to $\cos\psi$, $\cos\psi'$, of the well-known expressions for ${}_\psi Z_i$, ${}_{\psi'} Z_i$, and the integrals then got by direct integration; but the theorem

INTEGRATION OF ZONAL HARMONIC SERIES

expressed by (8) above, together with a supplementary theorem for the zonal harmonics of odd order, which we shall now prove, yields at once the indefinite integrals required.

Assume that

$$A \int \frac{\psi Z'_{2i-1}}{r^{2i+1}} dx = \frac{\partial^{2i-1} A_0}{\partial x^{2i-1}} \quad \ldots \quad (48)$$

Zonal Harmonic Terms of Series found by Differentiation.

where A is a constant. Then differentiating we find by (8)

$$A \frac{\psi Z'_{2i-1}}{r^{2i+1}} = \frac{\partial^{2i} A_0}{\partial x^{2i}} = -(2i-1)! \int \frac{\psi Z'_{2i}}{r^{2i+2}} dx$$

and therefore also

$$A\{(1-\mu^2)\,_\psi Z''_{2i-1} - (2i+1)\mu \cdot \,_\psi Z'_{2i-1}\} = -(2i-1)! \, a^2 \cdot \,_\psi Z'_{2i}. \quad (49)$$

The assumption made in (48) will be justified if the relation just found holds for a constant value of A. Now if Z_i denote a zonal harmonic of any order i, we have by the fundamental relations of zonal harmonics (writing μ for $\cos\psi$)

$$\mu Z_i - Z_{i-1} = -\frac{1}{i}(1-\mu^2)Z'_i$$

$$Z_i - \mu Z_{i-1} = -\frac{1}{i}(1-\mu^2)Z'_{i-1}.$$

Eliminating from these first Z_i, then Z_{i-1}, we find

$$Z_{i-1} = \frac{1}{i}(Z'_i - \mu Z'_{i-1}) \quad \ldots \quad (50)$$

$$Z_i = \frac{1}{i}(\mu Z'_i - Z'_{i-1}). \quad \ldots \quad (50')$$

Differentiating (50) and (50') with respect to μ, and eliminating Z''_i from the resulting equations, we obtain the relation

$$(1-\mu^2)Z''_{i-1} - \mu(i+1)Z'_{i-1} = -(i-1)Z'_i,$$

which with $2i$ written for i, agrees with (49), if we put

$$A = (2i-2)! \, a^2.$$

VOL. II.   T

Thus the assumption is justified, and we have

$$(2i - 2)! \frac{a^2}{r^{2i+1}} \cdot \psi Z'_{2i-1} = \frac{\partial^{2i} A_0}{\partial x^{2i}} \quad \ldots \quad (51)$$

Equations (8) and (51) may be combined in the single equation

$$(-1)^{i+1}(i-1)! \frac{a^2}{r^{i+2}} \cdot \psi Z'_i = \frac{\partial^{i+1} A_0}{\partial x^{i+1}} \quad \ldots \quad (52)$$

where i is any integer.

Theorem in Zonal Harmonics. Integrating, we get from (52)

$$(-1)^{i+1}(i-1)! \, a^2 \int \frac{\psi Z'_i}{r^{i+2}} \, dx = \frac{\partial^i A_0}{\partial x^i}$$

so that successive differentiation of $A_0 (= \sqrt{a^2 + x^2} - x)$ gives the first set of integrals required for (47). The differential coefficients as far as the 11th are given in the Appendix on Spherical Harmonics.

The second set of integrals is also very easily obtained by calculating $\psi Z'_i$ by (52), modified by putting ξ for x, α for a, and ρ for r. Or they may be found by differentiation of the zonal harmonic expressions given in terms of μ in the Appendix, substitution of ξ/ρ for μ in the result, and integration with respect to ξ. The factor ρ^{i-1} converts in each case the expression to be integrated into a rational integral function of ξ, so that the integration presents no difficulty.

Integrated Expression for Mutual Energy. Thus we obtain

$$T = \pi^2 nn' \gamma \gamma' a^2 \alpha^2 \{ K_1 k_1 \cdot {}_\phi Z_1 + K_2 k_2 \cdot {}_\phi Z_2 + K_3 k_3 \cdot {}_\phi Z_3 + . \} \quad (53)$$

where

$$K_1 = \frac{2}{a^2} \left(\frac{x_2}{r_2} - \frac{x_1}{r_1} \right), \qquad K_2 = -\left(\frac{1}{r_2^3} - \frac{1}{r_1^3} \right),$$

$$K_3 = -\frac{1}{2} \left(\frac{x_2}{r_2^5} - \frac{x_1}{r_1^5} \right),$$

$$K_4 = -\frac{1}{8} \left\{ \frac{1}{r_2^7} (4x_2^2 - a^2) - \frac{1}{r_1^7} (4x_1^2 - a^2) \right\},$$

$$K_5 = -\frac{1}{8} \left\{ \frac{x_2}{r_2^9} (4x_2^2 - 3a^2) - \frac{x_1}{r_1^9} (4x_1^2 - 3a^2) \right\},$$

$$K_6 = -\frac{1}{8}\left\{\frac{1}{r_2^{11}}(4x_2^4 - 6r_2^2a^2 + \tfrac{1}{2}a^4) - \frac{1}{r_1^{11}}(4x_1^4 - 6x_1^2a^2 + a^4)\right\},$$

$$K_7 = -\frac{1}{8}\left\{\frac{x_2}{r_2^{13}}(4x_2^4 - 10x_2^2a^2 + \tfrac{5}{2}a^4) - \frac{x_1}{r_1^{13}}(4x_1^4 - 10x_1^2a^2 + \tfrac{5}{2}a^4)\right\},$$

. . . .

$$k_1 = \xi_2 - \xi_1, \qquad k_2 = \xi_2^2 - \xi_1^2,$$

$$k_3 = 2\xi_2^3 - \tfrac{3}{2}\xi_2 a^2 - 2\xi_1^3 + \tfrac{3}{2}\xi_1 a^2, \quad k_4 = 2\xi_2^4 - 3\xi_2^2 a^2 - 2\xi_1^4 + 3\xi_1^2 a^2.$$

$$k_5 = 2\xi_2^5 - 5\xi_2^3 a^2 + \tfrac{5}{4}\xi_2 a^4 - 2\xi_1^5 + 5\xi_1^3 a^2 - \tfrac{5}{4}\xi_1 a^4,$$

$$k_6 = 2\xi_2^6 - \tfrac{15}{2}\xi_2^4 a^2 + \tfrac{15}{4}\xi_2^2 a^4 - 2\xi_1^6 + \tfrac{15}{2}\xi_1^4 a^2 - \tfrac{15}{4}\xi_1^2 a^4,$$

$$k_7 = 2\xi_2^7 - \tfrac{21}{2}\xi_2^5 a^2 + \tfrac{35}{4}\xi_2^3 a^4 - \tfrac{35}{32}\xi_2 a^6$$
$$\qquad\qquad - 2\xi_1^7 + \tfrac{21}{2}\xi_1^5 a^2 - \tfrac{35}{4}\xi_1^3 a^4 + \tfrac{35}{32}\xi_1 a^6.$$

. . . .

The moment Θ of the forces tending to turn either coil about the origin in the plane of their axes is $\partial T/\partial \phi$. Hence **Turning Moment on Either Coil.**

$$\Theta = -\pi^2 n n' \gamma \gamma' a^2 a'^2 \sin\phi \left\{K_1 k_1 . \phi Z'_1 + K_2 k_2 . \phi Z'_2 + \ldots\right\}. \quad (54)$$

If we examine the values of the quantities $K_1, K_2, \ldots k_1, k_2, \ldots$ we see that if one at least of the coils (say that of radius a) be placed so that its centre is at the intersection of the axes, the even terms in (53) and (54) will all vanish, since then $\xi_2 = -\xi_1$, $\rho_2 = \rho_1$. If besides being so placed this coil have its length $2\xi_2 = \sqrt{3} . a$, the third term will vanish; and the fifth term also disappears when the larger coil fulfils the same conditions. Further, if both coils are thus placed, the even terms, so to speak, doubly vanish, so that any little error in the placing of the coils can only insensibly affect the vanishing of the even terms.

With coils thus constructed and placed, the next term of the series in (53), (54), after the first is the seventh, and only the odd

276   MAGNETIC ACTION OF COILS

Vanishing of Terms of Series for Particular Arrangement and Construction of Coils. terms after that have any value, and this holds whatever the angle between the axes of the coils. The seventh term amounts to only about 1/27000 of the first if the ratio of the radii of the coils be 1/2, or to about 1/4700 if the ratio be 2/3. With the former ratio of radii, the error made in taking only the first term of the series amounts thus to a quantity quite inappreciable in the electrical measurements made with standard coils, and an important application of the result might be made in the construction of electro-dynamometers.

Application to Absolute Electro-Dynamometer. Two single-layer coils could be accurately made, of radii differing very considerably but each having the ratio $\sqrt{3}/1$ of length to radius, and of dimensions so great in each case as to be determinable with accuracy. If placed concentrically the fixed coil will act on the movable with a couple given by (54) in which the first term only need be taken. Thus for such an arrangement

$$\Theta = -8\pi^2 nn' \gamma \gamma' a^2 \frac{x_2 \xi_2}{r} \sin \phi \quad . \quad . \quad . \quad . \quad (55)$$

Absolute Galvanometer with Coil of one Layer A single-layer coil made as here of considerable length seems very suitable also for use as an absolute galvanometer. It has sufficient uniformity of field to render the very exact placing of the needle at the centre quite unessential, and it can be made sufficiently sensitive, so that it possesses most of the advantages of the Helmholtz double-coil arrangement, without the uncertainty which exists in the latter as to the distribution of the different turns of wire in the two multiple-layer bobbins, or requiring the correction terms which the bobbins involve on account of their finite cross-section.

Couple on Needle. We may find the couple acting on the needle of such a galvanometer as follows, provided the needle be suspended with its axis intersecting that of the coil. The suspended coil in the above discussion may be taken as a solenoidal magnet of magnetic moment $\pi a^2 n' \gamma'$ per unit of length, and therefore of total magnetic moment $M = \pi a^2 n' \gamma' (\xi_2 - \xi_1)$. Hence by (54)

$$\Theta = -\pi n \gamma a^2 M \sin \phi \frac{1}{\xi_2 - \xi_1} \left\{ K_1 k_1 \cdot \phi Z_1 + K_2 k_2 \cdot \phi Z_2 + \ldots \right\}. \quad (56)$$

from which by means of the values of $K_1, K_2, \ldots k_1, k_2, \ldots$ given above the value of Θ in the general case can be calculated.

If the coil and solenoidal magnet be concentric all the even terms vanish as before, and by making the length of the coil

$\sqrt{3}$ times its radius we can cause the fifth term to disappear. The couple therefore to the seventh term inclusive is given by

$$\Theta = -2\pi n \gamma a^2 M \sin\phi \left\{ \frac{a^2}{2}\frac{x_2}{r_2} - \frac{1}{4}\frac{x^2}{r_2^6}(4\xi_2^2 - 3a^2)\phi Z'_3 \right.$$
$$- \frac{1}{8}\frac{x_2}{r_2^{13}}\left(4r_2^4 - 10r_2^2 a^2 + \frac{5}{2}a^4\right)\left(2\xi_2^6 - \frac{21}{2}\xi_2^4 a^2 + \frac{35}{4}\xi_2^2 a^4\right.$$
$$\left.\left. - \frac{35}{32}a^6\right)\phi Z'_7 \right\} \quad . \quad . \quad . \quad . \quad (57)$$

Couple on Needle Concentric with Coil:

With an actual magnet it is impossible to set up any definite relation between ξ_2 and a; but by using a thin uniform needle it is possible to make $2a$, which is a quantity of the order of magnitude of the thickness, small compared with ξ_2 and therefore practically zero. Then by making ξ_2, which for a thin needle of uniform thickness is approximately its half-length, small in comparison with r_2, the second and third terms in (57) may be made quite negligible. For example, if a needle 1 cm. long be used in a coil of 20 cms. radius, and therefore of axial length 34·64 cms., and the value of ϕ be approximately 90°, the second term in (57) is only about 1/(6500) of the first.

Simplification for Short Needle.

We may notice here (though the subject of induction coefficients belongs to next section of this chapter), that in (53) $T/\gamma\gamma'$ is the coefficient of mutual induction of the two coils. Thus if two coils of considerably different radii, but each having its length $\sqrt{3}$ times its radius, be arranged concentrically, their mutual induction coefficient is given for any angle between their axes with great accuracy by the first term of (53). In this way standards of mutual inductance could be made with very considerable exactness.

Coefficient of Mutual Induction of Two Single-Layer Coils: Standards of Inductance.

By supposing the coils equal in every respect and coincident, we can calculate the self-induction coefficient of each, by taking the value of $T/\gamma\gamma'$ given by (53). In this case however the first term does not give a result nearer the truth than to about $\frac{1}{5}$ per cent., and it is necessary to take in at least one more term of the series.

With a certain amount of accuracy the single-layer coils discussed above might be replaced by coils consisting of several layers, the ends of the channel in each case being frustums of a cone having its vertex at the common centre and semi-vertical angle equal to $\tan^{-1} 2/\sqrt{3}$. This makes each layer (unless a whole number of turns cannot be made in each case to fulfil the relation), have its length equal to $\sqrt{3}$ times its radius. The

Multiple-Layer Coils, each Layer having Length = Radius $\times \sqrt{3}$.

278   MAGNETIC ACTION OF COILS

mutual energy and the action of one coil on the other can then be calculated by considering separately each pair of single-layer coils which can be formed by taking one layer in each coil. In such an arrangement, however, as in all multiple layer coils, the distribution of the wires would be in a certain extent uncertain.

We have now discussed for several different arrangements of circuits the potential and force at different points in their fields. As a final example we shall take the important case of a simple solenoid of uniform cross-section.

Simple Uniform Solenoid Defined. Such a solenoid may be defined as a tubular surface such, and so placed with reference to a directing curve in space, that every normal plane drawn to the directing curve gives a curve of section of the surface which is always of the same form, and makes with the point of intersection of the normal plane with the directing curve always the same geometrical arrangement or diagram. This directing curve is generally taken so that it passes through the mean point or centroid of each section, and it is then called the axis of the solenoid. The current in the solenoid is supposed to flow at every point along the curve of section at the point normal to the axis.

Solenoid Approximately Realised by Helix of Wire. Such an arrangement is approximated to closely by a coil consisting of a single layer of thin wire wound on the surface of a core, so that the wire forms a helix of step equal to the thickness of the wire. If the wire be thin the component of flow parallel to the axis is very small in comparison with that at right angles to the axis of the core, and may be neglected. Its effect may be annulled for points on the axis of a straight helix of circular section, and for points at a distance from the helix great in comparison with the radius of cross-section by bringing the wire back from one end to the other along a generating line of the cylindrical surface. If the wire be thin the effect of the confinement of the current to the cross-section of the wire, and its consequent want of uniformity of distribution over the surface of the solenoid, may also be neglected for points without and within at distances from the wire great in comparison with its radius.

Closed Solenoid with Circular Axis, or Ring Electro-Magnet. Considering then a pure solenoid, first of any form of cross-section, but having its directing axis a complete circle, and in a uniform medium which may be taken as of unit magnetic inductive capacity. (Fig. 62 shows a section of such a solenoid by a plane through the circular axis, the dotted circle.) If r be the radius of the circle formed by its axis, the length of the axis is $2\pi r$. Take any point within the solenoid and draw through it a circle in a plane parallel to the circular axis, and symmetrically situated with respect to its centre. Let H be the

CLOSED CIRCULAR SOLENOID

magnetic force at the point tangential to the circle, then by symmetry the value of **H** must be the same at every point of the circle. Let the position of the point be determined by coordinates x, y having their origin where the cross-section through the point meets the axis, x being drawn in the direction

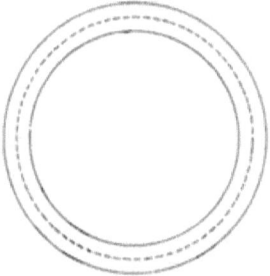

Fig. 62.

of the radius of the circular axis at the origin, y at right angles to the plane of the axis. The radius of the circle is $2\pi(r + x)$**H**. Hence, if γ' be the current in the solenoid at each point, taken per unit of length of the axis, the total current is $2\pi r \gamma'$, and we have by the theorem given at p. 108 above

$$4\pi \times 2\pi r \gamma' = 2\pi (r + x)\mathbf{H}$$

or

$$\mathbf{H} = \frac{4\pi r \gamma'}{r + x} \quad \ldots \ldots \quad (58)$$

In the case of a core lapped round by n turns of fine wire, in each of which flows a current of strength γ, so as to imitate this solenoid, this equation becomes

$$\mathbf{H} = \frac{2n\gamma}{r + x} \quad \ldots \ldots \quad (59)$$

Thus where x is zero, that is for all points within the solenoid which lie on a cylinder drawn through the axis at right angles to its plane, the value of **H** is the same, viz., $4\pi\gamma'$. At internal points outside this cylinder (that is, points for which x is positive) **H** is smaller, for points inside the cylinder, greater than $4\pi\gamma'$.

280 MAGNETIC ACTION OF COILS

Solenoid Considered as Lamellar Distribution of Magnetism.

The results already obtained might have been found by calculating, in the manner explained at p. 43 above, the potentials of magnetic induction Ω_i, Ω_e, for two infinitely near points, internal and external, to the solenoid. For we may regard the solenoid as a lamellar distribution of magnetism, of strength $d\phi/ds$, where ds is an element of length taken at right angles to a cross-section. The value of $d\phi/ds$ depends upon the radius, $r + x$, of the circle on which the point is taken, and clearly fulfils the condition

$$(r + x)\frac{d\phi}{ds} = C$$

where C is a constant, or if θ be the angle between the cross-section in which the point is taken and a cross-section of reference,

$$\frac{d\phi}{d\theta} = C = \gamma' r \quad \ldots \ldots \quad (60)$$

The value of ϕ is constant over each cross-section, and we may take that at the cross-section of reference as zero. Then

$$\phi = \int_0^\theta \frac{d\phi}{d\theta} d\theta = \gamma' r \theta \quad \ldots \ldots \quad (61)$$

Now by (69) p. 42

$$\Omega_e = \Omega_i + 4\pi\phi = \Omega_i + 4\pi\gamma' r \theta$$

where Ω_e, Ω_i, are the potentials of magnetic induction at infinitely near external and internal points, in the plane of the cross-section at which ϕ has the value $\gamma' r \theta$.

The value of Ω_e thus depends only on the value of θ and the assumed zero of reckoning of ϕ. It is further zero at an infinite distance, hence it is zero everywhere. We have therefore

$$\Omega_i = -4\pi\gamma' r \theta \quad \ldots \ldots \quad (62)$$

Induction everywhere Normal to Cross Section

This equation shows that there is no component of induction in the plane of any cross-section. Hence since magnetic induction, in space of unit magnetic inductive capacity, coincides in value with the magnetic force, we have within the solenoid the resultant magnetic induction $\mathbf{B} = \mathbf{H}$, and

$$\mathbf{B} = -\frac{1}{r+x}\frac{d\Omega_i}{d\theta} = \frac{4\pi\gamma'}{r+x} \quad \ldots \quad (63)$$

SOLENOID ENCLOSING DIFFERENT MEDIA

Since the value of Ω_e is zero there is no external action due to the solenoid.

The same method of proof can be applied to show that the external action of every closed solenoid is zero.

If the space within the solenoidal surface be filled with a uniform medium of magnetic inductive capacity μ, $\mathbf{B} = \mu \mathbf{H}$ and $\Omega_i = -4\pi\mu r\theta\gamma'$. Thus

$$\mathbf{B} = \frac{4\pi\mu r\gamma'}{r+x} \quad \ldots \ldots \quad (64)$$

External Action of closed solenoid is zero.

Again, let the internal space be filled (as shown in Fig. 62) by different media of magnetic inductive capacities μ_1, μ_2, μ_3, &c., occupying parts of the solenoidal space bounded by the current-surface and cross-sections of the solenoid, and extending over

Solenoid Enclosing Different Media.

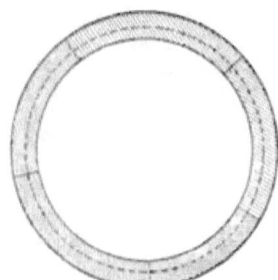

Fig. 63.

angles θ_1, θ_2, θ_3, &c., along the circular axis. Then if the effects of the magnetic forces due to the separating surfaces be neglected, the magnetic inductions must be taken constant throughout each space. Let its values for the respective media be denoted by $\mathbf{B}_1$, $\mathbf{B}_2$, $\mathbf{B}_3$, &c. We have

$$\mathbf{H} = \frac{\mathbf{B}_1}{\mu_1} = \frac{\mathbf{B}_2}{\mu_2} = \frac{\mathbf{B}_3}{\mu_3} = \&c. \quad \ldots \ldots \quad (65)$$

Thus (58) becomes

$$4\pi r\gamma' = (r+x)\left(\frac{\mathbf{B}_1\theta_1}{\mu_1} + \frac{\mathbf{B}_2\theta_2}{\mu_2} + \frac{\mathbf{B}_3\theta_3}{\mu_3} + \&c.\right) \quad . \quad (66)$$

or if we put $l_1 = (r + x)\theta_1$, $l_2 = (r + x)\theta_2$, &c.

$$4\pi r\gamma' = \frac{B_1 l_1}{\mu_1} + \frac{B_2 l_2}{\mu_2} + \frac{B_3 l_3}{\mu_3} + \&c. \quad \ldots \quad (67)$$

From (65) we find

$$B_1 = \frac{H}{\frac{1}{\mu_1}}, \quad B_2 = \frac{H}{\frac{1}{\mu_2}}, \&c. \quad \ldots \ldots \quad (68)$$

Notion of Magnetic Current and Magnetic Resistance. equations which are precisely similar in form to the (p. 195 above) equation $u = P/\rho$ for the current per unit of area at a point in an isotropic medium at which the resistance for unit of volume is ρ and the electromotive force P. Thus $1/\mu_1$, $1/\mu_2$, &c., might, if the term magnetic resistance is adopted generally, be called the specific magnetic resistances or *resistivities* of the media. The curve along which the induction is taken corresponds to the voltaic circuit, the inductions B_1, B_2, &c., to the values of u in its different parts, while (66) or (67) expresses that the line integral of H, or (as it has been called in this connection) the *magneto-motive* force round the circuit, is $4\pi r\gamma'$.

Magneto-Motive Force. Let the medium be such that B and H have the same direction at every point. The work done on a unit pole carried along in the direction of H in a closed path so as to thread through every turn of wire producing the magnetic force is numerically equal to the line-integral of the magnetic force, that is, to the magneto-motive force. But if N denote here the total number of turns of wire threaded through, and γ the current in each, the whole work done is $4\pi N\gamma$. Thus

$$\int \frac{B}{\mu} ds = 4\pi N\gamma \quad \ldots \ldots \quad (69)$$

where ds is an element of the closed path, and the integral is taken completely round the path.

Let dS be an element of a surface drawn at right angles to the tubes of induction, and B be the induction at that element: the total induction across the element is BdS. Now the integral in (69) is the same thing as the integral of $BdS.ds/\mu dS$, where dS is the cross-section of the tube at any element of length ds. But BdS by the property of a tube of induction is the same at every part of the tube. Hence supposing N the same for every tube

INDUCTION IN RING ELECTRO-MAGNETIC

$$\mathbf{B}dS \int \frac{ds}{\mu dS} = 4\pi N\gamma$$

or

$$\mathbf{B}dS = \frac{4\pi N\gamma}{\int \frac{ds}{\mu dS}} \quad \ldots \ldots \quad (69')$$

Let U be the ratio of $\mathbf{B}dS$ to the total magnetic induction $\mathbf{B}$ across a surface cutting all the tubes we have

$$\mathbf{B} = \frac{4\pi N\gamma}{U \int \frac{ds}{\mu dS}} \quad \ldots \ldots \quad (69'')$$

The denominator of the expressions on the right of (69') and (69'') have been called the magnetic resistances of the portions of the medium concerned. The name *magnetic reluctance* has also been proposed by Heaviside, who has pointed out that more properly the analogues of electric conductance and electric resistance, are given by the parallel relations of magnetic and electric forces explained at p. 201 above. *Magnetic Reluctance.*

It is of importance to know for an endless solenoid the total induction through a secondary circuit encircling it. Let the solenoid contain a core of magnetic permeability μ. The total induction (which we denote by the symbol $\mathbf{B}$) through a single turn of the secondary is given by *Total Induction through Solenoid.*

$$\mathbf{B} = \int \mathbf{B}dS + \int \mathbf{H}dS' \quad \ldots \ldots \quad (70)$$

where the first integral is taken over the cross-section of the core, and the second over the remainder of the cross-section of the solenoid.

Now taking the particular case of the ring electro-magnet we have by (64) *Total Induction in Ring Electro-Magnet.*

$$\mathbf{B} = 4\pi r\gamma' \left\{ \int \frac{\mu dS}{r+x} + \int \frac{dS'}{r+x} \right\} \quad \ldots \quad (71)$$

or, since $\mu = 1 + 4\pi\kappa$

$$\mathbf{B} = 4\pi r\gamma' \left\{ 4\pi \int \frac{\kappa dS}{r+x} + \int \frac{dS}{r+x} \right\} \quad \ldots \quad (71')$$

where the first integral is taken over the cross-section of the core, the second over that of the solenoid.

If now we suppose κ to be the same throughout the core of the solenoid, and the cross-sections to be circular, of radius ρ in the case of the solenoid, and ρ' in that of the core, and to have the same circular axis, then denoting $\sqrt{\rho^2 - x^2}$ by h, and $\sqrt{\rho'^2 - x^2}$ by h', we have

$$\mathsf{B} = 4\pi r \gamma' \left\{ \int_{-h}^{h} \int_{-\rho}^{\rho} \frac{dx\,dy}{r+x} + 4\pi\kappa \int_{-h'}^{h'} \int_{-\rho'}^{\rho'} \frac{dx\,dy}{r+x} \right\}$$

or after integration

$$\mathsf{B} = 8\pi^2 r \gamma' \left\{ r - \sqrt{r^2 - \rho^2} + 4\pi\kappa (r - \sqrt{r^2 - \rho'^2}) \right\} \quad . \quad (72)$$

If ρ may be taken as equal to ρ' this becomes

$$\mathsf{B} = 8\pi^2 \mu r \gamma' (r - \sqrt{r^2 - \rho^2}) \quad . \quad . \quad . \quad . \quad (73)$$

which gives the total induction through a single turn of the secondary coil when the magnetized core completely fills the solenoid.

To obtain the total induction through a secondary coil of n' turns, it is of course only necessary to multiply each of the preceding values of B by n'.

Induction in Straight Solenoid at Point distant from Ends.

If r be very great in comparison with ρ the last equation may be written

$$\mathsf{B} = 4\pi^2 \mu \rho^2 \gamma' \quad . \quad . \quad . \quad . \quad . \quad . \quad (74)$$

a result independent of r. In this case any portion of the solenoid, the length of which is of the same order of magnitude as ρ, may be regarded as straight. Thus we infer that in a long straight solenoid the total induction through a single turn of a secondary encircling it, at a place so distant from either end that the force due to the ends may be neglected, is given by (74).

Induction in Uniformly Wound Helix.

If the solenoid be lapped round uniformly and closely with n turns of fine wire per unit of length of r, and a current γ flow in each, then in the preceding equations we have simply to replace γ' by $n\gamma$ to obtain the corresponding expressions for the induction.

In particular from (74) we obtain for a straight solenoid at a cross-section the distance of which from either end fulfils the above condition

$$\mathsf{B} = 4\pi^2 \mu \rho^2 n \gamma \quad . \quad . \quad . \quad . \quad . \quad . \quad (75)$$

Thus the induction per unit area of the cross-section, or **B** is given by the equation

$$B = 4\pi\mu\gamma = \pi\mu n\gamma \quad \ldots \ldots \quad (76)$$

and hence for the magnetic force at any point of the cross-section we obtain

$$H = 4\pi n\gamma \quad \ldots \ldots \quad (77)$$

the result already given in equation (28''') above.

This is the intensity of the magnetic field within a straight solenoid at any point sufficiently remote from the ends, and is the result used at pp. 87, 179 above. It is remarkable that the intensity should be uniform over each cross-section, a result still very approximately true where the influence of the ends on the numerical value of **H** is quite sensible. The effect of the ends has already, p. 260 above, been investigated for a field of unit permeability.

Magnetic Field Intensity within Straight Helix.

The solution of the following important problem affords an interesting application of the theorem of p. 134 above, to the calculation of induction. A wire of iron or other highly magnetizable material, in the form of a right circular cylindric tube, carries a current of strength γ, uniform over the cross-section; it is required to find the influence of the resulting magnetization of the wire on the value of the coefficient of self-induction.

Induction in Magnetic Wire Carrying a Current. Effect on Self-Induction.

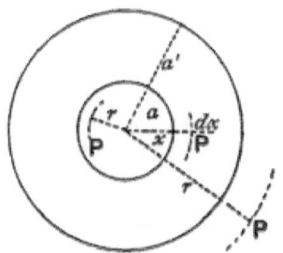

Fig. 64.

First we shall prove that the magnetic force at any point P external to a right cylindric tubular conductor, Fig. 64, or external to any part of such a conductor, is the same if the current be uniform over the cross-section, or symmetrically distributed round the axis, as if the current were confined to a filament coinciding with the axis of the tube. Let first P be external to the tube. Assuming that the current in each fila-

Cylindric Conductor Replaceable for External Points by Axial Filament.

ment produces the same magnetic force at each point as if the others did not exist (an assumption justified by experience and similar to those made in other parts of electrical theory), we see that the resultant magnetic force **H** at P is tangential to a circle in the plane of the cross-section through P, and concentric with the bounding circles of the section, and has the same numerical value at every point of that circle. Hence if r be the distance of P from the axis, the work done in carrying a unit pole round the circle is $2\pi r\mathbf{H}$. But by the theorem given at p. 107, if γ be the current in the circuit, $2\pi r H = 4\pi\gamma$, so that $H = 2\gamma/r$, which proves the statement made above.

Magnetic Force Zero within Hollow Cylindric Conductor.

Next let P be within the inner surface of the cylinder. By symmetry the magnetic force must be the same in numerical value at every point in a circle drawn through P in the cross-section, and having its centre on the axis. Further, by considering the effects of the two parts of the conductor on either side of a plane through the axis and P, imagined as divided into equal filaments, each producing magnetic force at P, we see at once that the force at P must be tangential to the circle. Hence for this case we have $2\pi r\mathbf{H}$ = current internal to $P + 4\pi = 0$. Hence $\mathbf{H} = 0$.

Magnetic Force within Substance of Cylinder.

It follows that if P be within the substance of the cylinder, the value of $\mathbf{H}$ depends only on the current internal to the circle described through P, concentric with the bounding circles of the cross-section.

Now consider any point P in the substance of the tube at distance x from the axis, as shown in Fig. 64, and let a, a' be the internal and external radii. Since the direction of $\mathbf{H}$ is tangential to the cross-section of the tube, and $\pi(x^2 - a^2)\gamma/\pi(a'^2 - a^2)$ or $\gamma(x^2 - a^2)/(a'^2 - a^2)$ is the current in the internal part of the conductor, we have

$$\mathbf{H} = 2\frac{x^2 - a^2}{a'^2 - a^2}\frac{\gamma}{x}$$

and if μ be the magnetic permeability of the material of the conductor

$$\mathbf{B} = \mu\mathbf{H} = 2\mu\frac{x^2 - a^2}{a'^2 - a^2}\frac{\gamma}{x}. \quad \ldots \ldots \quad (78)$$

Total Induction through Substance of Cylinder

The total induction across a strip of unit length, and breadth dx, of a radial plane of the cylinder through P, is therefore $2\mu\gamma(x^2 - a^2)dx/(a'^2 - a^2)x$, and the part of this due to the magnetization is, since $\mu = 1 + 4\pi\kappa$, $8\pi\kappa\gamma(x^2 - a^2)dx/(a'^2 - a^2)x$.

The energy due to the total induction across this area is equal

INDUCTANCE DUE TO MAGNETIZATION

to half the product of this induction by the current producing it, and is thus

$$\frac{2\mu\gamma(x^2-a^2)}{x(a'^2-a^2)}dx \times \frac{\gamma}{2}\frac{x^2-a^2}{a'^2-a^2} = \mu\gamma^2 \frac{(x^2-a^2)^2}{(a'^2-a^2)^2}\frac{dx}{x}.$$

Corresponding Energy.

Thus if T_c be the total energy depending on the induction in the wire itself

$$T_c = \frac{\mu\gamma^2}{(a'^2-a^2)^2}\int_a^{a'} \frac{(x^2-a^2)^2}{x} dx$$

$$= \frac{\mu\gamma^2}{(a'^2-a^2)^2}\left\{ \frac{a'^4-a^4}{4} - a^2(a'^2-a^2) + a^4\log\frac{a'}{a} \right\} \quad . \quad . \quad (79)$$

But if L_c be the coefficient of the self-induction which corresponds to T_c, $T_c = \tfrac{1}{2}L_c\gamma^2$, and therefore

$$L_c = \frac{2\mu}{a'^2-a^2}\left(\frac{a'^2+a^2}{4} - a^2 + \frac{a^4}{a'^2-a^2}\log\frac{a'}{a} \right) \quad . \quad . \quad (80)$$

The part of this, L_m say, depending on the magnetization of the wire, is $4\pi\kappa L_c/\mu$, since $\mu = 1 + 4\pi\kappa$. That is

Self-Inductance due to Magnetization of Conductor

$$L_m = \frac{8\pi\kappa}{a'^2-a^2}\left(\frac{a'^2+a^2}{4} - a^2 + \frac{a^4}{a'^2-a^2}\log\frac{a'}{a} \right) \quad . \quad . \quad (81)$$

If $a = o$, that is, if the conductor be solid throughout, the induction at P, or $\mathbf{B}$, is $2\mu\gamma x^2/a'^2$, and

$$T_c = \tfrac{1}{4}\mu\gamma^2, \quad L_c = \frac{\mu}{2}, \quad L_m = 2\pi\kappa \quad . \quad . \quad . \quad (82)$$

The corresponding expressions for the energy and the inductances, when a length l of the conductor is considered, are of course obtained by multiplying the values given above by l.

It may be noticed here that whatever the form of cross-section may be, the total induction through any finite area in the field of the conductor is finite, provided the area of cross-section carrying a finite current is not infinitely small.

Section II

CALCULATION OF COEFFICIENTS OF INDUCTION

Self-Inductance of Two Parallel Conductors.

We now consider the coefficient of self-induction of a circuit consisting of two long parallel wires of any form of cross-section, and carrying equal currents flowing in opposite directions. If the strength of the current be γ, and S_1, S_2 the areas of cross-section of A and B respectively, the current per unit area of cross-section in A is γ/S_1, in B γ/S_2. We may suppose each wire made up of the same number, n, of equal uniform filaments having their lengths parallel to the direction of flow, and each carrying a current γ/n. By making n large, the value γ/n may be made as small as we please.

The energy of the system per unit of length of the conductors can be calculated in the following manner. First we shall suppose, what is most frequently the case, that the magnetic permeability is everywhere unity; then take into account the permeability of the conductors when that is different from unity. The final result will thus be applicable to the important case of parallel conductors of iron, in a medium of unit magnetic permeability.

In dealing with this circuit we shall consider the parallel conductors as practically infinitely long, that is, such that the influence of the cross-conductors at the ends may be neglected for any point considered, and shall calculate only the induction and corresponding energy for the portion of the circuit intercepted between two parallel planes perpendicular to the conductors and at unit distance apart.

Let the distance between the conductors be measured in a plane at right angles to the two conductors, from a convenient point in one cross-section made by that plane to a convenient point in the other, and be denoted by b. Then let these points be taken as origins of rectangular coordinates (x, in the direction in which b is measured, and y at right angles to that direction) by which the position of the cross-section of any particular filament can be specified. We shall denote the coordinates of a filament in A by x_1, y_1, in B by x_2, y_2, and shall indicate a particular filament by its coordinates inclosed in brackets, thus (x_1, y_1), (x_2, y_2).

Now let (x'_1, y'_1) denote a second filament in A, dS'_1 its area, r' its distance from any point in the plane of its cross-section,

SELF INDUCTION OF TWO PARALLEL CONDUCTORS

r'_{11} its distance from the filament (x_1, y_1), and r'_{12} its distance from the filament (x_2, y_2) in B. The induction produced by (x'_1, y'_1) through unit length of the circuit formed by the two conductors (x_1, y_1) (x_2, y_2) is

Induction produced by Straight Current through Circuit of two Parallel Wires.

$$\frac{2\gamma dS_1}{S_1} \int_{r'_1}^{r'_2} \frac{dr'}{r'} = \frac{2\gamma dS_1}{S_1} (\log r'_{12} - \log r'_{11})$$

Hence the total induction $d\mathbf{B}_A$ per unit of length through this filamental circuit produced by the current in A is given by

$$d\mathbf{B}_A = \frac{2\gamma}{S_1} \int_A dS'_1 (\log r'_{12} - \log r'_{11}) \quad . \quad . \quad . \quad (83)$$

the integral being taken, as indicated, over the cross-section of A. To this is to be added the induction through this circuit due to the filaments of B. According to the principle of the notation adopted above we denote the distances of any filament (x'_2, y'_2) of B from (x_2, y_2) and (x_1, y_1) by r'_{22} and r'_{21}. Thus if B_B denote the induction specified

$$d\mathbf{B}_B = \frac{2\gamma}{S_2} \int_B dS'_2 (\log r'_{21} - \log r'_{22}) \quad . \quad . \quad . \quad (84)$$

The electrokinetic energy of the **circuit is half the** product of the induction by the current in the circuit. **The** value of the current may be written either $\gamma dS_1/S_1$ or $\gamma dS_2/S_2$. Using the second form in the first term of the integral in (83), and the second term of the integral in (84), and the first form in the remaining **two** terms, and denoting by dT the total electrokinetic **energy** depending on the filamental circuit considered, we get

$$dT = \gamma^2 \left\{ \frac{1}{S_1 S_2} dS_2 \int_A dS'_1 \log r'_{12} - \frac{1}{S_1^2} dS_1 \int_A dS'_1 \log r'_{11} \right.$$
$$\left. + \frac{1}{S_1 S_2} dS_1 \int_B dS'_2 \log r'_{21} - \frac{1}{S_2^2} dS_2 \int_B dS'_2 \log r'_{22} \right\} \quad (85)$$

Hence we get the total electrokinetic **energy** by finding **the** values of dT for all the circuits which **can be formed.** Thus we have only to integrate each of the terms of (27) over S_1 or S_2 as the case may be. Hence

$$T = \gamma^2 \left\{ \frac{2}{S_1 S_2} \int_B \int_A \log r_{12}\, dS_1 dS_2 - \frac{1}{S_1^2} \int_A \int_A \log r'_{11}\, dS_1 dS'_1 \right.$$
$$\left. - \frac{1}{S_2^2} \int_B \int_B \log r'_{22}\, dS_2 dS'_2 \right\} \quad \ldots \quad (86)$$

If we write

$$\left.\begin{array}{l} S_1 S_2 \log R_{12} = \int_B \int_A \log r_{12}\, dS_1 dS_2 \\[4pt] S_1^2 \log R_{11} = \int_A \int_A \log r'_{11}\, dS_1 dS'_1 \\[4pt] S_2^2 \log R_{22} = \int_B \int_B \log r'_{22}\, dS_2 dS'_2 \end{array}\right\} \quad \ldots \quad (87)$$

Geometric Mean Distance of Two Coplanar Areas.

then R_{12}, R_{11}, R_{22}, are called *geometric mean distances*, R_{12} of the area S_1 from S_2, R_{11} of S_1 from itself, and R_{22} of S_2 from itself.

The determination of the self-induction coefficient of a circuit composed of two long straight parallel wires, is thus reduced to the calculation of the geometric mean distances of the cross-sectional areas of the conductors from themselves, and from one another. The conductors may have any form of cross-section, and the calculation of their coefficient of self-induction is of course theoretically possible. Its evaluation, however, except in a few comparatively simple but important cases, is a tedious and troublesome operation. We shall consider these cases presently, in the meantime we can infer from electrical results already obtained the required geometrical mean distances for two right circular cylindric conductors, whether tubular or solid. For brevity we shall denote in the letter-press geometrical mean distances by G. M. D.

G.M.D. of Two Circles.

In the first place the G. M. D. of the conductors from one another is equal to the distance between their axes. As we have seen, the magnetic force at any point external to either of the conductors (say A) is the same as if the whole current were collected in a filament along the axis. Thus the induction through any external area may be found by supposing the conductor A replaced by an axial filament carrying the same current. The electromagnetic action of the current in A on unit length of an external parallel filament carrying unit current is $2\gamma/r$, if r be the distance of the filament from the axis of A. We infer therefore that the reaction of the filament on A is the same as would be exerted on the axial filament replacing the latter. Thus the total action of the conductor B on A is the same as if the conductors were replaced by filaments coinciding with their axes.

It follows from this that the expression for their *mutual* electrokinetic energy must be the same as if the conductors were replaced by axial filaments, that is, the G. M. D. between the conductors is equal to the distance between their axes. A direct analytical proof of this theorem will be given presently.

We can now find very simply the energy of the arrangement of two tubular wires, provided we can neglect the disturbance of the magnetic field produced by the magnetization of the conductors themselves, taking into account the magnetic permeabilities of the substance of the conductors A and B, which we suppose μ_1 and μ_2 respectively.

Energy of Opposite Currents in Parallel Tubular Conductors.

By (79) above, if a_1, a'_1, be the internal and external radii of the conductor A, the total amount T_1 of energy corresponding to the induction in A produced by its own current is given by the equation

$$T_1 = \frac{\mu_1 \gamma^2}{(a'_1{}^2 - a_1{}^2)^2} \left\{ \frac{a'_1{}^4 - a_1{}^4}{4} - a_1{}^2(a'_1{}^2 - a_1{}^2) + a_1{}^4 \log \frac{a'_1}{a_1} \right\} . \quad (88)$$

which if $a_1 = 0$, that is, if the conductor is a solid wire, reduces to

$$T_1 = \frac{1}{4} \mu_1 \gamma^2 \quad \ldots \ldots \ldots \quad (89)$$

Similarly for the energy corresponding to the induction in B produced by the current in B we obtain

$$T_2 = \frac{\mu_2 \gamma^2}{(a'_2{}^2 - a_2{}^2)^2} \left\{ \frac{a'_2{}^4 - a_2{}^4}{4} - a_2{}^2(a'_2{}^2 - a_2{}^2) + a_2{}^4 \log \frac{a'_2}{a_2} \right\} . \quad (90)$$

or, if $a_2 = 0$,

$$T_2 = \frac{1}{4} \mu_2 \gamma^2 . \quad \ldots \ldots \ldots \quad (91)$$

The value of **B** at distance r from the axis of A in the medium between the conductors is $2\mu\gamma/r$, if μ be the magnetic permeability of the medium. In the substance of B the magnetic induction due to the current in A is to be calculated as if B were replaced by the medium occupying the field, and similarly the induction in A due to the current in B is to be dealt with. For since B is isotropic, and the effect of the magnetization of the conductor is neglected, whatever of A's lines of induction pass through the substance of B, follow the direction of the magnetic force, that is, are along circles, the common axis of which is the axis of A. Thus, if Fig. 65 represent a section of the conductors

CALCULATION OF INDUCTION

by a plane at right angles to the conductor, so that a is a point on the axis of A, and β the corresponding point on the axis of B, the induction through the surface, the projection of which is $a'\beta'$, may be equally taken through $a'_1\beta'_1$ where $\beta'_1\beta'_1$ lie on a line of induction. But $a'_1\beta'_1$ lies entirely in the external medium and is equal in length to $a'\beta'$, so that it represents the same area. Hence the result stated above.

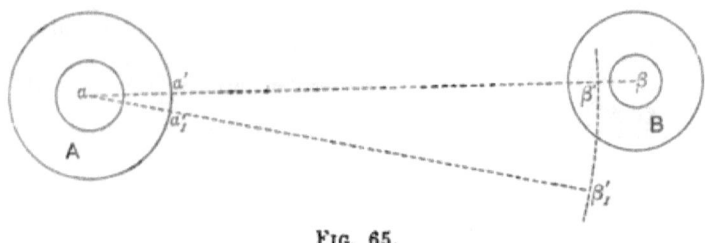

FIG. 65.

We thus have, by the theorem of the G. M. D. of B from A, for the total induction external to A the value

$$\int_{a_1}^{b}\frac{2\mu\gamma}{r}dr = 2\mu\gamma\log\frac{b}{a_1},$$

and for the corresponding part of the energy $\gamma^2\mu\log b/a_1$.

In the same way we find for the total induction external to B, produced by the current in B, and the corresponding energy, the values $2\mu\gamma\log b/a_2$, $\gamma^2\mu\log b/a_2$.

The total energy is therefore

$$T = \gamma_2\left[\frac{\mu_1}{(a'_1{}^2 - a_1{}^2)^2}\left\{\frac{a'_1{}^4 - a_1{}^4}{4} - a_1{}^2(a'_1{}^2 - a_1{}^2) + a_1{}^4\log\frac{a'_1}{a_1}\right\}\right.$$

$$+\frac{\mu_2}{(a'_2{}^2 - a_2{}^2)_2}\left\{\frac{a'_2{}^4 - a_2{}^4}{4} - a_2{}^2(a'_2{}^2 - a_2{}^2) + a_2{}^4\log\frac{a'_2}{a_2}\right\}$$

$$\left. + \mu\log\frac{b^2}{a_1 a_2}\right] \quad \ldots \ldots \ldots \quad (92)$$

Comparing with the expression $\tfrac{1}{2}L\gamma^2$ for the same quantity we get

$$L = \frac{2\mu_1}{(a'_1{}^2 - a_1{}^2)^2}\left\{\frac{a'_1{}^4 - a_1{}^4}{4} - a_1{}^2(a'_1{}^2 - a_1{}^2) + a_1{}^4 \log \frac{a'_1}{a_1}\right\}$$
$$+ \frac{2\mu_2}{(a'_2{}^2 - a_2{}^2)^2}\left\{\frac{a'_2{}^4 - a_2{}^4}{4} - a_2{}^2(a'_2{}^2 - a_2{}^2) + a_2{}^4 \log \frac{a'_2}{a_2}\right\}$$
$$+ 2\mu \log \frac{b^2}{a_1 a_2} \quad \ldots \ldots \quad (93)$$

L is therefore very **great** for wires of **very** small diameter, even if they be only a moderate distance apart. The least value which it can have is obtained when the two wires are put into contact, that is, when $b = a_1 + a_2$.

If the wires be solid (93) **reduces to**

$$L = 2\mu \log \frac{b^2}{a_1 a_2} + \frac{1}{2}(\mu_1 + \mu_2) \quad \ldots \quad (93')$$

For two solid wires in contact the last equation becomes

$$L = 2\mu \log \frac{(a_1 + a_2)^2}{a_1 a_2} + \frac{1}{2}(\mu_1 + \mu_2) \quad \ldots \quad (94)$$

This theory requires correction in the case of wires in which the current is not uniform over the cross-section, **which it never is when** its strength is rapidly varying. This **case will be considered** below in this chapter in the section **on Rapidly Varying and Alternating** Currents.

On account of the effect **of the magnetization of the con**ductors it is difficult to obtain a complete **solution except in the** case of coaxial **conductors.** The general case of **two parallel** cylindrical conductors, carrying steady currents, has been **worked** out by Mr. H. M. Macdonald, of Clare College, Cambridge (*Proc. Camb. Phil. Soc.* Vol. VII. Pt. V.). His result shows that when μ_1 (the common permeability of the conductors) is 100 and upwards, the part of L depending **on** the size of the conductors and their distance apart is only slightly affected by the permeability. Instead **of** (93') he obtains, putting $\lambda = (\mu_1 - \mu)/(\mu_1 + \mu)$,

$$L = \mu_1 + 2\mu\left[\log\frac{b^2}{a_1 a_2} + \lambda\log\frac{b^4}{(b^2 - a_1^2)(b^2 - a_2^2)} + 2\lambda^2\log\frac{b^2(b^2 - a_1^2 - a_2^2)}{(b^2 - a_1^2)(b^2 - a_2^2)}\right.$$
$$\left. + \lambda^3 \log\frac{b^4(b^2 - a_1^2 - a_2^2)}{(b^2 - a_1^2)(b^2 - a_2^2)\{(b^2 - a_1^2) - a_2^2 b^2\}\{(b^2 - a_2^2) - a_1^2 b^2\}} + \&c.\right]$$
$$(93'')$$

Calculation of Geometric Mean Distances.

We can now calculate the G. M. D. in some important cases. It is first to be noticed that if there exist any number of areas of extent A, B, &c., the G. M. D.'s of which, R_A, R_B, &c., from another area S are known, it follows from the definition that the G. M. D., R of their sum from S is given by the equation

$$\log R = \frac{A \log R_A + B \log R_B + \ldots}{A + B + \ldots} \quad . \quad . \quad (95)$$

G.M.D. of Point from Circular Area.

We consider first the G. M. D. of a circular area, annular or complete, from a point P, (1) external to the area and in its plane, (2) in the circular area itself. Let b be the distance of P from the centre of the circular area, a, a' the internal and external radii of the latter, x and $x + dx$ the radii of two intermediate circles very near to one another. Let two radii OR, OS

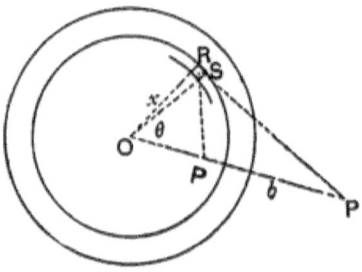

Fig. 66.

(Fig. 66) be drawn, making the angles θ, $\theta + d\theta$ with OP, so as to intercept the element of area $x\,dx\,d\theta$, on the annulus bounded by the circles of radii x and $x + dx$. The distance of P from the element, or r, is $\sqrt{b^2 + x^2 - 2bx \cos \theta}$. Hence the G.M.D. of the annulus from P is given by

$$2\pi x\,dx \log R = \tfrac{1}{2} x\,dx \int_0^{2\pi} \log (b^2 + x^2 - 2bx \cos \theta)\,d\theta.$$

This can be written

$$2\pi x\,dx \log R = \tfrac{1}{2} x\,dx \int_0^{2\pi} \log \left(1 + \frac{b^2}{x^2} - 2\frac{b}{x} \cos \theta\right) d\theta$$

$$+ x\,dx \log x \int_0^{2\pi} d\theta \quad . \quad . \quad . \quad . \quad . \quad (96)$$

GEOMETRIC MEAN DISTANCES

Now the first integral on the right of (36) is known to have the value $4\pi \log b/x$, if $b > x$, and o if $b < x$. Hence in case (1) for the annulus

$$2\pi x dx \log R = 2\pi x dx (\log b - \log x) + 2\pi x dx \log x$$

or
$$R = b \quad \ldots \ldots \ldots \quad (97)$$

On the other hand if P be within the inner boundary of the elementary annulus $b/x < 1$, and the first integral of (96) is zero. Hence we have for the annulus in case (2)

$$2\pi x \log R = 2\pi x \log x$$

or
$$R = x \quad \ldots \ldots \ldots \quad (98)$$

From these results it follows by (95) that the G. M. D. of any finite annulus from an external point P is simply the distance of the point from the centre. For the annulus is made up of elementary annuli, every one of which has the same G. M. D. from P. This includes, of course, as a particular case a complete circular area.

G.M.D. of Finite Annulus from External Point.

The G. M. D. of a finite annulus from a point within its inner bounding circle is now easily found. The area of the annulus is $\pi(a'^2 - a^2)$. Hence by (95) if R be the G. M. D. required

G.M.D. of Finite Annulus from Internal Point.

$$\pi(a'^2 - a^2) \log R = 2\pi \int_a^{a'} x \log x \, dx$$

$$= \pi \left(a'^2 \log a' - a^2 \log a - \frac{a'^2 - a^2}{2} \right),$$

that is

$$\log R = \frac{a'^2 \log a' - a^2 \log a}{a'^2 - a^2} - \frac{1}{2} \ldots \quad (99)$$

Lastly, if P be on the annulus at a distance b from the centre, the annulus divides into two parts, one internal and the other external to the concentric circle through P. Hence by (97), (99) and (95) if R now denote the G. M. D. for the whole area in this last case

$$\log R = \frac{a'^2 \log a' - b^2 \log b}{a'^2 - a^2} - \frac{1}{2} \frac{a'^2 - b^2}{a'^2 - a^2} + \frac{b^2 - a^2}{a'^2 - a^2} \log b,$$

or

$$\log R = \frac{a'^2 \log a' - a^2 \log b}{a'^2 - a^2} - \frac{1}{2} \frac{a'^2 - b^2}{a'^2 - a^2} \quad \ldots \quad (100)$$

The following corollaries follow at once from these results.

1. The G. M. D. from a circular area (complete or annular) of any area external to the circular area, and in the same plane, is equal to the G. M. D. of the figure from the centre of the circle. For the G. M. D. of every part of the area is its distance from the centre, and the result follows by (95).

2. The G. M. D. of any figure completely internal to an annular area from that area is the value of R given by (99). For R is the G. M. D. of every element.

3. The G. M. D. of a circular annulus of infinitesimal breadth from itself is simply its radius. For the G. M. D. of every point on it from the annulus is the radius.

G.M.D. of Finite Annulus from Itself.

4. The G. M. D. of the finite annulus from itself is given by

$$\log R = \log a' - \frac{a^4}{(a'^2 - a^2)^2} \log \frac{a'}{a} + \frac{1}{4} \frac{3a^2 - a'^2}{a'^2 - a^2}. \quad (101)$$

For consider the G. M. D. of the annulus from a point in it distant x from the centre. The G. M. D. of the internal part is x, the logarithm of the G. M. D. of the external part is $(a'^2 \log a' - x^2 \log x)/(a'^2 - x^2) - \frac{1}{2}$. Hence as found in (100) the G. M. D. of the whole area from the point is given by

$$\log R' = \frac{a'^2 \log a' - x^2 \log x}{a'^2 - a^2} - \frac{a'^2 - x^2}{2(a'^2 - a^2)} + \frac{x^2 - a^2}{a'^2 - a^2} \log x. \quad (102)$$

The G. M. D. of an infinitesimal annulus of breadth dx and radius x from the total area is thus R'. Hence by (95) the G. M. D. of the whole area from itself is to be found from

$$\pi(a'^2 - a^2) \log R = \int_a^{a'} 2\pi x dx \log R'.$$

Substituting the value of R' from (102) and integrating we obtain (101).

If $a = o$, the area is a complete circle, and (101) gives for that case

$$\log R = \log a' - \frac{1}{4},$$

or $\qquad R = a' \epsilon^{-\frac{1}{4}} = \cdot 7788 a' \quad \ldots \quad (103)$

Next consider the G. M. D. of a line from any point P. Let AB (Fig. 67) be the line, p the length of the perpendicular from

GEOMETRIC MEAN DISTANCES

P on the line, a and a' the lengths of the segments AO, OB into which the line is divided at O. Then the distance from P of any point Q in the line at distance x from O is $\sqrt{p^2 + x^2}$. Hence for the line

G.M.D. of Point from Straight Line.

$$(a + a') \log R = \int_{-a}^{a'} \log \sqrt{p^2 + x^2}\, dx$$
$$-\tfrac{1}{2}a' \log(a'^2 + p^2) + \tfrac{1}{2}a \log(a^2 + p^2) - (a + a')$$
$$+ p\left(\tan^{-1}\frac{a}{p} + \tan^{-1}\frac{a'}{p}\right). \quad \ldots \quad (104)$$

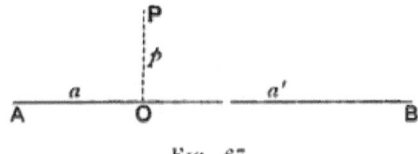

FIG. 67.

If O coincide with the centre of AB, $a = a' = \tfrac{1}{2}$ the length of AB, and

$$\log R = \tfrac{1}{2} \log(a^2 + p^2) - 1 + \frac{p}{2a}\left(\tan^{-1}\frac{a}{p} + \tan^{-1}\frac{a'}{p}\right) \quad (105)$$

If O coincide with B, $a = 0$, and

$$\log R = \tfrac{1}{2} \log(a'^2 + p^2) - 1 + \frac{p}{a'} \tan^{-1}\frac{a'}{p} \quad . . \quad (106)$$

where a' is now the whole length of AB.

From (105) we get at once the G. M. D. of four lines forming a rectangle from the centre. For let the length of the rectangle be a and its breadth b. Then since for the ends $p = \tfrac{1}{2}a$, and for the sides $p = \tfrac{1}{2}b$.

G.M.D. of Boundary of Rectangle from Centre.

$$2(a + b)\log R = (a + b)\log \frac{a^2 + b^2}{4} - 2(a + b)$$
$$+ 2a \tan^{-1}\frac{b}{a} + 2b \tan^{-1}\frac{a}{b}$$

or

$$\log R = \tfrac{1}{2} \log \frac{a^2 + b^2}{4} - 1 + \frac{a}{a+b} \tan^{-1} \frac{b}{a}$$
$$+ \frac{b}{a+b} \tan^{-1} \frac{a}{b} \quad \ldots \ldots \quad (107)$$

G.M.D. of Boundary of a Square from Centre.

If the rectangle be a square $a = b$, and thus

$$\log R = \log \left(\frac{a}{\sqrt{2}}\right) - 1 + \frac{\pi}{4}$$
$$= a\epsilon^{-\cdot 2146} / \sqrt{2} \quad \ldots \ldots \quad (108)$$

G.M.D. of Two Parallel Lines from one another.

The G. M. D. of two parallel lines from one another can now be found. This is an important case, as it enables the self-inductance of a circuit composed of two parallel thin sheets of conducting material to be calculated. Let AB, CD (Fig. 68) be

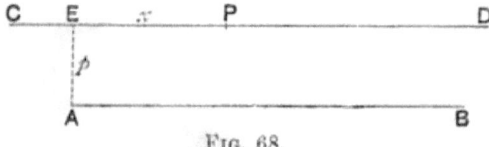

Fig. 68.

the lines, and E the foot of the perpendicular from A on CD. Let x be the distance of P from E, and p its distance from AB, a and β the distance of C and D from E, taken as positive quantities when measured from E to the right, and negative when measured the other way. The length of CD is thus $\beta - a$, and if a be the length of AB, we have to put in (104) x for a, and $a - x$ for a'. Thus multiplying the expression on the right of (104) by dx, and integrating from $x = a$ to $x = \beta$, we find by (95), if R now be the G. M. D. of CD from AB

$a(\beta - a) \log R$

$= \tfrac{1}{4} (\beta^2 + p^2) \log (\beta^2 + p^2) - \tfrac{1}{4} (a^2 + p^2) \log (a^2 + p^2)$

$- \tfrac{1}{4} \{(a - \beta)^2 + p^2\} \log \{(a - \beta)^2 + p^2\}$

$+ \tfrac{1}{4} \{(a - a)^2 + p^2\} \log \{(a - a)^2 + p^2\}$

GEOMETRIC MEAN DISTANCES

$$+ p\beta \tan^{-1}\frac{\beta}{p} - pa \tan^{-1}\frac{a}{p}$$

$$- p(a - \beta) \tan^{-1}\frac{a - \beta}{p} + p(a - a) \tan^{-1}\frac{a - a}{p}$$

$$- \tfrac{3}{2} a (\beta - a) \quad \ldots \quad \ldots \quad (109)$$

The value of R given by this equation may be used for the calculation of the self-induction of a circuit composed of two long thin strips of conducting material arranged with their lengths and planes parallel. The lines AB, CD represent the cross-sections of such an arrangement made by a plane at right angles to the conductors.

The G. M. D. of each line from itself can of course be found from (109) by putting $a = o$, $\beta = a$ (the length of the line considered), and $p = o$. We thus obtain

G.M.D. of Straight Line from Itself.

$$\log R = \log a - \tfrac{3}{2} \quad \ldots \quad \ldots \quad (110)$$

which can be verified at once by calculating directly for this particular case.

We can now find the G. M. D. of a given line from an area in the same plane. We shall consider first a given line and a parallel rectangle, and from the result for this case deduce the G. M. D. of two parallel coplanar rectangles from one another. The practically important arrangements are those in which the line and rectangle, or two rectangles, are symmetrical about the line passing through their centres, as shown in Figs. 69, 70, and 71.

G.M.D. of Straight Line from Parallel Rectangle.

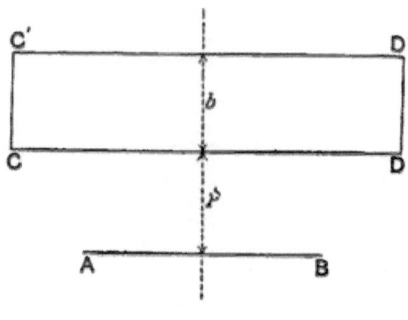

FIG. 69.

CALCULATION OF INDUCTION

G.M.D. of Straight Line from Parallel Rectangle.

Taking first the line and rectangle, as in Fig. (69), and putting a, a' for the lengths AB and CD, we suppose the rectangle to be generated by the motion of CD, at right angles to itself through a distance b, the breadth of the rectangle. We may thus suppose CD made up of parallel strips of area, each of infinitesimal breadth dp, and find the G. M. D. of the rectangle by multiplying the expression for $\log R$ in (109) (modified to suit the circumstances supposed) by dp, and integrating from p to $p+b$. The constant factors on the left will for simplicity be retained.

We have here $\alpha = \frac{1}{2}(a-a')$, $\beta = \frac{1}{2}(a+a')$ so that

$$a - \alpha = \frac{1}{2}(a+a') = \beta,$$
$$a - \beta = \frac{1}{2}(a-a') = \alpha.$$

Thus (109) becomes

$aa' \log R$

$$= \frac{1}{2}(\beta^2 + p^2)\log(\beta^2 + p^2) - \frac{1}{2}(\alpha^2 + p^2)\log(\alpha^2 + p^2)$$
$$+ 2p\beta \tan^{-1}\frac{\beta}{p} - 2p\alpha \tan^{-1}\frac{\alpha}{p}$$
$$- \frac{3}{2} aa'. \quad \ldots \ldots \ldots (111)$$

Multiplying by dp, integrating as stated above, and putting R now for the G. M. D. of the rectangle CD from the line AB, we obtain

$aa'b \log R$

$$= \frac{1}{2}(p+b)\left\{\beta^2 - \frac{(p+b)^2}{3}\right\}\log\{(p+b)^2 + \beta^2\}$$
$$- \frac{1}{2}p\left(\beta^2 - \frac{p^2}{3}\right)\log(p^2 + \beta^2)$$
$$+ \beta(p+b)^2 \tan^{-1}\frac{\beta}{p+b} - \beta p^2 \tan^{-1}\frac{\beta}{p}$$
$$+ \frac{1}{3}\beta^3 \tan^{-1}\frac{p+b}{\beta} - \frac{1}{3}\beta^3 \tan^{-1}\frac{p}{\beta}$$

– (the same series of terms with β replaced by α)

$$-\frac{11}{6} aa'b \quad \ldots \ldots \ldots (112)$$

GEOMETRIC MEAN DISTANCES   301

A rectangle of breadth b, might have been generated by moving the line AB away from CD (Fig. 70). We should have obtained the same expression for the G. M. D. of the latter rectangle from the line CD, as is given in (112) for the other case.

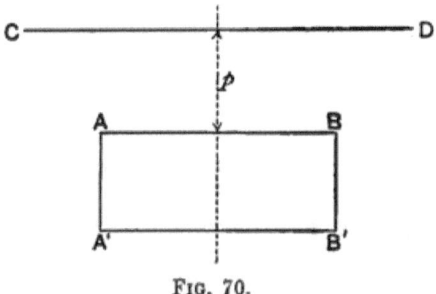

FIG. 70.

That these two G. M. D.'s are equal is easily seen from (111). Each rectangle may be divided into the same number of strips of equal breadth, and the G. M. D. of each strip in the rectangle CD' from AB is the same as the G. M. D. of each strip of AB' from CD, so that the result follows by (95).

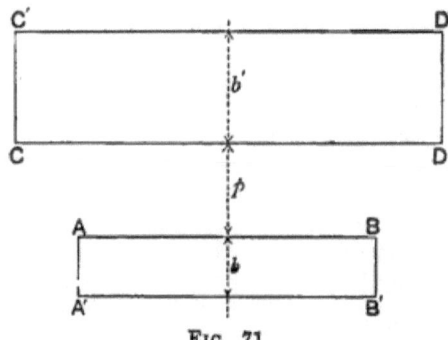

FIG. 71.

We can now find the G. M. D. in the important case of the two rectangles shown in Fig. 71. Multiplying the expression on the right of (112) by dp, integrating from p to $p + b'$ (so that p is the distance of AB from CD) and arranging the results

302   CALCULATION OF INDUCTION

G.M.D. of Two Parallel Rectangles from one another. we find for R, the G. M. D. of the rectangles from one another.

$$4aa'bb'\log R$$
$$=[(p+b+b')^2\{\beta^2-\tfrac{1}{6}(p+b+b')^2\}-\tfrac{1}{2}\beta^4]\log\{(p+b+b')^2-\beta^2\}$$
$$-[(p+b')^2\{\beta^2-\tfrac{1}{6}(p+b')^2\}-\tfrac{1}{2}\beta^4]\log\{(p+b')^2+\beta^2\}$$
$$-[(p+b)^2\{\beta^2-\tfrac{1}{6}(p+b)^2\}-\tfrac{1}{2}\beta^4]\log\{(p+b)^2+\beta^2\}$$
$$+\{p^2(\beta^2-\tfrac{1}{6}p^2)-\tfrac{1}{2}\beta^4\}\log(p^2+\beta^2)$$

$-$ (the same series of terms with β replaced by a)

$$+\tfrac{4}{3}\beta(p+b+b')\left\{(p+b+b')^2\tan^{-1}\frac{\beta}{p+b+b'}\right.$$
$$\left.+\beta^2\tan^{-1}\frac{p+b+b'}{\beta}\right\}$$
$$-\tfrac{4}{3}\beta(p+b')\left\{(p+b')^2\tan^{-1}\frac{\beta}{p+b'}+\beta^2\tan^{-1}\frac{p+b'}{\beta}\right\}$$
$$-\tfrac{4}{3}\beta(p+b)\left\{(p+b)^2\tan^{-1}\frac{\beta}{p+b}+\beta^2\tan^{-1}\frac{p+b}{\beta}\right\}$$
$$+\tfrac{4}{3}\beta p\left\{p^2\tan^{-1}\frac{\beta}{p}+\beta^2\tan^{-1}\frac{p}{\beta}\right\}$$

$-$ (the same series of trigonometrical terms with β replaced by a)

$$-\tfrac{1}{2}(\beta^2-a^2)\{(p+b+b')^2-(p+b')^2-(p+b)^2+p^2\}$$
$$-\tfrac{22}{3}aa'bb'.\quad\ldots\ldots\quad(113)$$

Here it is to be remembered that $\beta=\tfrac{1}{2}(a+a')$, $a=\tfrac{1}{2}(a-a')$.

G.M.D. of a Rectangle from Itself. The G. M. D. of either rectangle from itself can be found from (113) by putting $a=0$, $a=a'=\beta$, $b=b'$, $p+b=p+b'=0$. Hence for the G. M. D. from itself of a rectangle of length a and breadth b, we have the equation

$$\log R=\tfrac{1}{2}\log(a^2+b^2)-\frac{1}{12}\frac{b^2}{a^2}\log\left(1+\frac{a^2}{b^2}\right)-\frac{1}{12}\frac{a^2}{b^2}\log\left(1+\frac{b^2}{a^2}\right)$$
$$+\frac{2}{3}\frac{b}{a}\tan^{-1}\frac{a}{b}+\frac{2}{3}\frac{a}{b}\tan^{-1}\frac{b}{a}-\frac{25}{12}.\quad\ldots\quad(114)$$

If the rectangle is a square $a = b$, and *G.M.D. of Square from Itself.*

$$\log R = \log a + \frac{1}{3}\log 2 + \frac{\pi}{3} - \frac{25}{12}$$

or

$$R = \cdot 44705a \quad \ldots \ldots \quad (115)$$

The determination of the G. M. D. of the cross-section of a conductor is important in other cases than that of a long straight conductor. For example, if we have a circular coil of n turns each of radius great in comparison with any dimension of cross-section, it is easy to see that the coefficient of self-induction of the coil is very approximately equal to n^2 times the coefficient of mutual induction of two parallel coaxial circles, each of radius equal to the mean radius of the section, and at a distance apart equal to the G. M. D. of the cross-section from itself. For the coefficient of self-induction of a circuit is equal to the total magnetic induction through the circuit produced by unit current, and the coefficient of mutual induction of two circuits is the total induction through either produced by unit current in the other. Consider then the induction through a circle of reference A coaxial with the given circuit, and at a distance from the latter small in comparison with the radius. Let it be supposed as before that the current is of uniform density over the cross-section, so that the cross-section may be supposed divided into a very large number of parallel thin filaments each of cross-section dS. If S be the whole area of cross-section, and unit current flow in the conductor, the current in each filament is ndS/S. Let a cross-section of the whole system, including A, by a plane through the axis, be taken, and let r_a be the distance of the section of A from that of any one of the system of equal filaments, and r_m the distance between the section of the latter filament and any other of cross-section dS_m. The difference between the total induction produced by the assemblage of filaments through the circuit of this latter filament, and that which they produce through A is

Self-Inductance of Circular Coil of Large Radius

Found from Mutual Inductance of Two Parallel Circles,

$$2n\int \frac{dS_m}{S}(\log r_a - \log r_m) = 2n(\log R_a - \log R_m)$$

where R_a is the G. M. D. of the cross-section of the given conductor from that of A, and R_m is its G. M. D. from dS_m.

Now let A be composed of as many coincident filaments as there are imagined to be in the given circuit. Thus the induc-

at Distance apart equal to G.M.D. of Cross-Section from Itself.

tion through each filament of the conductor may be compared with that through a corresponding filament of A. Since the number of filaments is S/dS, we have for the total difference between the induction through A, and the sum of the inductions through each filament of the conductor due to the whole assemblage

$$2n\frac{S}{dS}\left(\log R_a - \int\frac{dS}{S}\log R_m\right) = 2n\frac{S}{dS}\left(\log R_a - \log R\right)$$

where R is now the G. M. D. of the cross-section of the conductor from itself.

The energy of the given system corresponding to this induction is half the product of the current ndS/S, in each filament into the expression just found, that is, it is $n^2(\log R_a - \log R)$. This vanishes when $R_a = R$, that is, when the G. M. D. of the cross-section of A from that of the conductor is equal to the G. M. D. of the latter from itself. The energy of the given system is then equal to half the product of the total current into the induction through A, that is, in other words, the self-induction coefficient of the given circuit is equal to that of mutual induction between the given circuit and A.

That the coefficient of mutual induction in the latter case is equal to that between A and an equal circuit B at a distance apart equal to R, if not evident, may be seen as follows. The induction through A due to the given circuit is for equal currents equal to that produced by A through the given circuit, and by the reasoning above, this is equal to the induction due to A through a circuit B replacing the given circuit at the distance R.

Mutual Induction of Two Close Coils of Large Radius.

The expression found above (113) for the G. M. D. between two symmetrically placed rectangles is applicable to the approximate calculation of the coefficient of mutual induction of two coils of which the cross-sections by a plane passing through the common axis are rectangles, provided the radius of either coil is great in comparison with every dimension of the sections, and with the distance between them. Clearly to find the total induction through coil B due to unit current in A, we may proceed by calculating (a) the total induction through each turn of A due to unit current in that turn ; (b) that part of each of these total inductions which does not pass through B. The difference between the sum of the results in (a) and the sum of those in (b) is the coefficient M of mutual induction. First we suppose the current in the coil A to be uniformly distributed over the cross-section, so that if S_1 be the area of the section, and there be n_1 turns each carrying unit current, the current per unit area will be n/S_1. Thus the current across an element of the cross-section dS_1 is $n_1 dS_1/S_1$.

Now consider, as before, the difference between that part of the total induction due to a filament of section dS_1, which escapes a filament of the other coil of section dS_2, and that part which escapes a near coaxial circular circuit of reference. Let r_{12}, r, be the distances from dS_1 to dS_2, and from dS_1 to the cross-section of the circle of reference. The difference of total inductions specified is then $n_1 dS_1/S_1 \cdot (\log r_{12} - \log r)$. Integrating over the whole area S_1 we get for the difference due to all the filaments into which S_1 can be divided the value

is Equal to Mutual Induction of Two Parallel Circles,

$$\frac{n_1}{S_1} \int dS_1 (\log r_{12} - \log r).$$

Now let the other circuit be divided into any convenient number n of circuits, each of the same small area dS_2. It is the difference between the total induction through one of these, and that through the circle of reference that has just been found. We have then $dS_2 = S_2/n$. Hence the result just obtained may be written

$$\frac{nn_1}{S_1 S_2} \int dS_1 dS_2 \log r_{12} - \frac{n_1}{S_1} \int dS_1 \log r.$$

Integrating now over both cross-sections we get for the total difference

$$nn_1 (\log R_{12} - \log R)$$

at Distance Apart equal to G.M.D. of Cross-Sections.

where R_{12} is the G. M. D. between the cross-sections, and R is that of the section S_1 from the circuit of reference.

If the number of turns in the second coil be n_2 instead of n, this result must be reduced in the ratio of n_2 to n, by multiplying it by n_2/n. For accuracy of course n_2 must be large. Hence for the final value of the difference of the total inductions we have

$$n_1 n_2 (\log R_{12} - \log R).$$

If $R = R_{12}$, that is, if the G. M. D. of the cross-section of the conductor of reference from A be equal to that of the cross-sections of A and B from one another, the total induction which escapes the conductor of reference is equal to that which escapes the coil; in other words, the coefficient of mutual induction of the two coils is equal to that of the coil A and a coaxial circular conductor, the cross-section of which by any plane through the axis is at a G. M. D. from that of the coil A, equal to that of the cross-sections of A and B from one another.

It must be possible to replace the coil A by a conductor of proper mean radius carrying the whole current of n units which flows in the coil, so that the total induction through it is equal to the sum of those through the coaxial filamental conductors into which the coil has been supposed divided. If the radius of any part of the coil be large in comparison with the dimensions of cross-section, this proper mean radius may be taken as the simple mean radius of the coil. The other coil can then be also supposed replaced by a coaxial circular conductor at a distance from the other equal to R_{12}. Thus the determination of the coefficient of mutual induction of two coaxial coils is reduced to the determination of that of two coaxial circles.

Positions of Equivalent Circles not Definite.

The relative positions of these two circles is not definite. If we consider the lines of force through a coil due to the current in it, we see that these are closed round the coil, and any closed circuit placed in its field will pass through certain lines of force. The circuit may be placed in any position or have any size consistent with passing through the same lines of force, and the coefficient of mutual induction of the coil and circuit will be the same for all. If, in the present case we suppose the primary circular conductor fixed, the other may be situated anywhere on the toroidal surface marked out by the circular lines of force, the radius of which is the G. M. D. of the cross-sections.

Mutual Inductance of Two Coaxial Circles.

We proceed now to calculate the coefficients of mutual induction of coaxial circular circuits and coils. Taking first the case of two coaxial circles of nearly equal radii, we see that if we can find their coefficient of mutual induction when the circles are in one plane, we can find that for the actual arrangement by calculating, in the manner described above, the portion of the total induction due to one which escapes passing through the other owing to the deviation from coplanarity.

Consider first two coaxial circles in the same plane. Let the radius of the outer circle be $a + c$, and of the inner a. Then if we take any element ds of the outer circle at A (Fig. 72), and let θ be the angle OAE between the diameter through ds and a line of length r drawn to an element E_1 of area $rd\theta dr$ in the inner circle, we have for the magnetic induction through that area the value $ds \cos\theta/r^2 \cdot rd\theta dr$. Hence for the total induction **B** through the inner circle we get

$$\mathbf{B} = 2\int ds \int_{r_1}^{r_2} \int_0^{\theta_1} \frac{\cos\theta \, d\theta \, dr}{r} \qquad (116)$$

TWO COAXIAL CIRCLES

where $r_1 = AB$, $r_2 = AC$, $\theta_1 = \sin^{-1} a/(a+c)$, and the final integral is taken round the outer circle. The distances r_1, r_2 are evidently the roots of the equation

Mutual Inductance of Two Coaxial Circles.

$$r^2 - 2r(a+c)\cos\theta + (a+c)^2 - a^2 = 0.$$

These roots are

$$\left.\begin{array}{c}r_1\\r_2\end{array}\right\} = (a+c)\cos\theta \mp \sqrt{(a+c)^2\cos^2\theta - c^2 - 2ac} \quad (117)$$

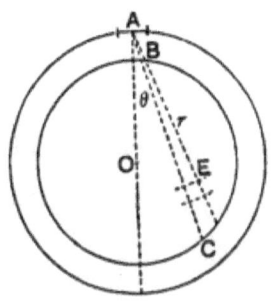

FIG. 72.

If c be very small then approximately

$$r_2 = 2a\cos\theta, \quad r_1 = c/\cos\theta.$$

Integrating then with respect to r we find

$$\mathbf{B} = 2\int ds \int_0^{\theta_1} \cos\theta \log\frac{r_2}{r_1} = 2\int ds \int_0^{\theta_1} \cos\theta \log\left(\frac{2a}{c}\cos^2\theta\right)d\theta$$

Now

$$\int_0^{\theta_1} \cos\theta \log\left(\frac{2a}{c}\cos^2\theta\right)d\theta = \sin\theta_1 \left\{\log\left(\frac{2a}{c}\cos^2\theta_1\right) - 2\right\}$$
$$+ \log\frac{1+\sin\theta_1}{1-\sin\theta_1}$$

which reduces approximately to

$$\log 4 - 2 + 2\log\sqrt{\frac{2a}{c}}.$$

CALCULATION OF INDUCTION

Hence to the same degree of approximation

$$M = 4\pi a \left(\log \frac{8a}{c} - 2\right) \quad \ldots \quad (118)$$

Now let the circle of radius a be carried out of the plane of the other circle a distance b while still remaining coaxial with the latter. The difference between the total inductions which escape from passing through the smaller circle in its two positions may be calculated as if the circles were straight. Putting now r for $\sqrt{b^2 + c^2}$, the shortest distance between the circular arcs, the difference of inductions per unit of length is approximately $2(\log r - \log c)$, and for the whole circle $4\pi a(\log r - \log c)$. Hence the coefficient M of mutual induction between the circles in the specified configuration is approximately given by

$$M = 4\pi a \left(\log \frac{8a}{r} - 2\right) \quad \ldots \quad (119)$$

Coil of Maximum Self-Inductance. From this result we can find approximately the relative dimensions of a coil of large radius, which for a given length and gauge of wire has a maximum coefficient of self-induction. By the theorem proved above (p. 304), the self-induction coefficient is equal to the coefficient of mutual induction between two equal coils each of the given mean radius, and at a distance apart equal to the G. M. D. of the cross-section from itself. Let the G. M. D. be R. Then by the preceding result if the number of turns be n

$$L = 4\pi n^2 a \left(\log \frac{8a}{R} - 2\right). \quad \ldots \quad (119')$$

Now for similar sections of different linear dimensions R varies as the dimensions, and since for a given thickness of wire the number of turns varies as the cross-section, we have $n = CR^2$ where C is a constant. Again the total length of wire l is $2\pi na$, so that we have the two conditions, $2\pi na = l$, $2\pi CR^2 a = l$, which give $dn/da = -n/a$, $dR/da = -R/2a$. Hence taking a as independent variable, differentiating the value of L, and substituting these values of dn/da, and dR/da we find

$$\frac{dL}{da} = 4\pi n^2 \left(\frac{7}{2} - \log \frac{8a}{R}\right)$$

which for a maximum gives

$$\log R = \log 8a - \frac{7}{2} \quad \ldots \quad (120)$$

CIRCLE AND COAXIAL HELIX

If the section of the coil is circular of radius ρ, then by (43) above $\log R = \log \rho - \frac{1}{4}$, and (60) gives

$$a = \frac{1}{8}\rho e^{\frac{1}{4}} = 3\cdot 224\, c \quad . \quad . \quad . \quad . \quad . \quad (121)$$

If the section of the coil is square, the value of R from (115) substituted in (120), gives

$$a = 1\cdot 838 s \quad . \quad . \quad . \quad . \quad . \quad . \quad (122)$$

if s is the side of the square. These dimensions are, however, too nearly equal to enable the approximate formula by which the relation is found to apply with accuracy, and the result can only be regarded as a rough rule to guide the experimenter in the construction of coils.

We proceed now to calculate the coefficient of mutual induction between a circular conductor and a coaxial helix. We shall employ equation (47) of Chap. IV., viz.:

Mutual Inductance of Circle and Coaxial Helix.

$$M = \iint \frac{\cos e}{r}\, ds\, ds'. \quad . \quad . \quad . \quad . \quad (123)$$

where ds, ds', are elements of the helix and circle respectively at a distance r apart, and inclined to one another at an angle e, and the integrals are taken along both curves.*

If a be the radius of the helix, a' that of the circle, and θ, θ', the angles which the radii drawn from the axis to the elements ds, ds', make with the plane through the axis and the initial radius of the helix, we have $\cos e = \cos(\theta' - \theta)$, $ds = a d\theta$, $ds' = a' d\theta'$. Also if p be the pitch of the helix, so that between two radii separated by an angle θ the distance parallel to the axis is $p\theta$, and the circle be, as we here suppose, in the plane of the initial radius of the helix, then

Direct Calculation of M by Integration along the Curves.

$$r^2 = a^2 + a'^2 - 2aa'\cos(\theta - \theta') + p^2\theta^2.$$

Thus instead of (123) we have

$$M = \int_0^{2\pi}\int_0^{\theta_1} \frac{aa'\cos(\theta' - \theta)\, d\theta\, d\theta'}{\{a^2 + a'^2 - 2aa'\cos(\theta' - \theta) + p^2\theta^2\}^{\frac{1}{2}}}. \quad (124)$$

---

* The method of calculation based on (126) below is due to Prof. J. Viriamu Jones. See *Phil. Mag.* Jan. 1889, and *Phil. Trans. R.S.* vol. 182 (1891) *A*.

where θ_1 is the superior limit of integration for θ, and is the angle between the terminal radii of the helix.

Now clearly we may take $\theta' - \theta$ as a new variable ϕ instead of θ', so that

$$M = \int_{-\theta}^{2\pi - \theta} \int_0^{\theta_1} \frac{aa' \cos\phi \, d\theta d\phi}{\sqrt{\alpha^2 + p^2\theta^2}} \quad \ldots \quad (125)$$

But it is obvious that we can find for every value of θ, that is, for every element of the helix, an element on the circle corresponding to a constant value of ϕ. Hence, if we integrate, first with respect to θ, keeping ϕ constant, and then with respect to ϕ from 0 to 2π, we shall obtain the integral sought.

$$M = \int_0^{\theta_1} \int_0^{2\pi} \frac{aa' \cos\phi \, d\phi d\theta}{\sqrt{\alpha^2 + p^2\theta^2}} \quad \ldots \quad (126)$$

where $\alpha^2 = a^2 + a'^2 - 2aa' \cos\phi$.

But

$$\int_0^{\theta_1} \frac{aa' \cos\phi \, d\theta}{\sqrt{\alpha^2 + p^2\theta^2}} = \frac{aa' \cos\phi}{p} \log(p\theta_1 + \sqrt{\alpha^2 + p^2\theta_1^2})$$

and therefore

$$M = \frac{aa'}{p} \int_0^{2\pi} \cos\phi \, d\phi \log(p\theta_1 + \sqrt{\alpha^2 + p^2\theta_1^2}) \quad . \quad (127)$$

By direct expansion of $\log(p\theta_1 + \sqrt{\alpha^2 + p^2\theta_1^2})$, or by expanding $1/\sqrt{\alpha^2 + p^2\theta^2}$ in (126) by the binomial theorem, and then integrating, we can if $\alpha > p\theta$ write

$$M = aa' \int_0^{2\pi} \cos\phi \, d\phi \left(\frac{\theta_1}{\alpha} - \frac{1}{2 \cdot 3} \frac{p^2\theta_1^3}{\alpha^3} + \frac{1 \cdot 3}{2 \cdot 4 \cdot 5} \frac{p^4\theta_1^5}{\alpha^5} - \ldots \right) (128)$$

Expression of M in a Series of Definite Integrals.

Now if we put $k^2 \equiv 4aa'/(a + a')^2$, $\psi = \pi/2 - \phi/2$, in the general term of this series we find

$$\int_0^{2\pi} \frac{\cos\phi \, d\phi}{\alpha^{2m+1}} = -\frac{4}{(a + a')^{2m+1}} \int_0^{\frac{\pi}{2}} \frac{(1 - 2\sin^2\psi) d\psi}{(1 - k^2 \sin^2\psi)^{(2m+1)/2}}.$$

Hence substituting in (128) we get

$$M = 4aa' \sum (-1)^{m+1} \frac{1.3\ldots(2m-1)}{2.4\ldots 2m} \frac{p^{2m}}{2m+1} \left(\frac{\theta_1}{a+a'}\right)^{2m+1} P_m \quad (129)$$

where

$$P_m = \int_0^{\frac{\pi}{2}} \frac{\cos 2\psi \, d\psi}{(1-k^2 \sin^2 \psi)^{(2m+1)/2}} \quad \ldots \quad (130)$$

If the actual coil extend on both sides of the plane of the circle, it will form two coils of axial lengths $p\theta_1$, $p\theta_2$, for each of which M must be calculated by (129) and the results added. If $\theta_1 = \theta_2$, numerically, the value of M is double that given by (129).

To calculate the values of P_m we proceed as follows. After a slight reduction we find

$$P_m = \left(1 - \frac{2}{k^2}\right) Q_m + \frac{2}{k^2} Q_{m-1}$$

if

$$Q_m = \int_0^{\frac{\pi}{2}} \frac{d\psi}{(1-k^2 \sin^2 \psi)^{(2m+1)/2}}$$

$\quad \ldots \quad (131)$

Now

$$Q_m = Q_{m-1} + \frac{k}{2m-1} \frac{d}{dk} Q_{m-1} \quad \ldots \quad (132)$$

and it is to be noticed that, $Q_0 = F$, $Q_{-1} = E$, where F and E are Legendre's complete elliptic integrals to modulus k.

Between these integrals are the following well-known and easily established relations   *Integrals expressed in terms of Auxiliary Functions.*

$$\left.\begin{array}{l} \dfrac{dF}{dk} = \dfrac{E}{k(1-k^2)} - \dfrac{F}{k} \\[6pt] \dfrac{dE}{dk} = \dfrac{E-F}{k} \end{array}\right\} \quad \ldots \quad (133)$$

Hence by (132)

$$Q_1 = F + k\frac{dF}{dk} = \frac{E}{1-k^2} \quad \ldots \quad (134)$$

Process of Computation of Integrals.

From (134) Q_2 can be found by (132), then Q_3 from the result and so on in succession. These used in (131) give the values of $P_0, P_1, P_2 \ldots$. Thus

$$P_0 = \frac{1}{k^2}\{(k^2-2)F + 2E\}$$

$$P_1 = \frac{1}{k^2(1-k^2)}\{(k^2-2)E + 2(1-k^2)F\}$$

$$P_2 = \frac{1}{3k^2(1-k^2)^2}\{-(k^2-2)F - 2(1-k^2+k^4)E\}$$

. . . .

Thus the value of P_m could be calculated for successive values of m, but the process is laborious. It is easier to proceed thus :—

By (134) we have, writing for brevity $Q'_1 \equiv dQ_1/dk$, $Q''_1 \equiv d^2Q_1/dk^2, \ldots$

$$\left.\begin{array}{l} k(1-k^2)Q'_1 - 2k^2 Q_1 = E - F \\ k(1-k^2)Q''_1 + (1-5k^2)Q'_1 - 3kQ_1 = 0 \\ k(1-k^2)Q'''_1 + 2(1-4k^2)Q''_1 - 13kQ'_1 - 3Q_1 = 0 \end{array}\right\} \quad . . \quad (135)$$

. . . .

from which $Q'_1, Q''_1, \ldots$ can be calculated numerically in succession from the given data.

Then the work is carried on as follows :—

(1) Q_2, Q'_2, Q''_2, Q'''_2, are calculated from the equations

$$\left.\begin{array}{l} Q_2 = Q_1 + \dfrac{k}{3}Q'_1 \\[4pt] Q'_2 = \dfrac{4}{3}Q'_1 + \dfrac{k}{3}Q''_1 \\[4pt] Q''_2 = \dfrac{5}{3}Q''_1 + \dfrac{k}{3}Q'''_1 \\[4pt] Q'''_2 = \dfrac{6}{3}Q'''_1 + \dfrac{k}{3}Q'''' \end{array}\right\} \quad \ldots \quad (136)$$

of which the last three are successive derivatives of the first. The quantities on the right are all known from the computations indicated in equation (135).

(2) Q_3, Q'_3, Q''_3 are calculated from these quantities by the equations

$$\left.\begin{aligned} Q_3 &= Q_2 + \frac{k}{5} Q'_2 \\ Q'_3 &= \frac{6}{5} Q'_2 + \frac{k}{5} Q''_2 \\ Q''_3 &= \frac{7}{5} Q''_2 + \frac{k}{5} Q'''_2 \end{aligned}\right\} \quad \ldots \quad (137)$$

of which the last two are successive derivatives of the first.

(3) Q_4, Q'_4 are calculated from the quantities found in (137) by the equations

$$\left.\begin{aligned} Q_4 &= Q_3 + \frac{k}{7} Q'_3 \\ Q'_4 &= \frac{8}{7} Q'_3 + \frac{k}{7} Q'' \end{aligned}\right\} \quad \ldots \quad (138)$$

the latter of which is the first derivative of the former.

(4) Q_5 is found from the values of Q_4, Q'_4, by the equation

$$Q_5 = Q_4 + \frac{k}{9} Q'_4 \quad \ldots \quad (139)$$

(5) P_0, P_1, ... P_5 are found from the values of E, F, and of Q_1, ... Q_5 [equations (134) ... (139)], by successive applications of (131).

Professor J. V. Jones thus calculated the coefficient of mutual induction of a helix and the edge of a coaxial disc in the mean plane of the helix, which he used in a very careful determination of the specific resistance of mercury in absolute measure.*

Results of Calculation in Actual Case.

The dimensions of the helix and disc, and the results as computed by him, are given in the following table :—

---

* *Phil. Trans. R.S.* vol. 182 (1891) A. See also Chap. X. below.

Axial length of Helix $= 2p\theta = 4\cdot 625$ inches
Mean radius of Helix $= a = 10\cdot 53774$ inches
Radius of Disk $= a' = 0\cdot 40493$ inches

$\log_{10} k = \overline{1}\cdot 9874084$
$E = 1\cdot 0065493$
$F = 2\cdot 8509609$

$Q_1 = 1\cdot 8931552 \times 10$
$Q_2 = 2\cdot 1977948 \times 10^2$
$Q_3 = 3\cdot 0951221 \times 10^3$
$Q_4 = 4\cdot 6957152 \times 10^4$
$Q_5 = 7\cdot 3990015 \times 10^5$

$P_0 = -\cdot 93092203$
$P_1 = -1\cdot 5149680 \times 10$
$P_2 = -2\cdot 0589783 \times 10^2$
$P_3 = -2\cdot 9988831 \times 10^3$
$P_4 = -4\cdot 6004090 \times 10^4$
$P_5 = -7\cdot 2872416 \times 10^5$

These gave

$$\sum (-1)^m \frac{1\cdot 3\cdot 5\ldots(2m-1)}{2\cdot 4\cdot 6\ldots 2m}\frac{1}{2m+1}\left(\frac{p\theta_1}{a+a'}\right)^{2m} P_m$$
$$= -\cdot 88891460$$

and

$$M = 89\cdot 7717 \times \text{number of turns in coil,}$$

or

$$M = 16606\cdot 6$$

since the number of turns was 185.

Application of Method to Case of Two Circles. It is to be noticed that if, instead of a circle and a helix, we have simply two circles of wire, the same mode of calculation is applicable. For putting b for the distance between the planes of the circles, we have for M equations (124) and (125), with $p^2\theta^2$ replaced by b^2. Expanding in ascending powers of b^2/a^2, and integrating with respect to θ between the limits 0 and 2π, we find

$$M = 2\pi a a' \int_0^{2\pi} \cos\phi\, d\phi \left(\frac{1}{a} - \frac{1}{2}\frac{b^2}{a^3} + \frac{1\cdot 3}{2\cdot 4}\frac{b^4}{a^5} - \&\text{c.}\right). \quad (140)$$

which is convergent so long as $b^2 < a^2$. Hence b must be less than the difference of the radii of the circles.

From (140) we find by the same process as before

$$M = 8\pi a a' \sum (-1)^{m+1} \frac{1\cdot 3\ldots(2m-1)}{2\cdot 4\ldots 2m}\left(\frac{b}{a+a'}\right)^{2m+1} P_m$$
$$(141)$$

where P_m has the value stated in (130), and may be calculated for $m = 0, 1, 2, 3 \ldots$ as described above.

Case of Two Circles of nearly Equal Radius.

This process will not be applicable when the circles are very nearly equal in radius; and in general it is more convenient to use the result (89) stated on p. 50 above. By the equivalence of two circular currents to two circular magnetic shells, the edges of which coincide with the circuit, we find from the equation referred to

$$M = 4\pi \sqrt{aa'} \left\{ \left(\frac{2}{k} - k\right) F - \frac{2}{k} E \right\} \quad . \quad . \quad . \quad (142)$$

M expressed in Elliptic Integrals.

where it is to be noticed the modulus k has the value $2\sqrt{aa'}/\sqrt{(a+a')^2 + b^2}$, and F and E are the complete elliptic integrals to that modulus. If in one circle there are n_1 turns, and in the other n_2 turns, the value of M must of course be multiplied by $n_1 n_2$.

When the radii are nearly equal, and are great in comparison with b, the modulus k is only a little less than unity, and the expansion of the elliptic integrals yields slowly convergent series. In such a case we can improve as much as we please the convergency of the series by employing a sufficient number of times in succession what is known as Landen's transformation, whereby the integrals are transformed to a new modulus $k_1(<k)$ defined by the equation $k = 2\sqrt{k_1}/(1 + k_1)$. If $F(k_1)$, $E(k_1)$ be the complete elliptic integrals to this modulus, $F(k)$, $E(k)$ those to the old modulus k,

Convergency of Elliptic Integral Expansions. Landen's Transformation.

$$\left. \begin{array}{l} F(k) = (1 + k_1) F(k_1) \\ E(k) = \dfrac{2E(k_1)}{1 + k_1} - (1 - k_1) F(k_1) \end{array} \right\} \quad . \quad . \quad . \quad (143)$$

These relations are easily proved thus. Draw (Fig. 73) a circle APB_1 from centre O with radius r, and join the point P with C (a point on the radius AO) and with A. Denote the angles PAO, PCO by θ, θ'_1, and the distance CO by r_1. Now if Q be a point on the circle adjacent to P, we have, since angle $POB = 2\theta$, $PQ/CP = 2r d\theta / \sqrt{r^2 + r_1^2 + 2r r_1 \cos 2\theta} = 2r d\theta / \{(r + r_1)\sqrt{1 - 4 r r_1 \sin^2\theta/(r + r_1)^2}\}$. But from the triangle CQP, $PQ/CP = d\theta_1 / \sin CQP = d\theta_1 / \sqrt{1 - \sin^2 OPC} = d\theta_1 / \sqrt{1 - r_1^2/r^2 \cdot \sin^2 \theta_1}$. Hence

$$\frac{2r}{r + r_1} \frac{d\theta}{\sqrt{1 - \dfrac{4 r r_1}{(r + r_1)^2} \sin^2\theta}} = \frac{d\theta_1}{\sqrt{1 - \dfrac{r_1^2}{r^2} \sin^2\theta}}$$

and, if $k = 2\sqrt{rr_1}/(r+r_1)$, $k_1 = r_1/r$, this gives

$$\frac{2}{1+k}\int_0^\theta \frac{d\theta}{\sqrt{1-k^2\sin^2\theta}} = \int_0^{\theta_1} \frac{d\theta_1}{\sqrt{1-k_1^2\sin^2\theta_1}}$$

which gives the first relation stated in (143), since when $\theta = \pi/2$, $\theta_1 = \pi$.

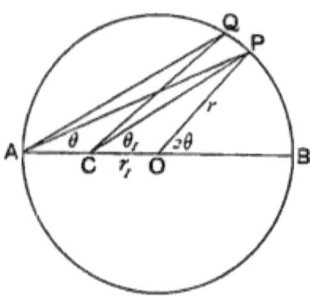

Fig. 73.

From the triangle CPO we have $\sin(2\theta - \theta_1)/\sin\theta_1 = r_1/r$, by which the limits of integration with respect to θ_1 can be fixed from those for θ.

Modular Relations of Elliptic Integrals. The second equation stated in (143) can be proved from the first. By differentiation of the latter we have

$$(1+k_1)\frac{dF(k_1)}{dk_1} + F(k_1) = \frac{dF(k)}{dk}\frac{dk}{dk_1}.$$

Substituting in this the values of the differential coefficients, namely, $-F(k_1)/k_1 + E(k_1)/k_1(1-k_1^2)$ for $dF(k_1)/dk_1$, $E(k)/k(1-k^2) - F(k)/k$ for $dF(k)/dk$, given by (133), and $(1-k_1)/\sqrt{k_1}(1+k_1)^2$ for dk/dk_1, and reducing by means of the first relation we arrive at the required result.*

---

* The same formulæ of transformation are available for the reduction of an elliptic integral the modulus of which is greater than unity to one less than unity. In this case the initial modulus is to be taken as k_1, and that to which the transformation is made as k. Whatever k_1 may be $2\sqrt{k_1}/(1+k_1)$ is less than 1, since it is the geometric mean of the two quantities k_1 and 1, divided by their arithmetic mean. Thus the new modulus is less than 1.

EXPANSION OF ELLIPTIC INTEGRALS

Substituting from (143) in (142) we get

$$M = 8\pi \sqrt{aa'} \cdot \frac{1}{\sqrt{k_1}} \{F(k_1) - E(k_1)\} \quad . \quad . \quad . \quad (144)$$

Expression of M in Elliptic Integrals Transformed to New Modulus.

a formula for M sometimes more convenient to use than that in (82).

Taking the old modulus k, as above, the new modulus k_1 is given by $2\sqrt{k_1}/(1 + k_1) = 2\sqrt{aa'}/\sqrt{(a + a')^2 + b^2}$, or what is the same thing $(1 - k_1)/(1 + k_1) = r_2/r_1$, where r_1, r_2 are the greatest and least distances from one circle to the other. Hence in this case

$$k_1 = \frac{r_1 - r_2}{r_1 + r_2} \quad . \quad . \quad . \quad . \quad . \quad (145)$$

Taking an angle $\gamma = \cos^{-1}(r_2/r_1)$ Maxwell* has calculated the values of $\log(M/4\pi\sqrt{aa'})$ for intervals of 6 minutes of angle for values of γ between 60 and 90°. This table is reproduced in the Appendix to this volume.

From the relations (133) above we can find differential equations which are satisfied by the complete elliptic integrals F, E; and by aid of these equations the integrals can be expanded in ascending powers of k' $(= \sqrt{1-k^2})$, the complementary modulus) when k is very nearly equal to unity, that is when k' is small.† The relations referred to can be written

Modular Differential Equations Satisfied by Elliptic Integrals.

$$(1-k^2)\left(k\frac{d}{dk} + 1\right)F = E$$

$$-F = \left(k\frac{d}{dk} - 1\right)E.$$

Hence operating on the first of these by $kd/dk - 1$, and subtracting the second we get after reduction

$$(1-k^2)\frac{d^2F}{dk^2} + \frac{1-3k^2}{k}\frac{dF}{dk} - F = 0 \quad . \quad . \quad . \quad (146)$$

Again operating on the second relation by $(1-k^2)(1 + kd/dk)$ adding to the other, and reducing, we get

---

* *El. and Mag.* vol. ii. chap. xiv. (second edition).
† The process followed here is practically that given in Cayley's *Elliptic Functions*, Art. 77.

$$(1 - k^2)\frac{d^2E}{dk^2} + \frac{1 - k^2}{k}\frac{dE}{dk} + E = 0 \ . \qquad . \quad (147)$$

The corresponding equations in terms of the complementary modulus are obtained from these by simply substituting k' for k everywhere, and in (147) changing the coefficient of the second term to $-(1 + k'^2)/k'$, as the reader may verify by means of the defining relation $k^2 + k'^2 = 1$.

Solutions of Differential Equations. Thus both $F(k)$ and $F(k')$ are solutions of (146) whether the independent variable be k or k'. The complete solution of (146) is thus $aF(k) + a'F(k')$ where a, a', are arbitrary constants. Equation (147) is satisfied by $E(k)$. But clearly if y be any solution of (146), (147) must by the fundamental relations be satisfied by $(1-k^2)(y + kdy/dk)$. Hence since the complete solution of (146) is $aF(k) + a'F(k')$, (147) is satisfied by the result obtained by substituting this value for y in the preceding expression, that is, $aE(k) + a'\{F(k') - E(k')\}$. The form of this is not altered by adding the particular solution $E(k)$, so that the complete solution of (147) is $\beta E(k) + \beta'\{F(k') - E(k')\}$ where β, β', are arbitrary constants.

Value of $F(k)$ when k is nearly unity. For the expansion we see in the first place that when k is nearly equal to 1, k' is very small, and we may write

$$F(k) = \int_0^{\frac{\pi}{2}-\epsilon} \frac{d\theta}{\sqrt{\cos^2\theta + k'^2 \sin^2\theta}} + \int_{\frac{\pi}{2}-\epsilon}^{\frac{\pi}{2}} \frac{d\theta}{\sqrt{\cos^2\theta + k'^2\sin^2\theta}}$$

where ϵ denotes a quantity very small in comparison with $\frac{1}{2}\pi$, but large in comparison with k'. The first integral is, very approximately

$$\int_0^{\frac{\pi}{2}-\epsilon} \frac{d\theta}{\cos\theta} = \log\tan\tfrac{1}{2}(\pi - \epsilon) = \log\frac{2}{\epsilon}.$$

The other integral may be written by putting $u = \frac{1}{2}\pi - \theta$,

$$\int_0^\epsilon \frac{du}{\sqrt{k'^2 + k^2 u^2}} = \frac{1}{k}\log\frac{k\epsilon + \sqrt{k'^2 + k^2\epsilon^2}}{k'} = \log\frac{2\epsilon}{k'}$$

very nearly, since we have supposed k' small in comparison with $k\epsilon$, and k is very nearly unity. Hence adding these two results we have

EXPANSION OF ELLIPTIC INTEGRALS

$$F = \log \frac{2\epsilon}{k'} + \log \frac{2}{\epsilon} = \log \frac{4}{k'} \quad \ldots \quad (148)$$

Thus in the limit when k' is very small, $\log 4 - \log k'$ is the value of $F(k)$. This value must also satisfy, under the same condition, the differential equation (146). If then, we can find a series for $F(k)$ which satisfies the differential equation, and approaches more and more nearly to $\log(4/k')$ as k' is made smaller and smaller, we shall have obtained the required development.

Expansion of Complete Elliptic Integrals.

Writing then

$$F(k) = P \log \frac{4}{k'} + Q$$

where P and Q are functions of k' to be determined, and substituting in (146) (written with k' instead of k) we obtain a result of the form

$$M \log \frac{4}{k'} + N = 0$$

from which by equating M and N separately to zero we derive

$$\left. \begin{array}{l} (1-k'^2)\dfrac{d^2 P}{dk'^2} + \dfrac{1-3k'^2}{k'}\dfrac{dP}{dk'} - P = 0 \\[2mm] (1-k'^2)\dfrac{d^2 Q}{dk'^2} + \dfrac{1-3k'^2}{k'}\dfrac{dQ}{dk'} - Q - \dfrac{2(1-k'^2)}{k'}\dfrac{dP}{dk'} + 2P = 0 \end{array} \right\} \quad (149)$$

The first of these is by (146) satisfied by $P = CF(k')$ where C is a constant.

Developing now the element $d\theta / \sqrt{1 - k'^2 \sin^2 \theta}$ of the integral in a series of ascending powers of k', integrating then each term from 0 to $\pi/2$, and using Wallis's theorem that

$$\int_0^{\frac{\pi}{2}} \sin^{2m}\theta\, d\theta = \frac{1.3.5\ldots(2m-1)}{2.4.6\ldots 2m}\frac{\pi}{2}$$

we get

$$P = C\frac{\pi}{2}\left\{1 + \left(\frac{1}{2}\right)^2 k'^2 + \left(\frac{1.3}{2.4}\right)^2 k'^4 + \ldots\right\}$$

In order that when $k' = 0$, the value of $F(k)$ may reduce to $\log(4/k')$ we must take $C = 2/\pi$, so that

$$P = 1 + \left(\frac{1}{2}\right)^2 k'^2 + \left(\frac{1.3}{2.4}\right)^2 k'^4 + \ldots$$

Calculating from this we find

$$2\frac{1-k'^2}{k'}\frac{dP}{dk'} - 2P = -\left\{1 + \frac{3}{2}\left(\frac{1}{2}\right)^2 k'^2 + \frac{5}{3}\left(\frac{1\cdot 3}{2\cdot 4}\right)^2 k'^4 + \ldots\right\} \quad (150)$$

and this is the value of the part of the second of (149) which involves Q. It is clear thus from the form of that expression that Q cannot involve odd powers of k', and as, on account of the limiting value of $F(k)$, it cannot contain any constant term, we may suppose

$$Q = -\left(\frac{1}{2}\right)^2 A_1 k'^2 - \left(\frac{1\cdot 3}{2\cdot 4}\right)^2 A_2 k'^4 - \left(\frac{1\cdot 3\cdot 5}{2\cdot 4\cdot 6}\right)^2 A_3 k'^6 - \ldots$$

Substituting from this in the second of (149) using (150), we get finally by equating to zero separately the coefficients of k'^0, k'^2, k'^4, ...

$$A_1 = \frac{2}{1\cdot 2}, \quad A_2 = \frac{2}{1\cdot 2} + \frac{2}{3\cdot 4}, \quad A_3 = \frac{2}{1\cdot 2} + \frac{2}{3\cdot 4} + \frac{2}{5\cdot 6}, \ldots$$

so that

Series for First Complete Elliptic Integral.

$$F(k) = \log\frac{4}{k'}$$

$$+ \frac{1^2}{2^2} k'^2 \left(\log\frac{4}{k'} - \frac{2}{1\cdot 2}\right)$$

$$+ \frac{1^2\cdot 3^2}{2^2\cdot 4^2} k'^4 \left(\log\frac{4}{k'} - \frac{2}{1\cdot 2} - \frac{2}{3\cdot 4}\right)$$

$$+ \frac{1^2\cdot 3^2\cdot 5^2}{2^2\cdot 4^2\cdot 6^2} k'^6 \left(\log\frac{4}{k'} - \frac{2}{1\cdot 2} - \frac{2}{3\cdot 4} - \frac{2}{5\cdot 6}\right)$$

$$+ \ldots \ldots \ldots \ldots \ldots \ldots \ldots \quad (151)$$

The expansion of $E(k)$ can be obtained from this by the first relation of (133) namely

$$\frac{dF(k)}{dk} = -\frac{F(k)}{k} + \frac{E(k)}{kk'^2}$$

or, what is the same thing,

$$E(k) = k^2 F(k) - k(1 - k'^2)\frac{dF(k)}{dk'}.$$

This gives

$$\begin{aligned}E(k) = 1 \\ + \frac{1}{2}k'^2\left(\log\frac{4}{k'} - \frac{1}{1.2}\right) \\ + \frac{1^2 \cdot 3}{2^2 \cdot 4}k'^4\left(\log\frac{4}{k'} - \frac{2}{1.2} - \frac{1}{3.4}\right) \\ + \frac{1^2 \cdot 3^2 \cdot 5}{2^2 \cdot 4^2 \cdot 6}k'^6\left(\log\frac{4}{k'} - \frac{2}{1.2} - \frac{2}{3.4} - \frac{1}{5.6}\right) \\ + \quad \cdots \quad \cdots \quad \cdots \quad \cdots \\ \cdots \quad \cdots \quad \cdots \quad \cdots \quad \cdots \quad (152)\end{aligned}$$

Series for Second Complete Elliptic Integral.

These equations are available for the calculation of M from equation (142) or (144), and are conveniently applicable even when k is not very nearly equal to 1. The computation is easy and expeditious if systematically carried out, and can be pushed to any degree of accuracy without trouble, as the law of formation of successive terms is known, and the series are very convergent. Since k is supposed exactly known numerically, $\log(4/k')$, and k'^2, &c., can be found numerically with sufficient exactness once for all, and then used in the numerical computation of the successive terms of the series. $F(k)$ and $E(k)$, having been thus found, are multiplied by the already known coefficients in the formula employed, and the calculation of the coefficient of mutual induction of the two coaxial circles is complete. Of course, when k is very nearly equal to unity, any uncertainty as to its value will render that of $\log(4/k')$ very uncertain, since a small error in k will be comparable with k'.

Use of Elliptic Integral Expansions.

If the coils are of finite breadth and depth of cross-section ($2\beta, 2\delta$, say, for the coil of radius a, and $2\beta', 2\delta'$ for the other), we can, as at p. 256, obtain the corrections for the finite cross-sections by denoting the uncorrected value of M in (142) by M_0, and calculating d^2M_0/da^2, d^2M_0/da'^2, d^2M_0/db^2, &c. Then to a first approximation we have

Correction of Elliptic Integral Formula for Cross-Section of Coils.

$$M = M_0 + \frac{1}{6}\left\{\delta^2\frac{d^2M_0}{da^2} + \delta'^2\frac{d^2M_0}{da'^2} + (\beta^2 + \beta'^2)\frac{d^2M_0}{db^2}\right\} + \&c.$$

VOL. II. Y

By differentiation we find, having regard to the fundamental modular relations, that in the important case in which $a=a'$,*

$$\frac{d^2 M_0}{da^2} = \pi \frac{k}{a} \left\{ (2-k^2)F - \left(2 - k^2 \frac{1-2k^2}{1-k^2}\right) E \right\}$$

$$\frac{d^2 M_0}{db^2} = \pi \frac{k^3}{a} \left(F - \frac{1-2k^2}{1-k^2} E \right).$$

A pair of coils of equal mean radius has been used by Rowland, Glazebrook and others in experiments on the absolute determination of resistances (see Chapter X.).

When E and F have been accurately calculated for any given case of two equal coaxial coils (and M_0 has thus been found), their values can be used with facility in these expressions to give the necessary correction for the dimensions of cross-section. It is to be noticed that, as in other cases, these corrections are calculated on the hypothesis that the coils are uniformly wound with wire sufficiently fine to enable the current to be regarded as uniform over the cross-section. Of course every term in the expression for M here given, must be multiplied by $n_1 n_2$, the product of the numbers of turns in the two coils.

<small>Maxwell's Expansion for M.</small>
Maxwell has given an expansion for M in terms of b and x which can be deduced from the series (151), (152) by substituting the algebraic expression of k' in terms of a, x, and b in the above expression, expanding by logarithmic and binomial series $\log(1+x/a)$, k'^2, k'^4, &c., and multiplying by the binomial expansions of the coefficients [see (142)] $2/k - k$, $2/k$ of $F(k)$, $E(k)$ respectively. The result does not seem so convenient for practical purposes of numerical computation as the above series derived analytically from the properties of the elliptic integrals.

Maxwell however obtained his series by a most instructive method depending directly on electrical principles.† On that account it is here sketched, the reader being left to fill in the details of calculation, and verify the result.

In the first place we have the following differential equation

---

* Here a is put equal to a' after differentiation, and $d^2 M_0/da^2$, must, when $\delta = \delta'$, be applied twice, once for each coil.

† *El. and Mag.*, vol. ii. (second edition), p. 315. The elliptic integral expansion has been here given in preference, even at the cost of some space devoted to purely analytical investigations, required to make the process of derivation intelligible to readers unacquainted with the properties of elliptic integrals.

MAXWELL'S THEOREM

which is satisfied by the total induction **B** through a circle in a homogeneous isotropic medium in presence of any system of magnets or conductors. If the radius of the circle be r, its axis be along z, and z be the distance of its centre from a fixed point on the axis, the equation is

Maxwell's Theorem of Total Induction through a Circle.

$$\frac{\partial^2 \mathbf{B}}{\partial r^2} + \frac{\partial^2 \mathbf{B}}{\partial z^2} - \frac{1}{r}\frac{\partial \mathbf{B}}{\partial r} = 0 \quad \ldots \quad (153)$$

To prove it let the radius (1) increase by an amount dr, (2) move along the axis a distance dz. The increase of **B** due to the first displacement is $d\mathbf{B}/dr.dr$, and due to the second $d\mathbf{B}/dz.dz$. Now let the components of magnetic force at any point of the circle in its undisplaced position be p, q, γ, p being taken along the radius outwards, q along the tangent, and γ parallel to z. If we assume, as we may without loss of generality, that the magnetic permeability of the medium is unity, the increase of total induction due to change of radius is plainly

Proof of Maxwell's Theorem.

$$dr . r \int_0^{2\pi} \gamma d\theta,$$

and therefore

$$\frac{\partial \mathbf{B}}{\partial r} = r \int_0^{2\pi} \gamma d\theta \quad \ldots \ldots \quad (154)$$

The number of induction-tubes swept over by the circle in the other displacement is

$$dz . r \int_0^{2\pi} p d\theta;$$

and it is easy to see, from the directions in which the tubes are cut by the conductor in the two cases, that the total induction through the circuit will be diminished by the latter amount. Thus we have

$$\frac{\partial \mathbf{B}}{\partial z} = -r \int_0^{2\pi} p d\theta \quad \ldots \ldots \quad (155)$$

Differentiating (154) with respect to r, and (155) with respect to z, we find

$$\frac{\partial^2 \mathbf{B}}{\partial r^2} + \frac{\partial^2 \mathbf{B}}{\partial z^2} = \int_0^{2\pi} \gamma d\theta + r \int_0^{2\pi} \left(\frac{\partial \gamma}{\partial r} - \frac{\partial p}{\partial z}\right) d\theta.$$

But since $p = -dV/dr$, $\gamma = -dV/dz$, where V is the magnetic potential, the integral on the right of the last equation is zero. Hence by (154) we obtain the differential equation stated in (153).

Derivation of Maxwell's Series for M.
Returning to the problem of the coefficient of mutual induction of two coaxial circles of radii a and $a + x$, let the distance between their planes be now denoted by z, so that $r = \sqrt{x^2 + z^2}$. We assume that instead of (119) we have

$$M = 4\pi \left(A \log \frac{8a}{r} - B \right),$$

where

$$A = a + A_1 x + A_2 \frac{x_2}{a} + A'_2 \frac{z_2}{a} + A_3 \frac{x^3}{a^2} + A'_3 \frac{xz^2}{a^2} + \ldots$$

$$B = -2a + B_1 x + B_2 \frac{x^2}{a} + B'_2 \frac{z^2}{a} + B_3 \frac{x^3}{a^2} + B'_3 \frac{xz^2}{a^2} + \ldots$$

To determine the coefficients in these series we have first by (101)

$$\frac{\partial^2 M}{\partial x^2} + \frac{\partial^2 M}{\partial z^2} - \frac{a}{a + x} \frac{\partial M}{\partial x} = 0,$$

which by differentiation of M and substitution enables one set of relations among the coefficients to be found. Secondly, since either circuit may be taken as the primary in which the current producing the induction flows, we must obtain the same result by putting everywhere in the above expressions $x = 0$, and substituting $a + x$ for a. The former substitution reduces the radius of the secondary from $a + x$ to a, and the latter enlarges that of the primary to $a + x$. The condition that these substitutions leave the value of M unaffected, gives another set of relations among the coefficients. From this set and the former together, as many equations are obtained as suffice to determine the coefficients. The final result is

$$M = 4\pi a \log \frac{8a}{r} \left\{ 1 + \frac{1}{2} \frac{x}{a} + \frac{x^2 + 3z^2}{16a^2} - \frac{x^3 + 3xz^2}{32a^3} + \ldots \right\}$$
$$+ 4\pi a \left\{ -2 - \frac{1}{2} \frac{x}{a} + \frac{3x^2 - z^2}{16a^2} - \frac{x^3 - 6xz^2}{48a^3} + \ldots \right\} \quad . \quad (156)$$

This formula may be used according to the principle explained above (p. 305) to find the coefficient of self-induction L of a coil of radius large in comparison with the dimensions of cross-section, and in this case r is the G.M.D. of the cross-section from itself. In ordinary cases it is probably more convenient to calculate the value of the G.M.D. once for all numerically, and then use the elliptic integral formula given by (142) when $4\pi a n^2$ is put for $4\pi \sqrt{aa'}$.

Here $k = 2a/\sqrt{4a^2 + r^2}$, since the radii are equal, and r, the distance between the planes of the circles, is the G.M.D. of the cross-section from itself. By means of the expansions of the elliptic integrals given in (151) and (152) the value of L can be obtained to any required degree of accuracy, provided the radius a is so large that the principle stated on p. 305 can be applied.

Weinstein has found (*Wied. Ann.* xxi. 329), from the series given in (156), that for a coil of axial breadth $2b$, and radial depth $2d$,

$$L = 4\pi n^2 (a\lambda + \mu) \quad . \quad . \quad . \quad . \quad . \quad (157)$$

where, if $x = b/d$,

$$\lambda = \log \frac{4a}{d} + \frac{1}{12} - \frac{\pi x}{3} - \left(\frac{1}{2} - \frac{1}{12x^2} - \frac{x^2}{12}\right)\log(1 + x^2)$$
$$- \frac{x^2}{12}\log x^2 + \frac{2}{3}\left(x - \frac{1}{x}\right)\tan^{-1} x,$$

$$\mu = \frac{d^2}{24a}\left\{\log\frac{4a}{d} - \left(\frac{1 + 3x^2}{2} + \frac{1}{10x^2} - \frac{x^4}{2}\right)\log(1 + x^2)\right.$$
$$\left. - \frac{x^4}{2}\log x^2 + 3 \cdot 45 x^2 + \frac{221}{60} - 1 \cdot 6\pi x^3 + 3 \cdot 2 x^3 \tan^{-1} x.\right.$$

A formula of approximation, used by Lord Rayleigh to find the coefficient of mutual induction of two coaxial circular coils of rectangular cross-section, will be given in next chapter.

Lord Rayleigh's Approximation Formula.

Section III

INDUCTION IN CONDUCTORS CARRYING VARYING OR ALTERNATING CURRENTS

We conclude this chapter with an investigation of the effect of the distribution of the current over the cross-section of a straight cylindrical conductor on the self-induction and effective resistance of the conductor. Take the conductor as a straight circular cylinder, with axis along z, then the density, w, of the

Varying Current in Cylindrical Conductor.

326   VARYING CURRENTS

current is a function of the time and the distance r of the point considered from the axis of the wire. Hence also the vector potential H at each part of the circuit is a function of the same variables. Assume that

$$H = S + T + T_1 r^2 + T_2 r^4 + \ldots + T_n r^{2n} + \ldots \quad (158)$$

where S, T, T_1, &c., are functions of the time.

Now we have

$$4\pi w = \frac{\partial \beta}{\partial x} - \frac{\partial a}{\partial y}$$

and

$$\mu a = \frac{\partial H}{\partial y} - \frac{\partial G}{\partial z} = \frac{\partial H}{\partial y}, \quad \mu \beta = \frac{\partial F}{\partial z} - \frac{\partial H}{\partial x} = -\frac{\partial H}{\partial x}.$$

Hence

$$\frac{\partial^2 H}{\partial x^2} + \frac{\partial^2 H}{\partial y^2} = -4\pi \mu w,$$

Current-Density at any Point of Cross-Section. or, if r be taken as a radial coordinate ($= \sqrt{x^2 + y^2}$),

$$\frac{\partial^2 H}{\partial r^2} + \frac{1}{r}\frac{\partial H}{\partial r} = -4\pi \mu w \quad \ldots \quad (159)$$

This gives by differentiation of (158)

$$-\pi \mu w = T_1 + 2^2 T_2 r^2 + 3^2 T_3 r^4 + \ldots + n^2 T_n r^{2n-2} + \ldots \quad (160)$$

If ρ be the specific resistance of the material, the component electromotive force in the direction of z is ρw at every point where the current is w. Hence by (5) of Chapter V,

$$\rho w = -\frac{dH}{dt} - \frac{\partial \Psi}{\partial z},$$

where Ψ is taken as the potential corresponding to that part of the electromotive force which is independent of current induction. This by (158) becomes

$$\rho w = -\frac{\partial \Psi}{\partial z} - \frac{dS}{dt} - \frac{dT}{dt} - r^2 \frac{dT}{dt} - \ldots \quad (161)$$

Comparing coefficients in (160) and (161) we find

$$\rho T_1 = \pi \mu \left(\frac{\partial \Psi}{\partial z} + \frac{dS}{dt} + \frac{dT}{dt} \right)$$

$$2^2 \rho T_2 = \pi \mu \frac{dT_1}{dt}, \ldots, n^2 \rho T_n = \pi \mu \frac{dT_{n-1}}{dt^{n-1}}, \ldots$$

Putting $dS/dt = -\partial\Psi/\partial z$, and reducing, we get

$$T_1 = \frac{\pi\mu}{\rho}\frac{dT}{dt},\ T_2 = \frac{1}{2^2}\frac{\pi^2\mu^2}{\rho^2}\frac{d^2T}{dt^2},\ \ldots,\ T_n = \frac{1}{(n!)^2}\frac{\pi^n\mu^n}{\rho^n}\frac{d^nT}{dt^n},\ \ldots$$

and therefore

$$-\pi\mu w = \frac{\pi\mu}{\rho}\frac{dT}{dt} + \frac{\pi^2\mu^2}{\rho^2}\frac{d^2T}{dt^2}r^2 + \ldots$$

$$+ \frac{1}{\{(n-1)!\}^2}\frac{\pi^n\mu^n}{\rho^n}\frac{d^nT}{dt^n}r^{2n-2} + \ldots \quad \ldots \quad (161')$$

If γ be the total current in the conductor

$$\gamma = 2\pi\int_0^a wr\,dr,$$

where a is the radius of the wire. Thus writing κ for $\pi a^2/\rho$, the conductivity of unit length of the wire, we get from the value of $\pi\mu w$ **Total Current in the Conductor.**

$$-\mu\gamma = \mu\kappa\frac{dT}{dt} + \frac{2\mu^2\kappa^2}{1^2 . 2^2}\frac{d^2T}{dt^2} + \ldots + \frac{n\mu^n\kappa^n}{(n!)^2}\frac{d^nT}{dt^n} + \ldots \quad (162)$$

Outside the wire the value of H does not depend on the distribution of the current in the wire (see p. 285 above), but only on the total current, γ. Hence at the surface of the wire we may put $H = A\gamma$ where A is a multiplier to be determined. Thus by (158) above **Lord Rayleigh's Calculation for Alternating Currents.**

$$A\gamma = S + T_0 + T_1 a^2 + \ldots + T_n a^n + \ldots$$

and therefore by the values of $T_1, T_2, \ldots$ formed above,

$$A\gamma - S = T + \mu\kappa\frac{dT}{dt} + \frac{\mu^2\kappa^2}{1^2 . 2^2}\frac{d^2T}{dt^2} + \ldots + \frac{\mu^n\kappa^n}{(n!)^2}\frac{d^nT}{dt^n} + \ldots$$

or if we write

$$\phi(x) = \sum_{n=0}^{n=\infty}\frac{x^n}{(n!)^2} \quad \ldots \quad \ldots \quad (163)$$

$$A\gamma - S = \phi\left(\mu\kappa\frac{d}{dt}\right)T.$$

Hence

$$\frac{dS}{dt} = A\frac{d\gamma}{dt} - \phi\left(\mu\kappa\frac{d}{dt}\right)\frac{dT}{dt}.$$

Equation (162) may also be written

$$\gamma = -\kappa\phi'\left(\mu\kappa\frac{d}{dt}\right)\frac{dT}{dt}.$$

Elimination of dT/dt between the two last equations gives

$$\frac{dS}{dt} = A\frac{d\gamma}{dt} + \frac{1}{\kappa}\frac{\phi\left(\mu\kappa\frac{d}{dt}\right)}{\phi'\left(\mu\kappa\frac{d}{dt}\right)}\gamma.$$

Expression for Non-Inductive E.M.F. But it has been assumed above that $-d\Psi/dz = dS/dt$, so that dS/dt is the part of the electromotive force at each point which does not depend on the inductive action of the current. This part of the electromotive force we suppose to be the same at every point, if the conductor be (as is here supposed) of uniform resistance, so that if E be its line integral, R the resistance, and l the length of the conductor considered, $E = l\, dS/dt$. Hence the last equation becomes

$$E = lA\frac{d\gamma}{dt} + \frac{l}{\kappa}\frac{\phi\left(\mu\kappa\frac{d}{dt}\right)}{\phi'\left(\mu\kappa\frac{d}{dt}\right)}\gamma \quad \ldots \ldots (164)$$

Now if the currents be periodic according to the simple harmonic law, every such current is (to a constant factor) represented by the real part of ϵ^{int}, where n denotes $2\pi/T$; and we may use this expression until it is convenient to separate its real from its imaginary part. We have then to replace in (163) d/dt by in, and the equation then becomes

$$E = inlA\gamma + \frac{l}{\kappa}\frac{\phi(i\mu\kappa n)}{\phi'(i\kappa\mu n)}\gamma \quad \ldots \ldots (164')$$

Development of Expression in Series. If κ be small we obtain by division

$$\frac{\phi(x)}{\phi'(x)} = 1 + \frac{1}{2}x - \frac{1}{12}x^2 + \frac{1}{48}x^3 - \frac{1}{180}x^4 + \frac{13}{8640}x - \ldots$$

and hence

$$\frac{\phi(i\kappa\mu n)}{\phi'(i\kappa\mu n)} = 1 + \frac{1}{12}\mu^2\kappa^2 n^2 - \frac{1}{180}\mu^4\kappa^4 n^4 + \ldots$$
$$+ i\left(\frac{1}{2}\mu\kappa n - \frac{1}{48}\mu^3\kappa^3 n^3 + \frac{13}{8640}\mu^5\kappa^5 n^5 - \ldots\right).$$

RESISTANCE AND INDUCTANCE

This converts (164') into

$$E = R(1 + \frac{1}{12}\frac{\mu^2 l^2 n^2}{R^2} - \frac{1}{180}\frac{\mu^4 l^4 n^4}{R^4} + \ldots)\gamma$$

$$+ inl\left\{A + \mu\left(\frac{1}{2} - \frac{1}{48}\frac{\mu^2 l^2 n^2}{R^2} + \frac{13}{8640}\frac{\mu^4 l^4 n^4}{R^4} - \ldots\right)\right\}\gamma \quad (165)$$

since $l/\kappa = R$, the resistance for steady currents.

This equation is of the form

$$E = R'\gamma + inL'\gamma = R'\gamma + L'\frac{d\gamma}{dt} \quad \ldots \quad (166)$$

Evaluation of Effective Resistance and Inductance.

where

$$R' = R(1 + \frac{1}{12}\frac{\mu^2 l^2 n^2}{R^2} - \frac{1}{180}\frac{\mu^4 l^4 n^4}{R^4} + \ldots)$$

$$L' = l\left\{A + \mu\left(\frac{1}{2} - \frac{1}{48}\frac{\mu^2 l^2 n^2}{R^2} + \frac{13}{8640}\frac{\mu^4 l^4 n^4}{R^4} - \ldots\right)\right\} \quad (167)$$

In consequence therefore of the variability of the current the wire behaves as if it had a resistance R', and a self-inductance L'. If the frequency of alternation be very small the resistance approximates to R and the self-inductance to $l(A + \frac{1}{2}\mu)$, the values for steady currents.

With increasing frequency the resistance increases without limit and the inductance diminishes towards the value lA. This result may be obtained from the theorem (see Note in Appendix) that when x is very great $\phi(x) = e^{2\sqrt{x}}/(2\sqrt{\pi}x^{\frac{1}{4}})$, so that in this case $\phi(x)/\phi'(x) = \sqrt{x}$, and therefore $\phi(i\kappa\mu n)/\phi'(i\kappa\mu n) = \sqrt{\frac{1}{2}\kappa\mu n}(1 + i)$, which by (164') gives

$$\left.\begin{array}{l} R' = \sqrt{\frac{1}{2}\mu l n R} \\ L' = l\left(A + \sqrt{\frac{\mu R}{2nl}}\right) \end{array}\right\} \quad \ldots \quad (167')$$

Taking unit length of an iron wire ·4cm. in diameter, and estimating the specific resistance of iron as 10^4 and its permeability as 300, we find that $\frac{1}{12}\mu^2 l^2 n^2/R^2$, is for a period of $\frac{1}{1000}$ sec. about 47, so that the resistance is vastly increased, and the self-inductance diminished by the rapid alternation.

The value of this term for copper, is if $\mu = 1$, and its specific resistance $= 1640$, about $3 \times \pi^2 n^2 a^4/10^8$. A frequency of 100 gives therefore $\cdot 12 a^4$. Thus the effect of alternation becomes very sensible when $a > 1$.

Going back to (164′) we see that in order to find the effective resistance of a cylindrical conductor in which is flowing an alternating current of frequency $2\pi/n$, we have to multiply the actual resistance $R \; (= l/\kappa)$ of a length l, by the real part of $\phi(i\kappa\mu n)/(\phi'(i\kappa\mu n))$ where ϕ is defined by equation (163). Now in the notation of Bessel's functions if x be any number

Expression in terms of Bessel's Functions.
$$J_0(ix) = 1 + \frac{x^2}{2^2} + \frac{x^4}{2^2 \cdot 4^2} + \frac{x^6}{2^2 \cdot 4^2 \cdot 6^2} + \ldots = \phi\left(\frac{x^2}{2^2}\right).$$

Hence, putting $x^2/2^2 = i\kappa\mu n$, we have
$$x = 2\sqrt{i\kappa\mu n} = p\sqrt{i}$$

if $p = 2\sqrt{\kappa\mu n}$. Thus
$$J_0(ip\sqrt{i}) = \phi(i\kappa\mu n)$$
$$= 1 + \frac{ip^2}{2^2} - \frac{p^4}{2^2 \cdot 4^2} - \frac{ip^6}{2^2 \cdot 4^2 \cdot 6^2} + \frac{p^8}{2^2 \cdot 4^2 \cdot 6^2 \cdot 8^2} - \ldots$$

The real part of this expression is called ber p, and the multiplier of i in the imaginary part bei p, by Sir William Thomson, so that
$$\phi(i\kappa\mu n) = \text{ber } p + i \text{ bei } p.$$

Hence in this notation

$$\frac{\phi(i\kappa\mu n)}{\phi'(i\kappa\mu n)} = \frac{\text{ber } p + i \text{ bei } p}{(\text{ber}' p + i \text{ bei}' p) \dfrac{dp}{d(i\kappa\mu n)}} = \frac{ip}{2} \frac{\text{ber } p + i \text{ bei } p}{\text{ber}' p + i \text{ bei}' p}$$

$$= \frac{p}{2}\left\{\frac{\text{ber } p \text{ bei}' p - \text{ber}' p \text{ bei } p}{(\text{ber}' p)^2 + (\text{bei}' p)^2} + i \frac{\text{ber}' p \text{ ber } p + \text{bei } p \text{ bei}' p}{(\text{ber}' p)^2 + (\text{bei}' p)^2}\right\}$$

From this by (164′) and (166) we get, writing R for l/κ,

$$\left.\begin{array}{l} R' = \dfrac{p}{2} \dfrac{\text{ber } p \text{ bei}' p - \text{ber}' p \text{ bei } p}{(\text{ber}' p)^2 + (\text{bei}' p)^2} R \\[1em] L' = \dfrac{pl}{2\kappa n} \dfrac{\text{ber}' p \text{ ber } p + \text{bei } p \text{ bei}' p}{(\text{ber}' p)^2 + (\text{bei}' p)^2} + lA \end{array}\right\} \quad . \quad (168)$$

CALCULATION OF EFFECTIVE RESISTANCE

Sir William Thomson, to whom this form of the value of R' is due,* has treated the subject in a somewhat different manner from that adopted by Lord Rayleigh, and a comparison of the two methods is instructive. Writing as before w for the component of current along z, we have, if the simple harmonic law holds,

Thomson's Method of Calculating Effective Resistance.

$$w = G_1 \cos nt + G_2 \sin nt \quad \ldots \quad (169)$$

where G_1, G_2, are functions of r, the distance of the point considered from the axis of the cylinder. The total current γ, flowing across the section at time t, is given by

$$\gamma = \cos nt \iint G_1 dS + \sin nt \iint G_2 dS \quad \ldots \quad (170)$$

if dS be an element of the cross-section including the point where the current is w.

But equations (12) and (6) of Chapter V. give

$$4\pi\mu w = \frac{1}{k}\nabla^2 w \quad \ldots \ldots \quad (171)$$

Differential Equations for Current.

if, as we suppose, there be no displacement current to be reckoned, and k be the conductivity of the material. Using the equation in (169) we find

$$4\pi\mu k n G_1 = -\nabla^2 G_2, \quad 4\pi\mu k n G_2 = \nabla^2 G_1 \quad \ldots \quad (172)$$

The first of these gives

$$4\pi\mu k n \iint G_1 dS = -\iint \nabla^2 G_2 dS = -\int \frac{dG_2}{d\nu} ds$$

by a well-known theorem, if ν be an outward-drawn normal to an element ds of the periphery of the cross-section, and the last integral be taken round the section. Similarly

$$4\pi\mu k n \iint G_2 dS = \int \frac{dG_1}{d\nu} ds.$$

Hence by substitution in (170) we get

$$4\pi\mu k n \gamma = -\cos nt \int \frac{dG_2}{d\nu} ds + \sin nt \int \frac{dG_1}{d\nu} ds. \quad . \quad (173)$$

Calculation of Mean Square of Current.

---

* See his Presidential Address to the Inst. El. Engineers, 1889, *Math. and Phys. Papers*, vol. iii. p. 492 *et seq.*

which, if the time average of γ^2 be denoted by $\overline{\gamma^2}$, gives by integration with respect to t over the period T

$$(4\pi\mu kn)^2\overline{\gamma^2} = \tfrac{1}{2}\left(\int\frac{dG_1}{d\nu}ds\right)^2 + \tfrac{1}{2}\left(\int\frac{dG_2}{d\nu}ds\right)^2 \quad . \quad . \quad (174)$$

For the activity **A** in the conductor per unit length at any instant we have

$$\mathbf{A} = \frac{1}{k}\int w^2 dS$$

$$= \frac{1}{k}\left(\cos^2 nt \int G_1^2 dS + \sin^2 nt \int G_2^2 dS + 2\cos nt \sin nt \int G_1 G_2 dS\right)$$

and hence for the time-average $\overline{\mathbf{A}}$ of $\mathbf{A}$

$$\overline{\mathbf{A}} = \frac{1}{T}\int \mathbf{A}\,dt = \frac{1}{2k}\int(G_1^2 + G_2^2)dS \quad . \quad . \quad . \quad (175)$$

The effective resistance R' of any length l is the ratio of the mean activity $l\overline{\mathbf{A}}$ to $\overline{\gamma^2}$. Hence

$$R' = (4\pi\mu n)^2 kl \frac{\int(G_1^2 + G_2^2)dS}{\left(\int\frac{dG_1}{d\nu}ds\right)^2 + \left(\int\frac{dG_2}{d\nu}ds\right)^2} \quad . \quad . \quad (176)$$

But by the relations (172) above

$$\iint(G_1^2 + G_2^2)dS = \frac{1}{4\pi\mu kn}\iint(G_2\nabla^2 G_1 - G_1\nabla^2 G_2)dS.$$

If for ∇^2 we substitute $\partial^2/\partial x^2 + \partial^2/\partial y^2$ and integrate by parts with respect to x and y, we get at once, since the surface integral obtained in the process vanishes identically,

$$\iint(G_1^2 + G_2^2)dS = \frac{1}{4\pi\mu kn}\int\left(G_2\frac{dG_1}{d\nu} - G_1\frac{dG_2}{d\nu}\right)ds \quad . \quad (177)$$

CALCULATION OF EFFECTIVE RESISTANCE

This result used in (176) gives, since $R = l/\pi a^2 k$,

Expression for Effective Resistance.

$$R' = 4\pi^2 \mu k n a^2 \frac{\int \left(G_2 \dfrac{dG_1}{d\nu} - G_1 \dfrac{dG_2}{d\nu}\right) ds}{\left(\int \dfrac{dG_1}{d\nu} ds\right)^2 + \left(\int \dfrac{dG_2}{d\nu} ds\right)^2} R . \quad (178)$$

which holds for a conductor of any form of cross-section, on the supposition that the direction of the current is parallel to the axis at every point.

If however the conductor be a right circular cylinder of radius a, and the arrangement be such as to make G_1, G_2 functions of r (the distance from the axis) only, $dG_1/d\nu$, $dG_2/d\nu$, become the values of dG_1/dr, dG_2/dr for $r = a$, and are the same for every cross-section. Then also $\int ds = 2\pi a$, and (178) becomes

$$R' = 2\pi \mu n k a \left\{ \frac{G_2 \dfrac{dG_1}{dr} - G_1 \dfrac{dG_2}{dr}}{\left(\dfrac{dG_1}{dr}\right)^2 + \left(\dfrac{dG_2}{dr}\right)^2} \right\} R . \quad (178')$$

$$r = a$$

The relations (172) give for G_1 the differential equation

Solution of Differential Equation

$$\nabla^4 G_1 + (4\pi \mu k n)^2 G_1 = 0,$$

or for the case under consideration

$$\left(\frac{d^2}{dr^2} + \frac{1}{r}\frac{d}{dr}\right)^2 G_1 + (4\pi \mu k n)^2 G_1 = 0 \quad . \quad . \quad . \quad (179)$$

A solution of this equation, sufficiently general for our present purpose, and expressed in the notation explained above, is

and Expression of Effective Resistance in Series.

$$G_1 = C_1 \operatorname{ber} q + C_2 \operatorname{bei} q \quad . \quad . \quad . \quad . \quad (180)$$

where $q = r\sqrt{4\pi \mu k n}$, and C_1, C_2, are arbitrary constants. This gives by (172)

$$G_2 = C_2 \operatorname{ber} q - C_1 \operatorname{bei} q \quad . \quad . \quad . \quad . \quad (180')$$

Thus since $dq/dr = \sqrt{4\pi\mu kn}$

$$\left.\begin{aligned} \frac{dG_1}{dr} &= \sqrt{4\pi\mu kn}\,(C_1\,\text{ber}'\,p + C_2\,\text{bei}'\,p) \\ \frac{dG_2}{dr} &= \sqrt{4\pi\mu kn}\,(C_2\,\text{ber}'\,p - C_1\,\text{bei}'\,p) \end{aligned}\right\} \quad \ldots \quad (181)$$

and instead of (178′) we find

$$R' = \frac{p}{2}\,\frac{\text{ber}\,p\,\text{bei}'\,p - \text{ber}'\,p\,\text{bei}\,p}{(\text{ber}'\,p)^2 + (\text{bei}'\,p)^2} \quad \ldots \quad (178'')$$

which is identical with (168).

An important table of values of ber q, bei q, ber $'q$, bei $'q$, &c., for different values of q, is given in Sir William Thomson's paper, and is printed in the Appendix to the present work.

Investigation of Effective Inductance. The corresponding expression for the effective self-induction L' is not given by Sir William Thomson in the paper above referred to, but may be found from the above equations as follows. It is shown below (Chapter XII.) that if the resistance and self-inductance of a circuit or part of a circuit be denoted by R', L', and a simple harmonic electromotive force of maximum value E be *applied* to it, the mean square of the current is equal to $\tfrac{1}{2}E^2/(R'^2 + n^2 L_1'^2)$. On the other hand if R be the resistance of the same part of the circuit for steady currents, the mean square of the simple harmonic current which would flow if there were no inductive action in the conductor is $\tfrac{1}{2}E^2/R^2$. But by (169) the electromotive force $\mathbf{E}$ at the point in the conductor where the current is w, being w/k is

$$\frac{1}{k}\,(G_1 \cos nt + G_2 \sin nt).$$

The electromotive force applied independently of induction is thus

$$\frac{1}{k}\,(G_1 \cos nt + G_2 \sin nt) + \dot{H},$$

since $-\dot{H}$ is the part of the actual electromotive force which depends upon induction. This must be the same at every part of the cross-section, and hence we may take as the values of G_1, G_2 those for the outside skin of the wire, with the corresponding value of $\dot{H}$, that is, as given above $A\dot{\gamma}$, if γ be the total

CALCULATION OF INDUCTANCE

current actually flowing at any instant. Thus if **E** be the electromotive force at a point on the surface of the conductor,

$$\mathbf{E} = \frac{1}{k}(G_1 \cos nt + G_2 \sin nt) + A\gamma$$
$$r = a.$$

But by (173)

$$\dot{\gamma} = \frac{n}{4\pi\mu k n}\left(\sin nt \int \frac{dG_2}{dr} ds + \cos nt \int \frac{dG_1}{dr} ds\right)$$

$$= \frac{a}{2\mu k}\left(\sin nt \frac{dG_2}{dr} + \cos nt \frac{dG_1}{dr}\right).$$

Therefore

$$\mathbf{E} = \frac{1}{k}\left\{\left(G_1 + \frac{Aa}{2\mu}\frac{dG_1}{dr}\right)\cos nt + \left(G_2 + \frac{Aa}{2\mu}\frac{dG_2}{dr}\right)\sin nt\right\}.$$

The maximum non-inductive electromotive force at each point is thus

$$\frac{1}{k^2}\left[G_1^2 + G_2^2 + \frac{A^2 a^2}{4\mu^2}\left\{\left(\frac{dG_1}{dr}\right)^2 + \left(\frac{dG_2}{dr}\right)^2\right\} + \frac{Aa}{\mu}\left(G_1\frac{dG_1}{dr} + G_2\frac{dG_2}{dr}\right)\right]$$
$$r = a.$$

Hence as stated above

$$R^2\gamma_s^2 = \tfrac{1}{2}l^2\mathbf{E}^2 = \frac{l^2}{2k^2}\left[G_1^2 + G_2^2 + \frac{A^2 a^2}{4\mu^2}\left\{\left(\frac{dG_1}{dr}\right)^2 + \left(\frac{dG_2}{dr}\right)^2\right\}\right.$$
$$\left. + \frac{Aa}{\mu}\left(G_1\frac{dG_1}{dr} + G_2\frac{dG_2}{dr}\right)\right]$$
$$r = a,$$

and therefore by (174) applied to a right circular cylinder

$$R^2 \frac{\gamma_s^2}{\gamma^2} (= R'^2 + n^2 L'^2) = l^2 n^2 \left\{ \frac{4\mu^2}{a^2} \frac{G_1^2 + G_2^2}{\left(\frac{dG_1}{dr}\right)^2 + \left(\frac{dG_2}{dr}\right)^2} \right.$$
$$\left. + A^2 + \frac{4\mu A}{a} \frac{G_1\frac{dG_1}{dr} + G_2\frac{dG_2}{dr}}{\left(\frac{dG_1}{dr}\right)^2 + \left(\frac{dG_2}{dr}\right)^2} \right\}$$
$$r = a.$$

Expression for Self-Inductance.

Substituting in this equation the value of R'^2 given by (178), (with $l^2/\pi^2 a^4 k^2$ put for R^2) and reducing we find

$$L' = \frac{2\mu l}{a} \frac{G_1 \frac{dG_1}{dr} + G_2 \frac{dG_2}{dr}}{\left(\frac{dG_1}{dr}\right)^2 + \left(\frac{dG_2}{dr}\right)^2} + lA,$$

which by substitution from (180), (180'), (181), becomes

$$L' = \frac{2\mu l}{p} \frac{\operatorname{ber} p \operatorname{ber}' p + \operatorname{bei} p \operatorname{bei}' p}{(\operatorname{ber}' p)^2 + (\operatorname{bei}' p)^2} + lA. \quad . \quad (182)$$

This agrees with Lord Rayleigh's result as expressed in (168).

Cylindrical Conductor of Large Radius of Curvature at Every Point.

Consider now, in a cylindrical conductor of any form of cross-section, so great a rapidity of alternation that the current is insensible at every point beyond a certain depth below the surface small in comparison with the radius of curvature of the surface at the point considered, and also in comparison with the distance of any part of the nearest insulated conductor.* Then for any point in this surface stratum the variation in the current density will be practically zero in the direction of the line drawn in the section at right angles to the radius of curvature. Let the distance of the point considered from the point in which the radius of curvature through it cuts the surface be x. At the point G_1, G_2 will be sensibly functions of x only. Hence instead of (172) we have

Modified Differential Equation.

$$4\pi\mu n k G_1 = -\frac{d^2 G_2}{dx^2}, \quad 4\pi\mu n k G_2 = \frac{d^2 G_1}{dx^2}.$$

If for brevity $\sqrt{2\pi\mu n k}$ be denoted by a, the harmonic solution of these equations is

$$G_1 = g\epsilon^{-ax} \cos ax, \quad G_2 = g\epsilon^{-ax} \sin ax.$$

These values of G_1, G_2 give by (169)

$$w = g\epsilon^{-ax} \cos(ax - nt).$$

---

* See Sir William Thomson's paper already cited, p. 331 above.

EFFECTIVE RESISTANCE OF A CYLINDER

Again

$$\frac{dG_1}{dx} = -aG_1 - ag\epsilon^{-ax} \sin ax = -a(G_1 + G_2)$$

$$\frac{dG_2}{dx} = -aG_2 + ag\epsilon^{-ax} \cos ax = -a(G_2 - G_1)$$

which (since $d/d\nu = -d/dx$) yield

$$G_2 \frac{dG_1}{d\nu} - G_1 \frac{dG_2}{d\nu} = a(G_1^2 + G_2^2) = g^2 a \epsilon^{-2ax}.$$

Hence by (175) and (177), since x may be taken as zero,

$$\overline{A} = \frac{1}{4ka} \int g^2 ds\,;$$

and by (174)

$$\overline{\gamma^2} = \frac{1}{4a^2} \left(\int g\, ds \right)^2$$

where ds is as before an element of the periphery of the cross-section.

Hence for the effective resistance R' of a length l we have

$$R' = \frac{l\overline{A}}{\overline{\gamma^2}} = \frac{la}{k} \frac{\int g^2 ds}{\left(\int g\, ds \right)^2} \quad \ldots \ldots \quad (183)$$

If the arrangement is such that g is constant

$$R' = \frac{la}{k} \frac{1}{2\pi a} = \frac{l}{kc\frac{1}{a}} \quad \ldots \ldots \quad (183')$$

if c denote the circumference of the conductor. The interpretation of this result is that the effective resistance is the same as the resistance which a surface stratum of the conductor of thickness $1/a$ ($= 1/\sqrt{2\pi\mu kn}$) would offer to a steady current. This thickness is given in the following table for copper, lead, and iron, and frequences of alternation 80, 120, 160, 200.

VOL. II.   Z

Depth of Effective Surface Stratum of Cylindrical Conductor.

Frequency of Alternation.	Copper.	Lead.	Iron ($\mu = 300$)
80	·719 cm.	2·49 cm.	·0976 cm.
120	·587 ,,	2·04 ,,	·0798 ,,
160	·509 ,,	1·76 ,,	·0691 ,,
200	·455 ,,	1·58 ,,	·0617 ,,

Two Parallel Plane Strips of Great Width.

Lord Rayleigh has also worked out the case of two coaxial cylindrical conductors, one carrying the outgoing, the other the return current. The conductors are taken of equal thickness and of radii differing by a small fraction of either, and so large that the influence of the curvature on the current, the magnetic force, &c., at any point may be neglected. This is the case then of two parallel infinitely long, infinitely broad, equally thick plane strips, facing one another and enclosing a stratum of uniform dielectric.

Supposing that there is no current at right angles to the length of the strips * and using the same notation as before, we have

$$\frac{w}{k} = -\frac{\partial \Psi}{\partial z} - \dot{H} = -\frac{\partial \Psi}{\partial z} - inH. \quad . \quad . \quad (184)$$

since the value of $\dot{H}$ is time periodic with period $2\pi/n$.

But as there is no component of magnetization, d, at right angles to the planes of the strips, we have $\partial \beta / \partial x = 4\pi w$, $\mu \beta = -\partial H / \partial x$, and therefore

$$w = \frac{1}{4\pi} \frac{\partial \beta}{\partial x} = -\frac{1}{4\pi\mu} \frac{\partial^2 H}{\partial x^2}.$$

Hence (184) becomes

$$\frac{\partial^2 H}{\partial x^2} - 4\pi\mu k \frac{\partial \Psi}{\partial z} - 4\pi\mu kin\, H = 0 \quad . \quad . \quad (185)$$

---

* This will be the case if there is no sensible condenser-action of the opposed strips. If there is such an action there will be a component of current at right angles to the planes of the strips, which must be taken into account in the theory.

TWO PARALLEL PLANE SHEETS

But $\partial \Psi / \partial z$ is independent of x, so that we may write

$$\frac{\partial^2 \left(H - \frac{i}{n}\frac{\partial \Psi}{\partial z} \right)}{\partial x^2} - 4\pi\mu k i n \left(H - \frac{i}{n}\frac{\partial \Psi}{\partial z} \right) = 0. \quad (186)$$

Differential Equation.

The general solution of this equation may be written

Solution.

$$H = \frac{i}{n}\frac{\partial \Psi}{\partial z} + A'\epsilon^{mx} + B'\epsilon^{-mx}$$

where $m = 2\sqrt{\pi\mu k i n}$, and A', B' are constants. It will be convenient to take the plane $x=0$, midway between the conductors, and to write the above result in the form

$$H = \frac{i}{n}\frac{\partial \Psi}{\partial z} + B\epsilon^{m(x-a)} + B'\epsilon^{-m(x-a)}. \quad (187)$$

where B, B', are another pair of constants, and a is half the thickness of dielectric between the strips.

This gives

$$\mu\beta = -\frac{\partial H}{\partial x} = m \{B\epsilon^{m(x-a)} - B'\epsilon^{-m(x-a)}\}. \quad (188)$$

If the thickness of each strip be b, then, (since one side of the arrangement of two parallel strips may be regarded as the outside, and the other as the inside of the system of two coaxial cylinders, where, in each case, the magnetic force is zero), we have $\beta=0$ when $x=a+b$, and therefore

$$B' = \frac{\epsilon^{mb}}{\epsilon^{-mb}} B \quad \ldots \quad (189)$$

Where $x=a$ let β be denoted by β_a, so that,

$$\mu\beta_a = m(B - B'). \quad \ldots \quad (190)$$

But if γ be the total current per unit of breadth in the strip

$$\gamma = \int_a^{a+b} u dx = \frac{1}{4\pi}\int_a^{a+b} \frac{\partial \beta}{\partial x} dx = -\frac{m}{4\pi\mu}(B - B').$$

Hence by (189)

$$B = \frac{4\pi\mu\gamma}{m} \frac{\epsilon^{-mb}}{\epsilon^{mb} - \epsilon^{-mb}}$$
$$B' = \frac{4\pi\mu\gamma}{m} \frac{\epsilon^{mb}}{\epsilon^{mb} - \epsilon^{-mb}} \qquad (191)$$

and (190) becomes

$$\beta_a = -4\pi\gamma a \qquad \qquad (192)$$

Within the insulator, at all points of the plane ($x=0$), the vector potential produced by the current in one strip is equal and opposite to that produced by the other, so that there $H=0$. Calling H_a the value of H for $x=a$, we have, since $\partial\beta/\partial x = 4\pi w = 0$ within the dielectric,

$$H_a = H_a - H_0 = -\int \beta dx = -\beta_a a$$

if the permeability of the dielectric be taken as unity.
Hence by (187) and (192) we have for $x=a$

$$-\frac{\partial \Psi}{\partial z} = in\left(4\pi a + \frac{4\pi\mu}{m} \frac{\epsilon^{mb} + \epsilon^{-mb}}{\epsilon^{mb} - \epsilon^{-mb}}\right)\gamma. \qquad (193)$$

Expressions for Resistance and Inductance.
But $m = 2\sqrt{\pi\mu k in} = \sqrt{2\pi\mu k n}(1+i)$ so that the second term in the brackets is partly real partly imaginary. Thus if R', iL', denote the real and imaginary parts of the coefficient of γ in the expression for $-ld\Psi/dz$, (l denoting as before a given length of the strip) we can write

$$-l\frac{\partial \Psi}{\partial z} = R'\gamma + L'in\gamma. \qquad (194)$$

so that R', L', are the resistance and self-induction of a strip of unit breadth and length l.

Comparison of Results for Slow and Rapid Alternations.
The chief interest of the problem lies in the comparison of the values of R', L', for very slow and very rapid alternations. When n is small we find by expansion of the exponentials in (193), and ordinary division

$$-l\frac{\partial \Psi}{\partial z} = \frac{l}{bk}\gamma + inl(4\pi a + \tfrac{4}{3}\pi\mu b)\gamma. \qquad (195)$$

Thus for this limiting case

$$R' = \frac{l}{bk}. \quad \ldots \ldots \quad (196)$$

the actual resistance for steady currents of a portion of the strip of length l and unit breadth.

If the strips be very close so that $a = o$, we have

$$-l\frac{\partial \Psi}{\partial z} = \frac{l}{bk}\gamma + \tfrac{4}{3}in\pi\mu bl\gamma \quad \ldots \quad (196')$$

and the resistance is the same as before. The self-inductance is $\tfrac{4}{3}\pi\mu bl$ for a strip of unit breadth, so that for a strip of breadth $2\pi r$ it is

$$L' = \tfrac{2}{3}\mu\frac{bl}{r}. \quad \ldots \ldots \quad (197)$$

which is half the result obtainable by direct calculation after the method of p. 291, for steady currents in two close coaxial cylinders at a distance apart small in comparison with their radii. This is as it should be, since there the self-inductance would be found for what in the present reckoning is a length $2l$, viz., a length l in the outgoing, and an equal length in the return strip.

Now let n be very great. Then $(\epsilon^{mb} + \epsilon^{-mb})/(\epsilon^{mb} - \epsilon^{-mb})$ $= (1 + \epsilon^{-2mb})/(1 - \epsilon^{-2mb}) = 1$, so that for this case

$$-l\frac{\partial \Psi}{\partial z} = \frac{l}{k}\sqrt{2\pi\mu nk}\,\gamma + inl\left(4\pi a + \sqrt{\frac{2\pi\mu}{nk}}\right)\gamma. \quad (198)$$

Thus

$$\left.\begin{array}{l} R' = \dfrac{l}{k}\sqrt{2\pi\mu nk} \\[2mm] L' = l\left(4\pi a + \sqrt{\dfrac{2\pi\mu}{nk}}\right) \end{array}\right\} \quad \ldots \quad (199)$$

We conclude that as n is increased the resistance is increased without limit, while L' diminishes towards the limit $4\pi al$. The result shows moreover that the thickness of the strip which would give the same resistance is $1/\sqrt{2\pi\mu nk}$, which agrees with the result obtained at p. 337 above. This may be taken as the effectual thickness of the conductor. It diminishes as $\sqrt{n}$ increases.

Thickness of Effective Surface Stratum.

When the thickness is great so that b may be regarded as infinite, we have by (188) since $-4\pi w = \partial \beta/\partial x$

$$w = m\epsilon^{-m(x-a)}\gamma$$
$$= a(1+i)\epsilon^{-a(1+i)(x-a)}$$

if (as at p. 336 above) we denote $\sqrt{2\pi\mu nk}$ by a. This may be written

$$w = \sqrt{2}a\gamma\epsilon^{-a(x-a)}[\cos\{\tfrac{1}{4}\pi - a(x-a)\} + i\sin\{\tfrac{1}{4}\pi - a(x-a)\}].$$

General Dynamical Theory of Effects of Constraints. Thomson's and Bertrand's Theorems. Lord Rayleigh[*] has shown that the restriction of a rapidly varying current to the outer strata of the conductor is a consequence of a general dynamical principle which regulates the effects of constraints on the motion of a material system. This principle is embodied in two general theorems due to Thomson and Bertrand respectively. Thomson's theorem asserts that if any material system given at rest be suddenly set in motion with any specified velocities (possible under the kinematical conditions of the system) imposed on certain parts of the system, the other parts being left free to take such velocities as result from the connections, the resulting motion is that for which the kinetic energy has the smallest possible value consistent with fulfilment of the prescribed velocity conditions. Bertrand's theorem, on the other hand, asserts that if the impulses applied to certain parts of the system be specified, the resulting motion is that for which the kinetic energy has the greatest value consistent with the prescribed condition as to impulses.

Taking any case in which we consider a system impulsively set into motion with a single specified velocity, or with a specified impulse of the same type. Let Φ denote the impulsive force, then the impulse is (p. 149 above) the time-integral

$$\int_0^\tau \Phi\, dt = \Phi$$

and the corresponding velocity generated is ϕ. The resulting kinetic energy T is $\tfrac{1}{2}\phi\Phi$.

Effect of Constraint according as Velocity or Impulse is Given. According to Thomson's theorem the introduction of any constraint limiting the freedom of the system causes an increase of T if ϕ be given. On the other hand, according to Bertrand's theorem, if Φ_1 be given, the effect of the constraint will be to

---

[*] *Phil. Mag.* May, 1886.

diminish T. In both cases the ratio $2T/\dot\phi^2$, or $\Phi/\dot\phi$, is increased, for in the former case Φ is increased, and in the latter $\dot\phi$ is diminished. Thus the effect of constraint is in each case to increase the generalized inertia-coefficient corresponding to the coordinate in question.

Consider now a system in which a force Φ_1 of type corresponding to the coordinate ϕ_1, and varying according to a simple harmonic function of the time is applied to the system. Suppose the system to have no potential energy, and to be subject to dissipative forces given (according to the rule on p. 159 above) by a dissipation function F, which is a homogeneous quadratic function of the generalized velocities of the system. Let, further, the remaining coordinates $\phi_2, \phi_3, \ldots \phi_m$ of the system be so chosen that no product of them enters into the expressions of T and F. *System under Time-Periodic Applied Force and subject to Dissipative Resistances.*

Then we have

$$\left.\begin{array}{l} T = \tfrac{1}{2}(a_{11}\dot\phi_1^2 + a_{22}\dot\phi_2^2 + \ldots + 2a_{12}\dot\phi_1\dot\phi_2 + 2a_{13}\dot\phi_1\dot\phi_3 + \ldots) \\ F = \tfrac{1}{2}(b_{11}\dot\phi_1^2 + b_{22}\dot\phi_2^2 + \ldots + 2b_{12}\dot\phi_1\dot\phi_2 + 2b_{13}\dot\phi_1\dot\phi_3 + \ldots) \end{array}\right\} \quad (200)$$

But by Lagrange's equations

$$\frac{d}{dt}\frac{\partial T}{\partial \dot\phi_1} + \frac{\partial F}{\partial \dot\phi_1} = \Phi_1$$

and hence by (200)

$$\left.\begin{array}{l} a_{11}\ddot\phi_1 + a_{12}\ddot\phi_2 + a_{13}\ddot\phi_3 + \ldots + b_{11}\dot\phi_1 + b_{12}\dot\phi_2 + \ldots = \Phi_1 \\ a_{12}\ddot\phi_1 + a_{22}\ddot\phi_2 + b_{12}\dot\phi_1 + b_{22}\dot\phi_2 = 0 \\ a_{13}\ddot\phi_1 + a_{33}\ddot\phi_3 + b_{13}\dot\phi_1 + b_{33}\dot\phi_3 = 0 \\ \ldots \ldots \ldots \ldots \ldots \ldots \ldots \ldots \end{array}\right\} \quad (201)$$

Let now the whole motion be simple harmonic in the period of the force Φ_1. Representing the latter by ϵ^{int} we get instead of (201)

$$\left.\begin{array}{l} (ina_{11}+b_{11})\phi_1 + (ina_{12}+b_{12})\phi_2 + (ina_{13}+b_{13})\phi_3 + \ldots = \Phi_1 \\ (ina_{12}+b_{12})\phi_1 + (ina_{22}+b_{22})\phi_2 = 0 \\ (ina_{13}+b_{13})\phi_1 + (ina_{33}+b_{33})\phi_3 = 0 \\ \ldots \ldots \ldots \ldots \ldots \ldots \ldots \end{array}\right\} \quad (201')$$

The second and following equations of (201') give ϕ_2, ϕ_3, &c., in terms of ϕ_1, and these values substituted in the first equation of (201') yield

344   ALTERNATING CURRENTS

$$\frac{\Phi_1}{\phi_1} = ina_{11} + b_{11} - \frac{(ina_{12} + b_{12})^2}{ina_{22} + b_{22}} - \frac{(ina_{13} + b_{13})^2}{ina_{33} + b_{33}} - \quad (202)$$

Since to a constant factor ϕ_1 is represented by ϵ^{int} it is clear that $(\partial F/\partial \phi)/\phi$ is the real part of Φ/ϕ_1 and therefore corresponds to the dissipative force. Calling this quantity R' and the other inL', we have instead of (169)

$$\Phi_1 = (R' + inL')\phi_1. \quad \quad (202')$$

Effective "Resistance," and "Inertia" (or "Inductance") of System.

R' may be called the resistance of the system, and L' the generalized inertia-coefficient, or what corresponds to the self-inductance in the electrical theory.

To calculate R' we have to find the real parts of the successive terms in (202). Now

$$\text{real part of } \frac{(ina_{12} + b_{12})^2}{ina_{22} + b_{22}} = \text{real part of } \frac{(ina_{12} + b_{12})^2(b_{22} - ina_{22})}{b_{22}^2 + n^2 a_{22}^2}$$

$$= \frac{b_{12}^2}{b_{22}} - n^2 \frac{(a_{12}b_{22} - a_{22}b_{12})^2}{b_{22}(b_{22}^2 + n^2 a_{22}^2)}$$

and similarly the real parts of the other terms may be found. Hence

$$R' = b_{11} - \sum_{i=2}^{i=m} \frac{b_{1i}^2}{b_{ii}} + n^2 \sum_{i=2}^{i=m} \frac{(a_{1i}b_{ii} - a_{ii}b_{1i})^2}{b_{ii}(b_{ii}^2 + n^2 a_{ii}^2)}. \quad (203)$$

Expression for "Resistance."

It is clear that each term of the second series in this expression increases as n increases, that is, as the frequency increases. It follows that the value of R' increases with the frequency. When n is very great R' approaches the limiting value

$$b_{11} - \sum_{i=2}^{i=m} \frac{b_{ii}^2}{b_{ii}^2} + \sum_{i=2}^{i=m} \frac{(a_{1i}b_{ii} - a_{ii}b_{1i})^2}{b_{ii}a_{ii}^2}$$

When n is small R' is approximately equal to the first two terms in (203), and is an absolute minimum for steady, and for continuous non-periodic motions.

The imaginary part of (202) is easily found to be

GENERAL DYNAMICAL THEORY 345

$$in \left\{ a_{11} - \sum_{i=2}^{i=m} \frac{a_{1i}^2}{a_{ii}^2} + \sum_{i=2}^{i=m} \frac{(a_{1i}b_{ii} - a_{ii}b_{1i})^2}{a_{ii}(b_{ii}^2 + n^2 a_{ii}^2)} \right\}$$

Expression for Inductance.

Hence

$$L' = a_{11} - \sum_{i=2}^{i=m} \frac{a_{ii}^2}{a_{ii}^2} + \sum_{i=2}^{i=m} \frac{(a_{1i}b_{ii} - a_{ii}b_{1i})^2}{a_{ii}(b_{ii}^2 + n^2 a_{ii}^2)} \quad . \quad . \quad (204)$$

Each term of the second series in (204) is positive, and continually diminishes as n increases. Hence as n increases L' approaches more and more nearly the value

$$a_{11} - \sum_{i=2}^{i=m} \frac{a_{ii}^2}{a_{ii}^2}$$

which is independent of dissipative terms.

In these results we have, as Lord Rayleigh has pointed out, an analogue to Thomson's theorem. In the absence of constraints, R' is great and L' a minimum when the vibrations are very rapid, and on the other hand L' is great and R' a minimum when the vibrations are very slow.

Analogue of Thomson's Theorem. Application of General Results to Electrical Problems.

The application of these results to electrical problems is obvious. Let us take the case, already treated at p. 185 above, of a primary and secondary circuit. If R_1, L_1 be the resistance and self-induction of the primary, L_2, R_2 those of the secondary, and M the mutual inductance, we have

1. Primary and Secondary Circuits.

$$a_{11} = L_1, \; a_{12} = M, \; a_{22} = L_2$$
$$b_{11} = R_1, \; b_{12} = 0, \; b_{22} = R_2$$

Then for the resultant resistance and self-inductance of the primary

$$\left. \begin{array}{l} R'_1 = R_1 + \dfrac{n^2 M^2 R_2}{R_2^2 + n^2 L_2^2} \\[6pt] L'_1 = L_1 - \dfrac{n^2 M^2 L_2}{R_2^2 + n^2 L_2^2} \end{array} \right\} \quad . \quad . \quad . \quad (205)$$

It follows that if the alternations be very slow, the secondary has no effect on the primary. On the other hand, if they are very rapid, R' approaches the limit $R_1 + M^2R_2/L_2^2$, and L' the limit $L_1 - M^2/L_2$.

The resultant resistance and self-inductance of the secondary will be given in the same way. An inductive electromotive force of frequency $n/2\pi$ acts in the secondary. The reaction of the primary will therefore give for the secondary

$$R'_2 = R_2 + \frac{n^2M^2R_1}{R_1^2 + n^2L_1^2}$$

$$L'_2 = L_1 - \frac{n^2M^2L_1}{R_1^2 + n^2L_1^2}$$

as given above.

2. System of Primary, Secondary, Tertiary, &c. Circuits. Another interesting application given by Lord Rayleigh is to a series of conductors forming primary, secondary, tertiary, &c., circuits, but such that no mutual induction exists except between the primary and the secondary, the secondary and the tertiary, &c. Taking, for example, four circuits in the series, the current in the fourth is due to the inductive action of the third. The reaction on the third causes the latter to have a resultant resistance R'_3, and self-inductance L'_3, at once calculable from (172) by substituting for R_1, L_1, R_2, L_2, M, the quantities $R_3, L_3, R_4, L_4, M_{34}$. If R'_3, L'_3 are used as the resistance and self-inductance of the third circuit, the fourth circuit may be ignored. Then the resultant resistance and self-inductance of the second circuit due to the action of the third can be found in the same way, and (the third then being also ignored) used to obtain those of the primary circuit. The effect on the primary is to increase its effective resistance and diminish its effective self-inductance in a degree which is greater the greater the frequency of alternation.

Phases of Currents in Primary, Secondary etc. It can be shown that the phases of the currents in the different circuits of the series depend in the case of very rapid alternation on the induction coefficients only, and differ successively by half a period.

A very important example, which will be given later in connection with the measurement of activity in alternating circuits is that of two conductors in multiple arc, and a related one which may be worked out easily by the reader from the general formulæ given above, is the case already treated at p. 186, of a number of conductors joining two points in parallel, but so arranged as not to exert on one another any mutual induction.

www.ingramcontent.com/pod-product-compliance
Lightning Source LLC
Chambersburg PA
CBHW021622250426
43672CB00037B/345